The Mathematics of Encryption

The Mathematics of Encryption

An Elementary Introduction

Margaret Cozzens
Steven J Miller

AMS
American Mathematical Society
Providence, Rhode Island

2010 *Mathematics Subject Classification*. Primary 94A60, 68P25, 01-01.

For additional information and updates on this book, visit
www.ams.org/bookpages/mawrld-29

Library of Congress Cataloging-in-Publication Data
Cozzens, Margaret B.
The mathematics of encryption : an elementary introduction / Margaret Cozzens, Steven J. Miller.
pages cm. — (Mathematical world ; 29)
Includes bibliographical references and index.
ISBN 978-1-4704-3733-6
1. Coding theory–Textbooks. 2. Cryptography–Textbooks. 3. Cryptography–Mathematics–Textbooks. 4. Cryptography–History–Textbooks. 5. Data encryption (Computer science)–Textbooks. I. Miller, Steven J., 1974– II. Title.

QA268.C697 2013
652'.80151—dc23 2013016920

Published by the American Mathematical Society
First Indian edition 2017
Reprinted 2017

Distributed by
Universities Press (India) Private Limited
3-6-747/1/A & 3-6-754/1, Himayatnagar, Hyderabad 500 029, Telangana, India.
info@universitiespress.com; www.universitiespress.com

For sale in India, Pakistan, Nepal, Bhutan, Bangladesh, Sri Lanka and the Maldives only. Export of this edition to any unauthorized territories is strictly prohibited.

Printed at
Yash Printographics, Noida

Margaret Cozzens dedicates to Tyler and Brad.

Steven Miller dedicates to Cameron and Kayla Miller, for their patience as the book was written and for hours of fun exploring the Caesar cipher.

This book is also dedicated to the men and women who have advanced freedom's cause through the centuries by devoting and sometimes giving up their lives to breaking codes and protecting these successes.

Contents

Preface

Many of the challenges and opportunities facing citizens in the twenty-first century require some level of mathematical proficiency. Some obvious ones are optimization problems in business, managing your household's budget, weighing the economic policies and proposals of political candidates, and of course the ever-important quest to build the best fantasy sports team possible and, if not winning your local NCAA basketball pool, at least doing well enough to avoid embarrassment! As important as these are, there are many other applications of mathematics going on quietly around us all the time. In this book we concentrate on issues arising from cryptography, which we'll see is far more than soldiers and terrorists trying to communicate in secret. We use this as the vehicle to introduce you to a lot of good, applicable mathematics; for much of the book all you need is high school algebra and some patience. These are not cookbook problems to help you perfect your math skills, but rather the basis of modern commerce and security! Equally important, you'll gain valuable experience in how to think about and approach difficult problems. This is a highly transferable skill and will serve you well in the years to come.

Cryptography is one of the oldest studies, and one of the most active and important. The word **cryptography** comes from two Greek words: *κρυτττὸςς* (kryptos), meaning secret, and *γρὰφω* (grapho), meaning to write. As these roots imply, it all began with the need for people to communicate securely. The basic setup is that there are two people, and they must be able to quickly, easily, and securely exchange information, often in the presence of an adversary who is actively attempting to intercept and decipher the messages.

In the public mind, the most commonly associated images involve the military. While war stories make for dramatic examples and are very important in both the development of the field and its applications, they are only part of the picture. It's not just a subject for soldiers on the battlefield. Whenever you make an online purchase, you're a player. This example has many of the key features.

The first issue is the most obvious. You need to authorize your credit card company or bank to transfer funds to the merchant; however, you're not face-to-face with the seller, and you have to send your information through a probably very insecure channel. It's imperative that no one is able to obtain your personal information and pretend to be you in future transactions!

There are, however, two other very important items. The process must be fast; people aren't willing to wait minutes to make sure an order has been confirmed. Also, there's always the problem of a message being corrupted. What if some of the message is mistransmitted or misread by the party on the other end? These questions lead us to the study of efficient algorithms and error detection and correction codes. These have found a wealth of applications not just in cryptography, but also in areas where the information is not secret.

Two great examples are streaming video and Universal Product Codes (UPC). In streaming video the information (everything from sports highlights to CSPAN debates) is often unprotected and deliberately meant to be freely available to all; what matters is being able to transmit it quickly and play it correctly on the other end. Fruits and vegetables are some of the few remaining items to resist getting a UPC barcode; these black and white patterns are on almost all products. It may shock you to realize how these are used. It's far more than helping the cashier charge you the proper amount; they're also used to help stores update their inventory in real time as well as correlate and analyze your purchases to better target you in the future! These are both wonderful examples of the need to detect and correct errors.

These examples illustrate that problems and solutions arising from cryptography often have applications in other disciplines. That's why we didn't title this book as an introduction to cryptography, but rather to encryption. Cryptography is of course important in the development of the field, but it's not the entire story.

The purpose of this book is to introduce just enough mathematics to explore these topics and to familiarize you with the issues and challenges of the field. Fortunately, basic algebra and some elementary number theory is enough to describe the systems and methods. This means you can read this book without knowing calculus or linear algebra; however, it's important to understand what "elementary" means. While we don't need to use powerful theorems from advanced mathematics, we do need to be very clever in combining our tools from algebra. Fortunately we're following the paths of giants, who have had numerous "aha moments" and have seen subtle connections between seemingly disparate subjects. We leisurely explore these paths, emphasizing the thought processes that led to these remarkable advances.

Below is a quick summary of what is covered in this book, which we follow with outlines for semester-long courses. Each chapter ends with a collection of problems. Some problems are straightforward applications of material from the text, while others are quite challenging and are introductions to more advanced topics. These problems are meant to supplement the text and to allow students of different levels and interests to explore the material in different ways. Instructors may contact the authors (either directly or through the AMS webpage) to request a complete solution key.

- Chapter 1 is a brief introduction to the history of cryptography. There is not much mathematics here. The purpose is to provide the exciting historical importance and background of cryptography, introduce the terminology, and describe some of the problems and uses.
- Chapter 2 deals with classical methods of encryption. For the most part we postpone the attacks and vulnerabilities of these methods for later chapters, concentrating instead on describing popular methods to encrypt and decrypt messages. Many of these methods involve procedures to replace the letters of a message with other letters. The main mathematical tool used here is modular arithmetic. This is a generalization of addition on a clock (if it's 10 o'clock now, then in five hours it's 3 o'clock), and this turns out to be a very convenient language for cryptography. The final section on the Hill cipher requires some basic linear algebra, but this section may safely be skipped or assigned as optional reading.
- Chapter 3 describes one of the most important encryption methods ever, the Enigma. It was used by the Germans in World War II and thought by them to be unbreakable due to the enormous number of possibilities provided. Fortunately for the Allies, through espionage and small mistakes by some operators, the Enigma was successfully broken. The analysis of the Enigma is a great introduction to some of the basic combinatorial functions and problems. We use these to completely analyze the Enigma's complexity, and we end with a brief discussion of Ultra, the Allied program that broke the unbreakable code.
- Chapters 4 and 5 are devoted to attacks on the classical ciphers. The most powerful of these is frequency analysis. We further develop the theory of modular arithmetic, generalizing a bit more operations on a clock. We end with a discussion of one-time pads. When used correctly, these offer perfect security; however, they require the correspondents to meet and securely exchange a secret. Exchanging a secret via insecure channels is one of the central problems of the subject, and that is the topic of Chapters 7 and 8.
- In Chapter 6 we begin our study of modern encryption methods. Several mathematical tools are developed, in particular binary expansions (which are similar to the more familiar decimal or base 10 expansions) and recurrence relations (which you may know from the Fibonacci numbers, which satisfy the recursion $F_{n+2} = F_{n+1} + F_n$).

We encounter a problem that we'll face again and again in later chapters: an encryption method which seems hard to break is actually vulnerable to a clever attack. All is not lost, however, as the very fast methods of this chapter can be used in tandem with the more powerful methods we discuss later.

- Chapters 7 and 8 bring us to the theoretical and practical high point of the book, a complete description of RSA (its name comes from the initials of the three people who described it publicly for the first time—Rivest, Shamir, and Aldeman). For years this was one of the most used encryption schemes. It allows two people who have never met to communicate quickly and securely. Before describing RSA, we first discuss several simpler methods. We dwell in detail on why they seem secure but are, alas, vulnerable to simple attacks. In the course of our analysis we'll see some ideas on how to improve these methods, which leads us to RSA. The mathematical content of these chapters is higher than earlier in the book. We first introduce some basic graph theory and then two gems of mathematics, the Euclidean algorithm and fast exponentiation. Both of these methods allow us to solve problems far faster than brute force suggests is possible, and they are the reason that RSA can be done in a reasonable amount of time. Our final needed mathematical ingredient is Fermat's little Theorem. Though it's usually encountered in a group theory course (as a special case of Lagrange's theorem), it's possible to prove it directly and elementarily. Fermat's result allows the recipient to decrypt the message efficiently; without it, we would be left with just a method for encryption, which of course is useless. In addition to describing how RSA works and proving why it works, we also explore some of the implementation issues. These range from transmitting messages quickly to verifying the identity of the sender.
- In Chapter 9 we discuss the need to detect and correct errors. Often the data is not encrypted, and we are just concerned with ensuring that we've updated our records correctly or received the correct file. We motivate these problems through some entertaining riddles. After exploring some natural candidates for error detecting and correcting codes, we see some elegant alternatives that are able to transmit a lot of information with enough redundancy to catch many errors. The general theory involves advanced group theory and lattices, but fortunately we can go quite far using elementary counting.
- We describe some of the complexities of modern cryptography in Chapter 10, such as quantum cryptography and steganography.
- Chapter 11 is on primality testing and factorization algorithms. In the RSA chapters we see the benefits of the mathematicalization of messages. To implement RSA, we need to be able to find two large

primes; for RSA to be secure, it should be very hard for someone to factor a given number (even if they're told it's just the product of two primes). Thus, this advanced chapter is a companion to the RSA chapter, but is not needed to understand the implementation of RSA. The mathematical requirements of the chapter grow as we progress further; the first algorithms are elementary, while the last is the only known modern, provably fast way to determine whether a number is prime. As there are many primality tests and factorization algorithms, there should be a compelling reason behind what we include and what we omit, and there is. For centuries people had unsuccessfully searched for a provably fast primality test; the mathematics community was shocked when Agrawal, Kayal, and Saxena found just such an algorithm. Our goal is not to prove why their algorithm works, but instead to explain the ideas and notation so that the interested reader can pick up the paper and follow the proof, as well as to remind the reader that just because a problem seems hard or impossible does not mean that it is! As much of cryptography is built around the assumption of the difficulty of solving certain problems, this is a lesson worth learning well.

Chapters 1–5 and 10 can be covered as a one semester course in mathematics for liberal arts or criminal justice majors, with little or no mathematics background. If time permits, parts of Chapters 9 and 11 can be included or sections from the RSA chapters (Chapters 7 and 8). For a semester course for mathematics, science, or engineering majors, most of the chapters can be covered in a week or two, which allows a variety of options to supplement the core material from the first few chapters.

A natural choice is to build the semester with the intention of describing RSA in complete detail and then supplementing as time allows with topics from Chapters 9 and 11. Depending on the length of the semester, some of the classical ciphers can safely be omitted (such as the permutation and the Hill ciphers), which shortens several of the first few chapters and lessens the mathematical prerequisites. Other options are to skip either the Enigma/Ultra chapter (Chapter 3) or the symmetric encryption chapter (Chapter 6) to have more time for other topics. Chapters 1 and 10 are less mathematical. These are meant to provide a broad overview of the past, present, and future of the subject and are thus good chapters for all to read.

Cryptography is a wonderful subject with lots of great applications. It's a terrific way to motivate some great mathematics. We hope you enjoy the journey ahead, and we end with some advice:

- `Wzr fdq nhhs d vhfuhw li rqh lv ghdg.`
- `Zh fdq idfwru wkh qxpehu iliwhhq zlwk txdqwxp frpsxwhuv. Zh fdq dovr idfwru wkh qxpehu iliwhhq zlwk d grj wudlqhg wr edun wkuhh wlphv.`
- `Jlyh xv wkh wrrov dqg zh zloo ilqlvk wkh mre.`

Acknowledgments

This book is the outgrowth of introductory cryptography courses for non-math majors taught at Rutgers University and Williams College. It is a pleasure to thank our colleagues and our students for many helpful conversations that have greatly improved the exposition and guided the emphasis, in particular Elliot Schrock (who helped write the Enigma chapter) and Zane Martin and Qiao Zhang (who were the TAs for the 2013 iteration at Williams College, Math 10: `Lqwurgxfwlrq wr Fubswrjudskb`, and who helped guide the class in writing the solutions manual for teachers).

We are especially grateful to Wesley Pegden, who generously shared his notes from versions he taught at Rutgers, and who provided numerous, detailed comments. We also thank our editor, Ed Dunne, for all his help, comments and advice throughout the project, Barbara Beeton and Jennifer Sharp at the AMS for their excellent work in creating the final version of the text, Teresa Levy for designing the cover, and the anonymous referees for their suggestions. We wish to thank a number of people who have read and commented on the book, including especially Katrina Palmer at Appalachian State University, Andrew Baxter at Penn State, and Robert Wilson at Rutgers University. Several members of the NSA community kindly shared general references as well as allowed one of the authors to use an Enigma machine.

Some of this book was inspired by work done with the CCICADA Center at Rutgers, a Department of Homeland Security University Center. Miller was partially supported by NSF grants DMS0600848 and DMS0970067 during the writing of this book, and gratefully acknowledges their support.

Chapter 1

Historical Introduction

Cryptology, the process of concealing messages, has been used for the last 4,000 years. It started at least as long ago as the Egyptians, and continues today and into the foreseeable future. The term **cryptology** is from the Greek $\kappa\rho\upsilon\pi\tau\omega$ or **kryptós**, meaning secret or hidden, and $\lambda o\gamma o\zeta$ or **logós**, meaning science. The term cryptology has come to encompass **encryption** (**cryptography**, which conceals a message) and **decryption** (revelation by **cryptanalysis**).

In this chapter we give a quick introduction to the main terms and goals of cryptology. Our intent here is not to delve deeply into the mathematics; we'll do that in later chapters. Instead, the purpose here is to give a broad overview using historical examples to motivate the issues and themes. Thus the definitions are less formal than later in the book. As this is a cryptography book, we of course highlight the contributions of the field and individuals in the stories below, though of course this cannot be the entire story. For example, even if you know the enemy's plan of attack, men and women must still meet them in the field of battle and must still fight gallantly. No history can be complete without recalling and appreciating the sacrifices many made.

Below we provide a brief introduction to the history of cryptography; there are many excellent sources (such as [**45**]) which the interested reader can consult for additional details. Later chapters will pick up some of these historical themes as they develop the mathematics of encryption and decryption. This chapter is independent of the rest of the book and is meant to be an entertaining introduction to the subject; the later chapters are mostly mathematical, with a few relevant stories.

For the most part, we only need some elementary number theory and high school algebra to *describe* the problems and techniques. This allows us to cast a beautiful and important theory in accessible terms. It's impossible to live in a technologically complex society without encountering such issues, which range from the obvious (such as military codes and deciphering terrorist intentions) to more subtle ones (such as protecting information for online purchases or scanning purchases at a store to get the correct price

FIGURE 1.1. Hieroglyph on Papyrus of Ani. (Image from Wikipedia Commons.)

and update inventories in real time). After reading this book, you'll have a good idea of the origins of the subject and the problems and the applications. To describe modern attacks in detail is well beyond the scope of the book and requires advanced courses in everything from number theory to quantum mechanics. For further reading about these and related issues, see [**5, 6, 57**].

1.1. Ancient Times

The first practice of cryptography dates at least as far back as ancient Egypt, where scribes recorded various pieces of information as **hieroglyphs** on monuments and tombs to distinguish them from the commonly used characters of the time and give them more importance (see Figure 1.1). These hieroglyphics included symbols and pictures, and were translated by the hierarchy of the country to suit themselves. Thus, the hieroglyphs served the purpose of taking something in writing and masking the text in secrecy. s

The Egyptian hieroglyphs were initially done on stone as carvings and then later on papyrus. The Babylonians and others at about the same time used cuneiform tablets for their writing. One such tablet contained the secret formula for a glaze for pottery, where the figures defining the ingredients were purposefully jumbled so people couldn't steal the secret recipe for the pottery glaze. This is the oldest known surviving example of encryption.

As important as pottery is to some, when cryptography is mentioned people think of spies and military secrets, not pottery glazes. The first documented use of secret writing by spies occurred in India around 500 BCE. The Indians used such techniques as interchanging vowels and consonants, reversing letters and aligning them with one another, and writings placed at odd angles. Women were expected to understand concealed writings as an important skill included in the *Kama Sutra*.

The Old Testament of the Bible includes an account of Daniel. He was a captive of Babylon's King Nebuchadnezzar around 600 BCE and had won promotion with successfully interpreting one of the king's dreams. He saw a message "Mene, Mene, Tekel, Parsin" written on a wall (Daniel 5:5–28) and interpreted it as *Mene* meaning "God Hath numbered thy kingdom and finished it"; *Tekel* as "Thou art weighed in the balances and art found wanting"; and *Parsin* as "Thy kingdom is divided and given to the Medes and Persians". The king was killed that very night and Babylon fell to the Persians. Other passages of the Old Testament allude to passwords required for entry into various places. Very few knew the passwords, or keys as they were often called.

As time progressed and conflict became more prevalent and important to the spread of boundaries, the need for concealed messages grew. This was also at a time when written records began to be collected. Both the Greeks and the Persians used simple encryption techniques to convey battle plans to their troops in the fifth century BCE. For example, one technique was to wrap a missive written on parchment around rods of specific sizes and with writing down the length of the rod. When unwrapped the letters were not in the right order, but wound around the right size rod they were. Another example is the Greek use of wooden tablets covered with wax to make them appear blank (a steganographic technique discussed in Chapter 10), which were then decrypted by melting the wax to expose the written letters.

Various transmission systems were developed to send messages in the period between 400 and 300 BCE, including the use of fire signals for navigation around enemy lines. Polybius, the historian and cryptographer, advanced signaling and cipher-making based on an idea of the philosopher Democritus. He used various torch signals to represent the letters of the Greek alphabet, and he created a true alphabet-based system based on a 5×5 grid, called the Polybius checkerboard. This is the first known system to transform numbers to an alphabet, which was easy to use. Table 1.1 shows a Polybius checkerboard (note that `i` and `j` are indistinguishable):

Each letter is coded by its row and column in that order; for example `s` is coded as 43. The word "spy" would be coded by 43, 35, 54, while "Abe is a spy" is 11, 12, 15, 24, 43, 11, 43, 35, 54. It's easy to decode: all we have to

TABLE 1.1. The Polybius checkerboard.

	1	2	3	4	5
1	a	b	c	d	e
2	f	g	h	ij	k
3	l	m	n	o	p
4	q	r	s	t	u
5	v	w	x	y	z

do is look in the appropriate table entry to get the letter (remembering, of course, that 24 can be either an `i` or a `j`). For example, 22, 42, 15, 15, 43, 25, 11, 42, 15, 13, 34, 32, 24, 33, 22 decodes to either "Greeks are coming" or "Greeks are comjng"; it's clear from context that the first phrase is what's meant.

A **cipher** is a method of concealment in which the primary unit is a letter. Letters in a message are replaced by other letters, numbers, or symbols, or they are moved around to hide the order of the letters. The word **cipher** is derived from the Arabic **sifr**, meaning nothing, and it dates back to the seventh century BCE. We also use the word code, often interchangeably with cipher, though there are differences. A **code**, from the Latin **codex**, is a method of concealment that uses words, numbers, or syllables to replace original words or phases. Codes were not used until much later. As the Arabic culture spread throughout much of the western world during this time, mathematics flourished and so too did secret writing and decryption. This is when frequency analysis was first used to break ciphers (messages). **Frequency analysis** uses the frequency of letters in an alphabet as a way of guessing what the cipher is. For example, `e` and `t` are the two most commonly used letters in English, whereas `a` and `k` are the two most commonly used letters in Arabic. Thus, "native language" makes a difference. Chapters 4 and 5 include many examples of how frequency analysis can decrypt messages.

Abu Yusef Ya'qab ibn 'Ishaq as-Sabbah al-Kindi (Alkindus to contemporary Europeans) was a Muslim mathematician, who lived in what is now modern day Iraq between 801 and 873 AD. He was a prolific philosopher and mathematician and was known by his contemporaries as "the Second Teacher", the first one being Aristotle [**55**]. An early introduction to work at the House of Wisdom, the intellectual hub of the Golden Age of Islam, brought him into contact with thousands of historical documents that were to be translated into Arabic, setting him on a path of scientific inquiry few were exposed to in that time [**46**].

Al-Kindi was the first known mathematician to develop and utilize the **frequency attack**, a way of decrypting messages based on the relative rarity of letters in a given language. The total of his work in this field was published in his work *On Deciphering Cryptographic Messages* in 750 AD, one of over 290 texts published in his lifetime [**50**]. This book forms the first mention of cryptology in an empirical sense, predating all other known references by 300 years [**28**]. The focus of this work was the application of probability theory (predating Fermat and Pascal by nearly 800 years!) to letters, and is now called **frequency analysis** [**41**].

The roots of al-Kindi's insight into frequency analysis began while he was studying the Koran. Theologians at the time had been trying to piece together the exact order in which the Koran was assembled by counting the number of certain words in each *sura*. After continual examination it became clear that a few words appeared much more often in comparison to the rest and, after even closer study in phonetics, it became more evident that letters themselves appeared at set frequencies also. In his treatise on cryptanalysis, al-Kindi wrote in [**50**]:

> One way to solve an encrypted message, if we know its language, is to find a different plaintext of the same language long enough to fill one sheet or so, and then we count the occurrences of each letter. We call the most frequently occurring letter the "first", the next most occurring letter the "second", the following most occurring letter the "third", and so on, until we account for all the different letters in the plaintext sample. Then we look at the cipher text we want to solve and we also classify its symbols. We find the most occurring symbol and change it to the form of the "first" letter of the plaintext sample, the next most common symbol is changed to the form of the "second" letter, and the following most common symbol is changed to the form of the "third" letter, and so on, until we account for all symbols of the cryptogram we want to solve.

Interest in and support for cryptology faded away after the fall of the Roman Empire and during the Dark Ages. Suspicion of anything intellectual caused suffering and violence, and intellectualism was often labeled as mysticism or magic. The fourteenth century revival of intellectual interests became the Renaissance, or rebirth, and allowed for the opening and use of the old libraries, which provided access to documents containing the ancient ciphers and their solutions and other secret writings. Secret writing was at first banned in many places, but then restored and supported. **Nomenclators** (from the Latin *nomen* for name and *calator* for caller) were used until the nineteenth century for concealment. These were pairs of letters used to refer to names, words, syllables, and lists of cipher alphabets.

It's easy to create your own nomenclator for your own code. Write a list of the words you most frequently use in correspondence. Create codewords or symbols for each one and record them in a list. Then create an alphabet, which you will use for all the words that are not included in your list. Try

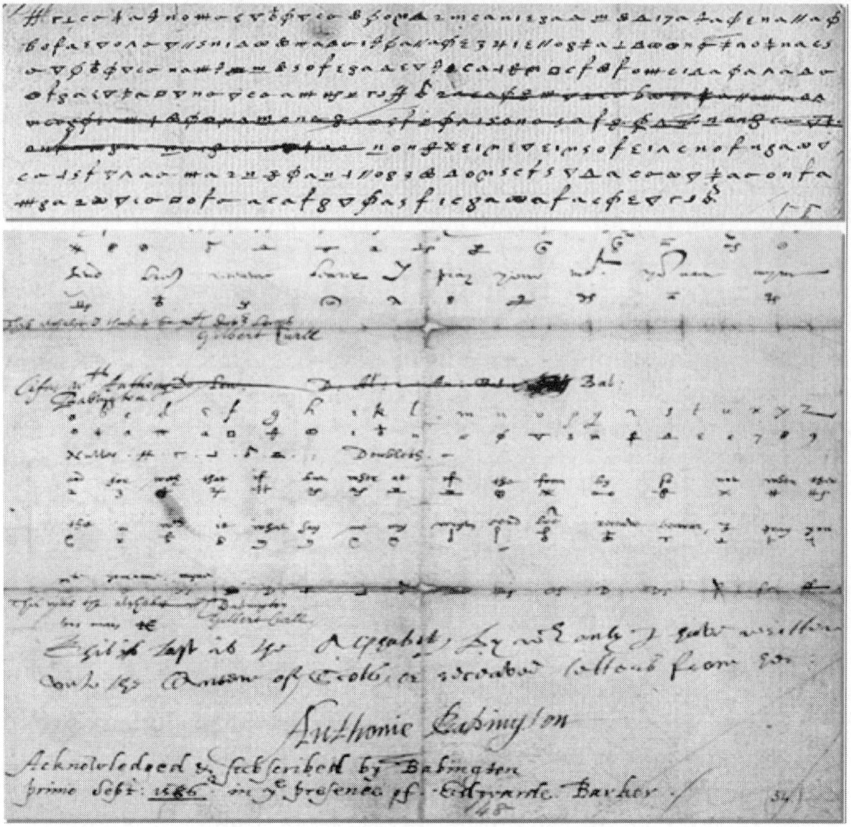

FIGURE 1.2. A forged nomenclator used in the Babington Plot in 1585. (Image from Wikipedia Commons.)

constructing a message to a friend by substituting the codeword for each word in the message that is on your list, and for those not in the list, use the alphabet you created. This should sound quite familiar to those who are used to texting. The difference here is that this uses your own codewords and alphabet, rather than commonly used phrases such as "lol" and "ttyl".

It wasn't until the seventeenth century that the French realized that listing the codewords in alphabetical order as well as the nomenclator alphabet in alphabetical order made the code more readily breakable. Figure 1.2 shows a portion of a fifteenth century nomenclator.

The Renaissance was a period of substantial advances in cryptography by such pioneer cryptographers, mostly mathematicians, as Leon Alberti, Johannes Trithemius, Giovanni Porta, Geirlamo Cardano, and Blaise de Vigenère. Cryptography moved from simple substitutions and the use of symbols to the use of keys (see Chapters 2 to 5) and decryption using probability.

Secrets were kept and divulged to serve many different purposes. Secret messages were passed in many ways, including being wrapped in leather and then placed in a corked tube in the stoppers of beer barrels for Mary Stuart, Queen of Scots. Anthony Babington plotted to kill Queen Elizabeth I. He used beer barrels to conceal his message, telling Mary Stuart of the plot and his intent to place her, Mary, on the throne. He demanded a personal reply. In doing so, Mary implicated herself when the barrels were confiscated long enough to copy the message. They decrypted the message using letter frequency techniques (see Table 4.1 of §4.1). Mary Stuart was subsequently charged with treason and beheaded.

Double agents began to be widespread, especially during the American Revolution. Indeed, the infamous Benedict Arnold used a particular code called a **book code**. Because he was trusted, his correspondence was never checked and thus never tested. Not knowing whether that would continue to be true, he often used invisible ink to further hide his code.

Aaron Burr, who had at one time worked for Arnold, got caught up in his own scandal after Thomas Jefferson was elected president. Burr had been elected vice president, and he was ambitious and wanted to advance to the presidency. Alexander Hamilton learned of a plot to have New England and New York secede and publicly linked Burr to the plot. This led to the famous Hamilton–Burr duel, where Hamilton was killed. People turned against Burr as a result, and he, in turn, developed an elaborate scheme to get rid of Jefferson. The scheme included ciphers to link all of the many parts and people, some from Spain and England. Despite eventual evidence of deciphered messages, Burr was not convicted of treason.

Telegraphy and various ciphers played key roles during the Civil War. The **Stager cipher** was particularly amenable to telegraphy because it was a simple word transposition. The message was written in lines and transcribed using the columns that the lines formed. Secrecy was further secured by throwing in extraneous symbols and creating mazes through the columns. Consider the following simple example:

j	o	e	i	s
a	n	t	t	o
s	o	r	o	n
o	n	a	r	t

Most likely this would be read as “Joe is ant [antithetical] to soron [General Soron] on art”. But the intent is to read it as “Jason traitor”.

Women have always been directly involved in cryptography. An interesting example occurred during the Battle of Bull Run. A woman called Rebel Rose Greenhow sent messages to the Confederate defenders about Union troop movements and numbers. She used everything from pockets hidden in her clothing to coded designs embroidered into her dresses. She was so effective that the Federal authorities began counterespionage missions and tracked leaks to party and parlor gossip. Greenhow’s chief nemesis turned out to be Allan Pinkerton, the famous detective. He eventually trapped her and had her imprisoned; however, even from her cell she managed to create new networks and methods of secret communication. In the end, the cryptographic efforts of the South were not as advanced and effective as those of the North. Despite the variety of codes and ciphers applied during the Civil War, none affected the outcome of the war as much as telegraphy did. Telegraphy and Morse code enabled Grant to use broad strategies on many fronts, contributing to Lee's surrender in 1865.

1.2. Cryptography During the Two World Wars

1.2.1. World War I

Cryptography has played an important role in the outcome of wars. The inadequacy of the cryptographic techniques at the beginning of World War I probably contributed to the loss of early potential Allied victories. Early attempts by the Russians, who far outnumbered the Germans, failed because the Russians sent messages in unprotected text that were picked up by German eavesdroppers, who then foiled the attacks.

The Allies were no better at intelligence gathering. Even though they intercepted a radio message from the German warship, *Goben*, in 1914 and deciphered the message, it was too late to prevent the shelling of Russian ports which ultimately caused Turkey to ally with the Germans. In general, decrypted messages were not generally trusted.

It was the hard work of the military and the intelligence gathering of the Allies that initially brought the plot of Zimmerman to the attention of the U.S. During the First World War, British naval intelligence began intercepting German radio messages. They amassed a group of scholars whose job was to decipher these German communications. With the help of the Allied forces and some good luck, they were able to come across German code books. Armed with their knowledge and hard work, the British cryptographers of what became known as **Room 40** decoded a message, called the **Zimmerman telegram**, from the German Foreign Minister Zimmerman.

It described German plans first sent to the German ambassador in the U.S. and then to the German ambassador in Mexico City. The message indicated that Germany was about to engage in submarine warfare against neutral shipping. Zimmerman, fearing that the U.S. would join England, proposed an alliance with Mexico. If the U.S. and Germany were to go to war with each other, Mexico would join forces with Germany, who would support Mexico regaining the land it lost to America in the Mexican-American War of 1846 to 1848. Room 40 analysts intercepted the telegram, deciphered it, and kept it secret for a while. It was then released to the Associated Press. The exposé shocked the U.S. into joining the war as an ally of the British.

1.2.2. Native Americans and Code Talkers in World War I and II

A group of **Choctaw Indians** were coincidentally assigned to the same battalion early in World War I, at a time when the Germans were wiretapping and listening to conversations whenever and wherever possible. It thus became critically important for the Americans to send coded messages.

As the Choctaws were overheard in conversation in the command posts, officers thought about using the Choctaw native tongue to send coded messages. They tried successfully using the Choctaw language with two battalions and found no surprise attacks. The officials now knew that this linguistic system could work. For the most part these messages were sent as natural communications without additional coding. There were some issues, as some words were not in the Chocktaw vocabulary. This led to codewords being substituted, such as "big gun" for artillery, "stone" for grenade, and "little gun shoot fast" for machine gun. Telephone and radio were the most efficient means of communication, yet were highly susceptible to penetration; however, the use of the Choctaw language baffled the Germans, who were unable to decipher the language or the coded vocabulary. Some coded written messages in Choctaw were given to runners to protect their secrecy from the Germans, who often captured Americans to steal the valuable information.

The most famous group of **code talkers** were the **Navajos**, who were used in the Pacific during World War II (see Figure 1.3). It all began with an older gentleman, a WWI veteran himself, reading a paper on the massive death tolls encountered by the Americans and their efforts to create a safe encryption code. Philip Johnston was a missionary's son who grew up playing with Navajo children and learned their language as a boy. He was perhaps one of merely 30 non-Navajos who could understand their language. He knew that the U.S. had befuddled the Germans in World War I by using Choctaws to transmit messages in their own language on field phones. Thus, in combination with his war experience and with his intricate knowledge of the Navajo language, he realized that this could be the key to an unbreakable code. The Navajo marines and the few others who understood the language trained like all other marines; their desert and rough lifestyle actually benefited them during rigorous training. But in addition they were trained for radio communications and were tasked to create a unique code that would soon be used on the battlefield. Their language was very complex, which helped the security of their encrypted messages. For example, the Navajo language has at least ten different verbs for different kinds of carrying, depending on the shape and physical properties of the thing being carried. Also, depending on the tone or pitch of the speaker's voice, the same word could have a multitude of meanings. Even prefixes can be added to a verb, as many as ten different ones, to the point where one word in Navajo can take the place of a whole sentence in English.

Although their language seemed quite uninterpretable in its natural form, they took it a step further. To further encrypt the messages, they created the code that would be utilized on the front lines. The Navajo code initially consisted of a 234-word vocabulary, which over the course of WWII grew to some 450 words. Some military terms not found in the Navajo

FIGURE 1.3. Newton's Photograph from the Smithsonian Exhibit on American Indian Code Talkers. `http://www.sites.si.edu/images/exhibits/Code\%20Talkers/pages/privates_jpg.htm`

language were given specific code names, while others were spelled out. For example, "dive bomber" became "*gini*" (the Navajo word for chicken hawk). Even when they would spell words out, the word remained complex. Each English letter was assigned a corresponding English word to represent it and then that word was translated into Navajo. For example, z became "zinc" which then became "*besh-do-gliz*", and those letters that were frequently used were given three word variations so that a pattern, if decrypted by the enemy, could not easily be found. As an indication of its complexity, consider the code in a message sent in 1944: "*A-woh Tkin Ts-a Yeh-hes Wola-chee A-chen Al-tah-je-jay Khut*", which translated means, "Tooth Ice Needle Itch Ant Nose Attack Ready or now" corresponding to the decrypted message, TINIAN Attack Ready.

The Navajo code talkers could take a three-line English message and encode, transmit, and decode it in twenty seconds. A machine would take thirty minutes. Their unique skills were an important asset in the victories in WWII. Some Japanese thought it could be a tribal language, and there were cases where Navajo soldiers in POW camps were tortured and forced to listen to these encrypted messages. But all they could tell was that it was just incoherent jumbled words in Navajo. In order to decode the transmission, one had to be fluent in English, Navajo, and know the secret code. It was never broken, and it wasn't until 1968 that the existence of these codes was released to the public, only after they had become obsolete.

1.2.3. World War II

Winston Churchill became Prime Minister of Great Britain seven months after the start of World War II. As a communications, intelligence, and security specialist in World War I, he was very aware of the importance of breaking German codes and ciphers. To respond to this need, he created a small group of decryption specialists, along with the Government Code and Cipher School at **Bletchley Park**, an estate 45 miles outside of London. Other linguists and mathematicians joined them in subsequent months to break the German encryptions, especially those generated by the **Enigma**. The Enigma, a rotor-based encryption device developed by the Germans, had the potential to create an immense number of electrically generated alphabets. Bletchley staff gave the code name **Ultra** to their deciphering efforts. Ultra was helped by French and Polish sources who had access to the Enigma's workings. The whole of Chapter 3 is devoted to the Enigma and the Ultra efforts.

The U.S. isolationist policies after World War I directed people away from the warning signs of trouble overseas, including some missed opportunities to detect the bombing of Pearl Harbor in December 1941. U.S. cryptographic units were blamed for not reading the signs. The Hypo Center in Hawaii did not have the decipherments of the "J" series of transposition ciphers used by Japan's consulate, despite the fact that one of the Japanese consulates was very near the U.S. naval base at Pearl Harbor. Had the Navy had access to the messages at the Hypo Center, history might have been different. In addition, the information filtering through the cryptoanalysts from the Japanese cipher machine **Purple** was not disseminated widely. They had broken the cipher, Red, from one of the Japanese cipher machines, but Purple was a complicated polyalphabetic machine that could encipher English letters and create substitutions numbering in the hundreds.

Dorothy Edgars, a former resident of Japan and an American linguist and Japanese specialist, noticed something significant in one of the decrypted messages put on her desk and mentioned it to her superior. He, however, was working on the decryption of messages from Purple and ignored her. She had actually found what is called the "lights message", a cable from the Japanese consul in Hawaii to Tokyo concerning an agent in Pearl Harbor, and the use of light signals on the beach sent to a Japanese submarine. After the shocking losses at Pearl Harbor, the U.S. leaders no longer put their faith in an honor code where ambassadors politely overlooked each other's communications. The U.S. went to war once again.

Naval battles became paramount, and cryptoanalysts played a key role in determining the locations of Tokyo's naval and air squadrons. The Navy relied heavily on Australian cryptoanalysts who knew the geography best.

General Douglas MacArthur commanded an Allied Intelligence Unit formed from Australian, British, Dutch, and U.S. units. They contributed to decisive Allied victories by successfully discovering Japan's critical military locations and their intended battles, such as Midway.

Traitors and counterespionage efforts continued to exist through the rest of the war. For example, the attaché Frank Fellers gave too-frequent and detailed reports about the British actions in North Africa, and German eavesdroppers snared various reports, reencrypted them and distributed them to Rommel. However, Fellers' activities were discovered, and Rommel was ultimately defeated after this source of information ceased.

Another aspect of cryptography is misdirection. The end of World War II was expedited through the transmission of codes and ciphers intended to be intercepted by German intelligence. Various tricks were employed to communicate false information and mislead them into believing something else was going on. They even had vessels sent to these bogus locations to give the appearance of an impending battle. We'll discuss some of these in greater detail in Chapter 3.

1.3. Postwar Cryptography, Computers, and Security

After World War II came the Cold War, which many feared could flare into an active war between the Soviets and the U.S. and her allies. It was a time of spies and counterspies, and people who played both sides of the fence. The damage to U.S. intelligence from activities of people like Andrew Lee and Christopher Boyce, the Falcon and the Snowman, was irreparable. They sold vital information to Soviet agents in California and Mexico, including top-secret cipher lists and satellite reconnaissance data in the 1970s. As a result, the Russians began protecting their launches and ballistic missile tests with better encrypted telemetry signals.

Another spy operated in the 1980s, John Walker. He was a Navy radio operator who used the KL-47, a mainstay of naval communications. It was an electronic rotor machine more advanced than the Enigma machine. He provided the Russians with wiring diagrams, and they were able to reconstruct the circuitry and determine with computer searches the millions of possible encrypted variations and read the encrypted messages.

Jewels was the codename for the carefully guarded cipher machines in Moscow used by the CIA and NSA cipher clerks. Many precautions were taken to protect the computer's CPU, and the cipher machines were state of the art with key numbers and magnetic strips that changed daily. Messages were double encrypted; however the Soviets managed to "clean" the power line to the machines so that electronic filters could be bypassed. The results of the subsequent leaks revealed many CIA agents who were then expelled, as well as revealing U.S. negotiating positions.

One of the more famous recent spies was identified in 1994 as Aldrich Ames, a CIA analyst, whose father Carleton had also been a CIA counterspy in the 1950s. Aldridge Ames had been divulging secrets for at least ten years and had been in contact with many Russians as a CIA recruiter. He applied cryptographic techniques to conceal his schemes, some as simple as B meaning meet in Bogota, Columbia, while others involved a series of chalk-marked mailboxes with codenames like "north" and "smile", signaling brief commands like "travel on". At the time of this writing, he is serving a life sentence in prison for treason.

Cryptology continued to use codes and ciphers but was intensified, and it became more sophisticated with the improvements in computer technology. Horse Feistel of IBM in the 1970s developed a process of computer enhanced transposition of numbers using binary digits. It began as a demonstration cipher. Known as **Demon**, and then **Lucifer**, this DES cipher is a complicated encrypting procedure built upon groups of 64 plaintext bits, six of which were parity bits to guarantee accuracy. Simultaneously, Professor Martin Hellman and students Whitfield Diffie and Ralph Merkle collaborated to present the public key as a solution to the problem of distributing individual keys. This system had a primary basis of two keys. One was published and the other was kept private (see §8.5). For a while this system proved unbreakable, but in 1982 a trio of mathematicians from MIT broke it. They, Leonard Adleman, Ronald Rivest, and Adi Shamir, created another two-key procedure based on prime numbers. Their public key version is called **RSA**, and it is discussed in Chapter 8. RSA is slower to implement than DES because of its many computations, but is useful in networks where there are many communicants and the exchange of keys is a problem.

Today, matters of security are ever present as Social Security numbers, bank account numbers, employment data, and others are digitized on a daily basis. Some of the alphanumeric components used include door openers, passwords, health plan numbers, PIN numbers, and many more. Even though these are not intended as encryptions, they are nonetheless to be kept hidden for privacy and security reasons. The U.S. government became obsessed with a system developed in the 1990's called **Pretty Good Privacy (PGP)** for email, because they could not access emails when they thought they needed to. PGP has since been replaced by a system not nearly as good. A system called **key escrow** involved sending and receiving equipment that electronically chose algorithms from millions of available keys to encrypt conversations or data exchanges. The keys were to be held by two secure agencies of the federal government and required court-approved permission to access. It never gained public approval.

As computer technology improves, new codes and ciphers are developed for encryption, and attempts are made at decryption, often successfully. In some cases, old techniques, such as **steganography**, are made even better. Steganography is the technique of passing a message in a way that even the existence of the message is unknown. The term is derived from the Greek *steganos* (which means covered) and *graphein* (to write). In the past, it was often used interchangeably with cryptography, but by 1967 it became

FIGURE 1.4. An embedded digital image that says "Boss says we should blow up the bridge".

used exclusively to describe processes that conceal the presence of a secret message, which may or may not be additionally protected by a cipher or code. The content of the message is not altered through the process of disguising it. The use of wax tablets discussed in §1.1 is an example of ancient steganography. Modern steganography, discussed in Chapter 10, not only conceals the content of messages, but hides them in plain sight in digital images, music, and other digitized media. The computer has provided a modern day invisible ink as these messages are not discernable by the naked eye or ear (see Figure 1.4).

Quantum computing has made **quantum cryptography** possible. Quantum cryptography uses quantum mechanical effects, in particular in quantum communication and computation, to perform encryption and decryption tasks. One of the earliest and best known uses of quantum cryptography is in the exchange of a key, called **quantum key distribution**. Earlier cryptology used mathematical theorems to protect the keys to messages from possible eavesdroppers, such as the RSA key encryption system discussed in Chapter 8. The advantage of quantum cryptography is that it allows fast completion of various tasks that are seemingly impractical using only classical methods, and it holds forth the possibility of algorithms to do the seemingly impossible, though so far such algorithms have not been found. Chapter 10 includes a longer discussion of quantum cryptography and the mathematics and physics behind it.

1.4. Summary

In this chapter we encountered many of the issues and key ideas of the subject (see [**12**] for an entertaining history of the subject). The first are various reasons requiring information protection. The case of Mary Stuart, Queen of Scots, and Anthony Babington show the grave consequences when ciphers are broken. While the effects here are confined to individuals, in

Chapter 3 we'll see similar issues on a larger scale when we explore Enigma and Ultra.

Another important takeaway is the need for speed and efficiency. In a battle situation, one does not have thirty minutes to leisurely communicate with headquarters. Decisions need to be made in real time. It's precisely for such reasons that the Navajo code talkers played such an important role, as they allowed U.S. units the ability to quickly communicate under fire. Of course, this code was designed to communicate very specific information. In modern applications we have a very rich set of information we want to encode and protect, and it becomes critically important to have efficient ways to both encrypt and decrypt.

Another theme, which will play a central role throughout much of this book, is replacing message text with numbers. We saw a simple recipe in the work of Polybius; we'll see more involved methods later. The monumental advances in the subject allow us to use advanced mathematical methods and results in cryptography.

We end with one last comment. Though there are many threads which we'll pursue later, an absolutely essential point comes from the Soviet efforts to read our ciphers. Even though the cipher machines in Moscow used double encryption, the Soviets were able to circumvent electronic filters by "cleaning" the power lines. This story serves as a powerful warning: in cryptography you have to defend against all possible attacks, and not just the expected ones. We'll see several schemes that appear safe and secure, only to see how a little more mathematics and a different method of attack are able to quickly break them.

1.5. Problems

EXERCISE 1.5.1. *Use the Polybius checkerboard to encode:*
(a) Men coming from the south.
(b) King has called a cease fire.

EXERCISE 1.5.2. *Use the Polybius checkerboard to encode:*
(a) Fire when ready.
(b) Luke, I am your father.

EXERCISE 1.5.3. *Use the Polybius checkerboard to decode:*
(a) $13, 54, 13, 32, 11, 14, 15, 44$.
(b) $33, 45, 35, 32, 54, 33, 41, 51, 44$.
(c) $23, 15, 32, 15, 34, 35, 21, 45, 43, 35, 54$.

EXERCISE 1.5.4. *Use the Polybius checkerboard to decode:*
(a) $43, 44, 15, 11, 31, 23, 34, 32, 15$.
(b) $35, 34, 31, 54, 12, 24, 45, 43$.

EXERCISE 1.5.5. *Use the Polybius checkerboard to decode*
$23, 22, 22, 22, 33, 25, 43$.

EXERCISE 1.5.6. *Come up with two messages that encrypt to the same text under the Polybius checkerboard but have different meanings; each message should make sense. Note there are not too many possibilities as almost all letters have a unique decryption.*

EXERCISE 1.5.7. *One difficulty in using the Polybius checkerboard is that it only has* 25 *squares, but there are* 26 *letters in the English alphabet. Show how we can overcome this by either increasing the size of the board or by considering a cube. What is the smallest cube that works?*

EXERCISE 1.5.8. *Create a nomenclator code book and alphabet, and use it to encrypt the message: "Meet me at our favorite restaurant at* 6 *PM."*

EXERCISE 1.5.9. *Using a Stager cipher, encode the message "Do you believe in miracles?"*

EXERCISE 1.5.10. *Using a Stager cipher, encode the message "It was early spring, warm and sultry glowed the afternoon." (*Note: *there is an interesting history to this quote, which can be uncovered by an internet search.)*

EXERCISE 1.5.11. *The following message was encrypted with a Stager cipher; what is it?*

d	*f*	*n*	*o*	*t*	*i*	*f*	*r*
o	*i*	*t*	*u*	*h*	*t*	*t*	*e*
n	*r*	*i*	*s*	*e*	*e*	*h*	*y*
o	*e*	*l*	*e*	*w*	*s*	*e*	*e*
t	*u*	*y*	*e*	*h*	*o*	*i*	*s*

EXERCISE 1.5.12. *The following message was encrypted with a Stager cipher; what is it?*

a	*s*	*l*	*l*	*u*	*p*	*o*
s	*i*	*i*	*l*	*t*	*t*	*r*
k	*x*	*o*	*a*	*a*	*f*	*m*
f	*m*	*n*	*r*	*c*	*o*	*o*
o	*i*	*d*	*s*	*c*	*u*	*r*
r	*l*	*o*	*b*	*e*	*r*	*e*

EXERCISE 1.5.13. *Deciphering the code in Exercises* 1.5.11 *and* 1.5.12 *is fairly easy* if *you know to read it in columns. We can increase the security by hiding the number of columns and writing it as* `d f n o t i f r o i t u h t t e n r i s e e h y o e l e w s e e t u y e h o i s`. *While this initially masks the number of columns, assuming we have at least two columns and at least two rows, show there are only six possibilities.*

EXERCISE 1.5.14. *Suppose your friend is considering encrypting a message to you through a Stager cipher. Having done Exercise* 1.5.13, *she knows that it would be a mistake to write the message in columns, as then it can be readily deciphered. She therefore decides to write it as a string of letters, and only the two of you will know the number of rows and columns. If there* *are r rows and c columns, this means she can send a message of rc letters. In terms of security and frustrating an attacker, which of the following is the best choice for rc and why:* $1331, 1369, 1800$, *or* 10201?

EXERCISE 1.5.15. *Research and write a brief description about one of the following:*

- *The Black Chamber.*
- *The technological treason in the Falcon and the Snowman case.*
- *Cryptography during Prohibition and the role of Elizabeth Smith Friedman.*
- *Echelon.*
- *The Kryptos sculpture at the NSA.*

EXERCISE 1.5.16. *A major theme of this book is the need to do computations quickly. The Babylonians worked base* 60*; this meant they needed to know multiplication tables from* 0×0 *all the way to* 59×59*, far more than we learn today (since we work base* 10*, we only go up to* 9×9*).*

(a) Calculate how many multiplications Babylonians must memorize or write down.

(b) The number in part (a) can almost be cut in half, as $xy = yx$*. Using this observation, how many multiplications must be memorized or written down?*

(c) As it is painful and expensive to lug clay tablets around, there was a pressing need to trim these numbers as much as possible. The Babylonians made the remarkable observation that

$$xy \ = \ \frac{(x+y)^2 - x^2 - y^2}{2}.$$

Show this formula is true, and thus reduces multiplication to squaring, subtracting, and division by 2*.*

REMARK. The above formula shows that the Babylonians need only learn the squares and can deduce the remaining products by elementary operations. This is an early example of a "look-up table", where some calculations are done and stored, then used to deduce the rest. This exercise shows that the *standard* way to do a problem is sometimes not the most practical.

Chapter 2

Classical Cryptology: Methods

In this chapter we explore in great detail one of the oldest methods of encryption, the Caesar cipher, and some of its descendants. This chapter is mostly devoted to describing how to use these methods; we discuss how to attack and break them in Chapters 4 and 5.

While this ordering of topics provides a nice way to mix a historical tour with an introduction to the needed mathematics, we could have instead presented each method and then immediately discussed attacks on it, and then moved on to a new method designed in response to these attacks. We chose this approach for several reasons. First, even if we are not aware of a successful attack on our system it seems natural to search for more and more complicated encryption methods. While there is not a perfect correlation between the size of the keyspace and the difficulty of cracking the code, there is frequently a relation and thus there is a compelling motivation to search for more complex systems. Second, this allows us to introduce the new mathematical ideas when we describe the methods and then revisit them when we discuss the attacks. When learning new concepts and material, it often helps to see the material again and again, from slightly different perspectives.

Remarkably, variants of a system that began almost 2000 years ago are still in use today, with many of the changes due to the availability of cheap and powerful computing. The main idea of many of these methods is a **letter swap**, where we're given rules to replace each letter with another letter, or possibly blocks of letters with another block of letters.

These encryptions are very easy to implement but face two serious drawbacks. First, many of these schemes are vulnerable to certain types of attack. Sometimes these issues can be addressed by adding layer upon layer of complication, but there's always the danger of forgetting to defend against an approach. We'll see a spectacular failure (or success, depending on whether or not you're trying to successfully encrypt a message or break a code) when we discuss Enigma and Ultra in Chapter 3. The Enigma is one of the most famous encryption methods of all time. It was used by the Germans in World War II. If done correctly, it should have offered complete security and dazzling efficiency. It was not. Ultra, the successful Allied effort to crack the code, was instrumental in the defeat of the Nazis; estimates on its worth range from shortening the war by two years to preventing the Axis powers from triumphing.

The second disadvantage of these methods is that they require the two parties to meet and agree upon a secret ahead of time. In other words, there must be a code book or some meeting where the encryption scheme is agreed upon. There are many drawbacks with this, ranging from complete disaster if a code book is captured by the enemy to having to know ahead of time who might need to communicate with whom and arranging for all those secrets to be exchanged (and remembered!). For example, with millions of consumers making constant online transactions, individual meetings just aren't feasible. Fortunately there are ways to exchange secrets in public; in fact, the ability to exchange secrets in public allows us to use a modification of these early ciphers with enormous security. We discuss these issues in Chapter 8.

2.1. Ancient Cryptography

For thousands of years, individuals and groups in all civilizations have needed ways to secretly and securely transmit information. Many ingenuous ways were found. Since this is a *mathematics* book, we concentrate on mathematical methods, but we would be remiss if we didn't at least mention some methods from the ancient world. A particularly clever one is due to the Greeks. All it required was one slave, a barber, and some ink. The slave's head would be shaved, providing a surface for the message. One then writes the message, waits for the hair to grow back, and then sends the slave off! This particular type of cryptography is called **steganography**, where we hide even the existence of the message. We discuss this in great detail in Chapter 10.

This is a terrific example for illustrating the key issues in **cryptography**, the science of encoding, securely transmitting, and decoding messages, often in the presence of attackers who are trying to intercept and decipher the message. Here are some of the major goals of cryptography, ones which make a valid encryption scheme. As you read below, try and think if there are any others.

We call an encryption scheme `valid` if it satisfies the the following properties:

- It should be easy to **encrypt** the message. Encryption is the process of converting the original message to a new message, which should be unreadable by anyone except the intended recipient (note the sender doesn't have to be able to read the enciphered message!)
- It should be easy to **transmit** the message. We need to quickly and correctly get the message from the sender to the recipient.
- It should be easy to **decode** the message. Once the message arrives, it shouldn't be hard to figure out what it is.
- If someone intercepts or eavesdrops on the message, it should be very hard for them to **decipher** it.

So, how does the example of the slave-barber method do relative to the described properties?

For the most part, it's easy to encrypt. There are some limitations due to the amount of space on a head, but it's pretty easy to shave and write. Similarly, decryption is easy—it's just a haircut away!

The problem with this method, of course, is that it fails the other two points. If the slave is intercepted, it's very easy to get the message. No special knowledge is required. Also, transmission is horrible. Hairfinder.com states that on average hair grows about 1.25cm or half an inch per month. Thus it's a very slow process, as it can take a long time for enough hair to grow back to cover the message. Further, we now have to send the slave to the intended recipient, which could involve a perilous journey.

There's another issue with the slave-barber method. Imagine you're the commander of a city allied with Athens in the Peloponnesian War, and you're surrounded by a vastly superior Spartan force. Do you fight to the death, leading to the probable slaughter of everyone, or do you surrender as your position is untenable? You ask the leaders in Athens for advice. A slave soon arrives. Yes, soon. Athens was intelligent and shaved many heads months ago, and wrote a variety of messages so they'd be ready at a moment's notice. You shave the head, and read that your position is not essential to the war effort, and there is no dishonor in surrendering. You send emissaries to the sieging Spartans, and soon thereafter you open your gates. The problem is, unbeknownst to you, the Spartans intercepted the Athenian slave, and replaced him with one of their own with a very different message!

If we're Athens, we've just learned an important lesson the hard way: we need a way to verify the **authenticity** of a message. In other words, we need to add a **signature** to our message to convince the recipient that the message truly does come from us. Thus, we need to add the following to our goals for a good cryptosystem:

An additional property of a `valid encryption scheme`:

- The source of the message must be easily verifiable. This means a third party cannot replace the intended message with their own and convince the receiver of its legitimacy.

Though authenticity is important, we won't discuss it again until Chapter 8, as none of the classical systems we study allow the recipient to verify the sender's identity. This is a severe defect. Fortunately there are other systems that are verifiable; if there weren't, e-commerce would be crippled.

Returning to our slave-barber system, clearly we need a better system! Writing on people just won't scale to our times in our fast-paced world, although, in fairness to it, there is a terrific feature which is easy to overlook (see Exercise 2.11.1). Before we can describe these better methods, we first need to set some notation.

Plaintext and ciphertext:

- **Plaintext:** The plaintext is the message we wish to send. For example, it might be `DO NOT FIRE UNTIL YOU SEE THE WHITES OF THEIR EYES`.
- **Ciphertext:** The ciphertext is the result of encrypting the plaintext and what we transmit to the recipient. For example, the above message might be encrypted to `QM HFN YOVV MJBTW GOG KRC NYY PNMKWO WQ EPEUJ RWYJ`.

2.2. Substitution Alphabet Ciphers

More than 2000 years ago, the military secrets of the Roman empire were protected with the help of cryptography. The **Caesar cipher**, as it's now called, was used by Julius Caesar to encrypt messages by "shifting" letters alphabetically.

Before we describe the mechanics of the Caesar cipher, it's worthwhile to place it in the context of **substitution alphabet ciphers**. We first set some terminology.

Substitution alphabet ciphers: In a substitution alphabet cipher, each letter of the alphabet is sent to another letter, and no two letters are sent to the same letter.

The way these ciphers work is that they permute the alphabet. One way to record a permutation is to write two rows. The first row is the alphabet, and the second row is what each letter is mapped to. For example, one possibility is

```
A B C D E F G H I J K L M N O P Q R S T U V W X Y Z
↕ ↕ ↕ ↕ ↕ ↕ ↕ ↕ ↕ ↕ ↕ ↕ ↕ ↕ ↕ ↕ ↕ ↕ ↕ ↕ ↕ ↕ ↕ ↕ ↕ ↕
Q W E R T Y U I O P A S D F G H J K L Z X C V B N M
```

(this choice was inspired by looking at a keyboard). If we want to encode a word, we see what letter is below our alphabet row, and that tells us what each letter is sent to. For example, to encode the word `cat`, we would see that `c` is sent to `e`, `a` is sent to `q`, and `t` is sent to `z`. Thus, we would send the word `eqz`. To decode, all we do is reverse direction. We start off with the code row and look and see what letter is above. Thus, if we receive the word `eqtlqk`, then the intended message was the word `caesar`.

How many alphabet codes are there? It's natural to guess there are 26^{26} possibilities, as there are 26 choices for each letter; however, not all of these choices are *simultaneously* possible. Remember, no two letters may be sent to the same letter. Thus, once we determine that `A` will be sent to `Q`, then there are only 25 choices remaining for `B`. If we then choose to send `B` to `W`, there are now only 24 choices left for `C`, and so on. The answer is therefore not 26^{26}, but rather

$$26 \cdot 25 \cdot 24 \cdot 23 \cdot 22 \cdot 21 \cdot 20 \cdot 19 \cdot 18 \cdot 17 \cdot 16 \cdot 15 \cdot 14 \cdot 13 \cdot 12 \cdot 11 \cdot 10 \cdot 9 \cdot 8 \cdot 7 \cdot 6 \cdot 5 \cdot 4 \cdot 3 \cdot 2 \cdot 1.$$

In the interest of brevity, we define the **factorial function** to compactly express such products. The factorial function is a specific example of a recursive function. We use the term **recursion** or **recursive** on and off in this book. Formally, we define **recursion** as a method where the solution to a problem depends on solutions to smaller instances of the same problem, or for functions when $f(n+1)$ depends on the values $f(1)$ to $f(n)$ for n a positive integer. For example $f(n+1) = 2f(n)$ and $f(1) = 10$ implies that the iterations of the function are 10, 20, 40, 80, ... Another example is the Fibonacci Sequence, obtained by taking $f(n+1) = f(n) + f(n-1)$ and $f(1) = 1$ and $f(2) = 1$.

The factorial function: Let n be a nonnegative integer. We define the factorial of n, written $n!$, recursively by setting $0! = 1$ and $n! = n \cdot (n-1)$ for $n \geq 1$. Thus $2! = 2 \cdot 1 = 2$, $3! = 3 \cdot 2 \cdot 1 = 6$, and in general $n! = n \cdot (n-1) \cdots 2 \cdot 1$.

The factorial function has a nice combinatorial interpretation: $n!$ is the number of ways to order n objects when order matters. We may interpret $0! = 1$ as saying there is one way to order no elements! We'll meet the factorial function again when we discuss the Enigma in Chapter 3 (see in particular §3.2).

If we plug 26! into a computer or calculator, we see it's approximately $4.03 \cdot 10^{26}$. This means that there are over 10^{26} *distinct* ciphers that can be created simply by switching the order of the alphabet. This is both a boon and a curse. A clear benefit is that we have a huge number of possible encryption schemes, but the cost is that we and our intended recipient have to find a way to agree upon the choice. Typically this means agreeing to a reordering of the alphabet, which requires agents and operatives to memorize a string of 26 letters. In the next section we discuss the Caesar cipher, which is a simple case of the substitution alphabet cipher and only requires participants to remember one letter!

2.3. The Caesar Cipher

In the last section we discussed substitution alphabet ciphers. Each of the approximately 10^{26} possible alphabet swaps are uniquely determined by a string of 26 letters, namely how we reorder the alphabet. As it's not easy to remember the agreed-upon ordering of the 26 letters, Julius Caesar always used a simple rule: "shift" all letters by 3. This way, people only needed to remember one piece of information: the shift.

A	B	C	D	E	F	G	H	I	J	K	L	M	N	O	P	Q	R	S	T	U	V	W	X	Y	Z
↕	↕	↕	↕	↕	↕	↕	↕	↕	↕	↕	↕	↕	↕	↕	↕	↕	↕	↕	↕	↕	↕	↕	↕	↕	↕
D	E	F	G	H	I	J	K	L	M	N	O	P	Q	R	S	T	U	V	W	X	Y	Z	A	B	C

What we're doing here is taking each letter and moving them down three places in the alphabet. Thus, A is sent to D while B is sent to E, and so on. Everything works well up to W, which is sent to Z. What about X? If we try to move three letters from X, we exceed the alphabet. The solution is to wrap around, and say that the letter following Z is A. Thus shifting X by three places moves us to A (and then similarly Y is sent to B and Z is sent to C). Instead of viewing the alphabet on two rows, it's easier to view them on a circle, which makes the wrapping clearer.

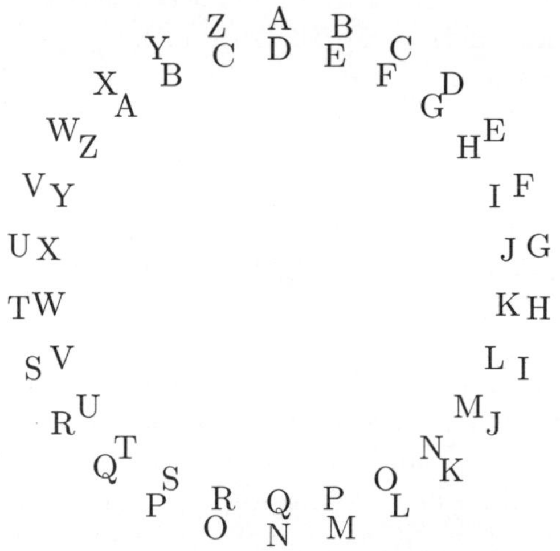

Using this method, the message MEET AT TEN is encrypted to PHHW DW WHQ. Remember, we call the original message **plaintext** and the encrypted message **ciphertext**, and we say the message was encrypted by a **Caesar cipher** with a shift of 3. As the Caesar cipher is a simple substitution alphabet cipher, decrypting a message is easy: all we have to do is reverse the arrows, and read up from the bottom row to the alphabet.

Of course, there's nothing special about using a shift of 3. While we can shift by any number, a little calculation shows that shifting by 1 is the same

TABLE 2.1. Number code for letters.

A	B	C	D	E	F	G	H	I	J	K	L	M	N	O	P	Q	R	S	T	U	V	W	X	Y	Z
0	1	2	3	4	5	6	7	8	9	10	11	12	13	14	15	16	17	18	19	20	21	22	23	24	25

as shifting by 27, or by 53, or by 79. What's happening is that every time we shift by 26, we return to where we started. Thus, there are really only 26 possibilities: we can shift by 0, by 1, by 2, and so on up to a shift of 25. These are all the *distinct* shifts possible.

As it's a little awkward to write relations such as $\texttt{B} + 3 = \texttt{E}$, we replace letters with numbers. In this case it's $1 + 3 = 4$. It's convenient to start with 0, and thus A becomes 0, B becomes 1 and so on; we record these values in Table 2.1. While it may seem a waste of time to replace numbers with letters, this is a very important advance in cryptography. As we progress in the book, we'll see more and more mathematical tools. We cannot directly apply these to letters, but we can to numbers.

To encrypt a message, we simply convert its letters to numbers, add 3 to each number, and then convert back into letters.

	M	E	E	T		A	T		T	E	N
	12	4	4	19		0	19		19	4	13
add 3:	15	7	7	22		3	22		22	7	16
	P	H	H	W		D	W		W	H	Q

The recipient decrypts PHHW DW WZR by shifting the letters *back* by 3. This corresponds to subtracting three when we convert to numbers.

	P	H	H	W		D	W		W	H	Q
	15	7	7	22		3	22		22	7	16
subtract 3:	12	4	4	19		0	19		19	4	13
	M	E	E	T		A	T		T	E	N

This lets them decrypt the ciphertext and recover the original message (the plaintext).

When Caesar used the cipher, he always shifted by 3, but there's no reason for us to stick with this convention. For example, we could have encrypted the message MEET AT TEN by shifting the letters by 5 instead of 3.

	M	E	E	T		A	T		T	E	N
	12	4	4	19		0	19		19	4	13
add 5:	17	9	9	24		5	24		24	9	18
	R	J	J	Y		F	Y		Y	J	S

The plaintext is still MEET AT TEN, but the ciphertext is now RJJY FY YJS. We need to tell our recipient how much to subtract. This is called the **key**, and in our most recent example it would be 5. Just like before, they would decrypt RJJY FY YJS by subtracting.

	R	J	J	Y	F	Y	Y	J	S
	17	9	9	24	5	24	24	9	21
subtract 5:	12	4	4	19	0	19	19	4	16
	M	E	E	T	A	T	T	E	N

2.4. Modular Arithmetic

There's a subtlety to the Caesar cipher that hasn't surfaced with our examples yet. Its analysis leads to clock arithmetic, which in addition to the applications here is a key ingredient in many modern encryption schemes (such as RSA).

Let's encrypt our original message to MEET AT TWO, and use 5 as the key.

	M	E	E	T	A	T	T	W	O
	12	4	4	19	0	19	19	22	14
add 5:	17	9	9	24	5	24	24	27	19
	R	J	J	Y	F	Y	Y	(?)	T

What should go in the place of the question mark? It doesn't seem like there is a letter corresponding to the number 27. Or is there? Such a letter would be two places *past* the letter Z. This is exactly the issue we faced earlier in the section, when we wrote down the shifted alphabet underneath the original alphabet. We solved the problem by wrapping the alphabet around, and thus the letter immediately following Z is A, and thus two letters after Z would be B. The encrypted message becomes RJJY FY YBT.

This is the same way we add when we're talking about time: what time will it be 5 hours after 10 o'clock? The answer isn't 15 o'clock (unless you're using 24-hour time): it is simply 3 o'clock.

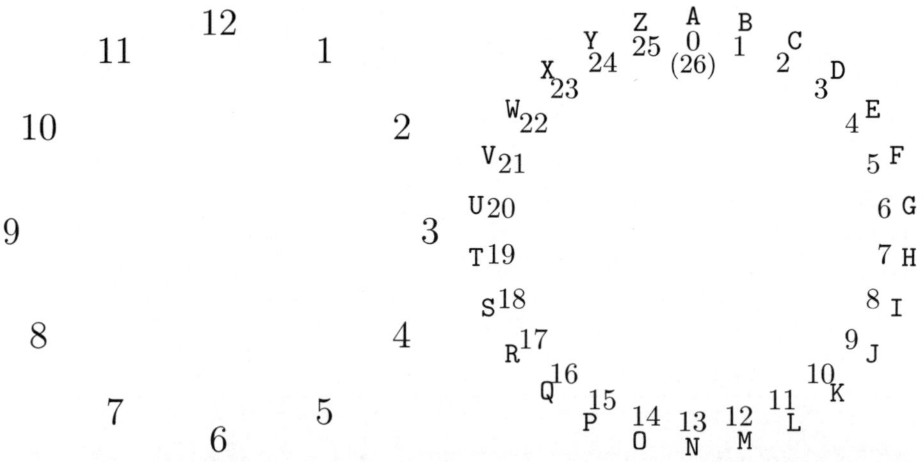

The rings above can be used to add for time and for the Caesar cipher, respectively. What time is it 10 hours after 10 o'clock? Count 10 places past 10 on the left wheel, and you get 8. What letter does S encrypt to using the Caesar cipher with key 10? Count 10 places past S on the wheel to the right, and get C.

Counting places on the wheel quickly becomes annoying; fortunately, we don't have to do that. In the case of the clock, observe that $10 + 10 = 20$. This is 8 more than 12 (which is one complete run of the clock). We write this fact as $20 = 8 \pmod{12}$, which is read as "20 is **equivalent** to 8 modulo 12" (some authors write **congruent** instead of equivalent). Similarly, we have that the letter S corresponds to the number 18, and $18 + 10 = 28$, which is 2 more than 26 (which is one complete turn of the letter wheel, since there are 26 letters). We write this $28 = 2 \pmod{26}$. Note we get the same answer by counting on the wheel, since 2 corresponds to the letter C. If we add big enough numbers, we can go around the wheels multiple times. For example, what time is it 21 hours after 9 o'clock? We have $9 + 21 = 30$, which is 6 hours past two complete runs of the clock (24 hours). Thus it'll be 6 o'clock.

These operations turn out to be very important in both cryptography and modern mathematics, and merit names and notation. We call this method of addition **clock arithmetic** or **modulo arithmetic**. On a clock the **modulus** is 12, while for a Caesar cipher it's 26. We first state the key definitions and results, and then discuss why they hold.

Clock arithmetic or modulo arithmetic: Given a positive integer m, we say two integers x and y are **equivalent modulo** m, if their difference is a multiple of m, and we write this as $x = y \pmod{m}$ or $x =_m y$. We sometimes omit writing the modulus when it's clear what it is. In general, $x = y \pmod{m}$ means there is some integer n (which may be positive, negative or zero) such that $x = y + n \cdot m$.
Every integer x is equivalent modulo m to some $y \in \{0, 1, 2, \ldots, m-1\}$. We call y the **reduction** of x modulo m, or say that x **reduces** to y. We use the notation MOD to indicate this reduction modulo m.

For now, we concentrate *only* on adding numbers on a clock; multiplication is important as well, and will be a major theme later. It is important to note that we could also go "backwards" on our clock. Thus, -4 is equivalent to 8 modulo 12, as $-4 = 8 + (-1) \cdot 12$.

Returning to the example of the letter S (corresponding to the number 18) being encrypted by the Caesar cipher using the key 10, we already pointed out that $18 + 10 = 2 \pmod{26}$. Thus the letter S encrypts to the letter C (since A is 0, we find B is 1 and hence C is 2). If you think about it though, $18 + 10 = 54 \pmod{26}$ is also true, since $28 = 54 + (-52)$, and -52 is a multiple of 26. In fact, there are infinitely many numbers that 28 is equivalent to modulo 26. For the purposes of encrypting the letter S, however, we don't use any of these other congruences, since they don't give numbers between 0 and 25. In general, given any problem of the form $a = \_ \pmod{m}$ there is exactly *one* solution from 0 to $m - 1$. This is extremely important, as it ensures that we have a unique encryption (and decryption) procedure.

How can we find that number? If our number is positive, we keep subtracting m until we have a number between 0 and $m - 1$. Since we are subtracting m each time, we can't go from a number greater than $m - 1$ to a number less than 0 in one subtraction, and thus we hit one of 0, 1, …, $m - 1$. If instead a is negative, we just keep adding m until we again land between 0 and $m - 1$. The result is the reduction of a modulo m, and we say that a reduces to that number. We denote reduction by using the notation MOD, so 28 MOD 26 = 2. Notice the difference between the problems 28 = __ (mod 26) and 28 MOD 26 = __. The first question has infinitely many correct answers (such as 2, 28, 54, -24, etc.), while the second question has only one correct answer, 2.

Armed with this new modular arithmetic, let's return to the Caesar cipher. Consider encryption of the phrase `THEY COME BY SEA` using the Caesar cipher with a key of 18. As before, we first translate letters into numbers.

	T	H	E	Y	C	O	M	E	B	Y	S	E	A
	19	7	4	24	2	14	12	4	1	24	18	4	0

Then we add the key (18 in this case) and reduce the results modulo 26.

	T	H	E	Y	C	O	M	E	B	Y	S	E	A
	19	7	4	24	2	14	12	4	1	24	18	4	0
add 18:	37	25	22	42	20	32	30	22	19	42	36	22	18
MOD 26:	11	25	22	16	20	6	4	22	19	16	10	22	18

Finally, we convert back to letters to get the ciphertext.

	T	H	E	Y	C	O	M	E	B	Y	S	E	A
	19	7	4	24	2	14	12	4	1	24	18	4	0
add 18:	37	25	22	42	20	32	30	22	19	42	36	22	18
MOD 26:	11	25	22	16	20	6	4	22	19	16	10	22	18
	L	Z	W	Q	U	G	E	W	T	Q	K	W	S

We thus send the message `LZWQ UGEW TQ KWS`. If the receiving party knows that the key is 18, they recover the original message by subtracting 18 and reducing modulo 26.

	L	Z	W	Q	U	G	E	W	T	Q	K	W	S
	11	25	22	16	20	6	4	22	19	16	10	22	18
subtract 18:	−7	7	4	−2	2	−12	−14	4	1	−2	−8	4	0
MOD 26:	19	7	4	24	2	14	12	4	1	24	18	4	0
	T	H	E	Y	C	O	M	E	B	Y	S	E	A

2.5. Number Theory Notation

As divisibility issues play a key role in cryptography problems, it's worth introducing some terminology before we return to encryption schemes in §2.6.

Definitions: factors, divisors, unit, composite, prime, and relatively prime.

- **Factors, divisors:** Let x and y be two positive integers. We say x divides y (or x is a factor of y) if y/x is an integer. We often write $x|y$. Note this means there is a positive integer d such that $y = dx$. It is convenient to say every integer is a factor of 0.
- **Proper divisor, nontrivial proper divisor:** If x is a positive integer, a proper divisor is any factor of x that is strictly less than x. If the factor is strictly between 1 and n, then we say it is a nontrivial factor or a nontrivial proper divisor.
- **Prime, composite, unit:** A positive integer n greater than 1 is prime if its only proper divisor is 1 (alternatively, its only factors are 1 and itself). If a positive integer n greater than 1 is divisible by a proper divisor greater than 1, we say n is composite. If $n = 1$ we say n is a unit.
- **Greatest common divisor (gcd):** The greatest common divisor of two positive integers is the largest integer dividing both.
- **Relatively prime:** Two integers are relatively prime if the only positive integer dividing both of them is 1. In other words, they have no common factors. Note this is the same as saying their greatest common divisor is 1.

For example, 26 is composite as it is divisible by 2 and 13, but 29 is prime as it is only divisible by 1 and 29. The greatest common divisor of 26 and 42 is 2, while the greatest common divisor of 12 and 30 is 6. As the greatest common divisor of 26 and 29 is 1, these two numbers are relatively prime. The first few primes are 2, 3, 5, 7, 11, 13, 17 and 19. The primes are the building blocks of integers. The **Fundamental Theorem of Arithmetic** asserts that every integer can be written uniquely as a product of powers of primes where the primes are in increasing order. Thus $12 = 2^2 \cdot 3$ and $30 = 2 \cdot 3 \cdot 5$, and there are no other ways of writing these numbers as a product of primes in increasing order (though we can write 12 as $3 \cdot 2^2$, this is the same factorization as before, just written in a different order). In fact, the reason 1 is declared to be a unit and not a prime is precisely to ensure each positive integer has a unique factorization into products of prime powers. If 1 were a prime, we could also write $12 = 1^{2013} \cdot 2^2 \cdot 3$.

Though we won't do too much with these concepts now, we will return to them later. Both primes and greatest common divisors play key roles in cryptography in general, and RSA (one of the most important systems ever) in particular.

It is one thing to define a concept, it is quite another to be able to use it. The definition of greatest common divisor is fairly clear: find the largest number dividing the two given numbers. We can find the greatest common divisor of x and y by starting with the smaller of x and y, and working down to 1. The first number dividing both is the greatest common divisor. Unfortunately, this definition is inadequate for computation. For large numbers, it would take far too long to find.

Similarly, it's easy to write down a method to check to see if a number is prime. All we need to do is try all numbers less than it; if none of them divide our number, then it is prime. Again, this method is far too slow to be of practical use. A large part of Chapters 7–9 is devoted to finding efficient algorithms for these and other problems. We discuss the Euclidean algorithm, a fast way to find greatest common divisors, in Chapter 8, and discuss fast primality testing in Chapter 11.

2.6. The Affine Cipher

While this chapter is mostly about describing various cryptosystems (with Chapters 4 and 5 devoted to the attacks), we have to say a few words about the security of the Caesar cipher as its insecurity led to the development of numerous other systems. We assume our sender is intelligent enough not to do a shift of zero, as that wouldn't encode the message! Thus a determined attacker can crack it in at most 25 attempts, as there are only 25 possible shifts. When most people couldn't even read, a small level of security sufficed; however, we clearly need a more powerful method than the Caesar cipher. There are several simple ways to generalize the Caesar cipher. In this section we discuss the affine cipher. It's a natural improvement, but unfortunately it doesn't improve the security significantly. Thus more powerful generalizations are needed. We discuss one of these in the next section, the Vigenère cipher.

Remember the Caesar cipher was a special case of a substitution alphabet cipher. There are $26! \approx 4.03 \cdot 10^{26}$ substitution alphabet ciphers, and only 26 of them are a Caesar cipher. This means there are still approximately $4.03 \cdot 10^{26}$ other substitution alphabet ciphers we could use! We want a simple cipher where we don't have to remember too much. The general substitution alphabet cipher requires 26 pieces of information (the total reordering of the alphabet), while the Caesar cipher requires just one (the shift). It's thus natural to look for something slightly more complicated than the Caesar cipher. The logical thing to try is a cipher which requires *two* pieces of information.

It turns out that the Caesar cipher is a special case of a cipher with *two* free parameters. What this means is that there is a family of encryption schemes that depend on *two* pieces of information, and if we set one of these values to 1 and the other to k, then we get a Caesar cipher with a shift of k. We now discuss how to find these generalizations of the Caesar cipher.

We start by writing the alphabet in two rows, and then we shift the bottom row by a fixed amount. There's another way to view this. We write the alphabet in the first row, and then underneath the `A` we write some letter, and then from that point on in the second row we just write the rest of the alphabet, wrapping around when needed.

When studying a complicated problem, it's often a good idea to concentrate on a simpler case first to build intuition. Therefore, let's simplify the world and imagine that we have just a six letter alphabet. We'll explore what happens here, and then generalize. Writing `C` under `A` gives

$$\begin{array}{l} \texttt{A B C D E F} \\ \updownarrow \\ \texttt{C} \end{array} \qquad \text{becomes} \qquad \begin{array}{cccccc} \texttt{A} & \texttt{B} & \texttt{C} & \texttt{D} & \texttt{E} & \texttt{F} \\ \updownarrow & \updownarrow & \updownarrow & \updownarrow & \updownarrow & \updownarrow \\ \texttt{C} & \texttt{D} & \texttt{E} & \texttt{F} & \texttt{A} & \texttt{B} \end{array}$$

While it seems that once we choose the letter to write under `A`, then the rest of the second row is uniquely determined, this is only true because we've implicitly made an assumption. When we write the letters in the first row, we might as well increment by one as we go along; however, why not increment by a different amount in the second row? What if we increment by 2, or 3, or any integer other than 1? Let's compare what happens when we increment each letter on the bottom by 1 versus what happens when we increment by 2.

$$\begin{array}{ccccccccccc} \texttt{A} & +1 \Rightarrow & \texttt{B} & +1 \Rightarrow & \texttt{C} & +1 \Rightarrow & \texttt{D} & +1 \Rightarrow & \texttt{E} & +1 \Rightarrow & \texttt{F} \\ \updownarrow & & \updownarrow & & \updownarrow & & \updownarrow & & \updownarrow & & \updownarrow \\ \texttt{C} & +1 \Rightarrow & \texttt{D} & +1 \Rightarrow & \texttt{E} & +1 \Rightarrow & \texttt{F} & +1 \Rightarrow & \texttt{A} & +1 \Rightarrow & \texttt{B} \end{array}$$

$$\begin{array}{ccccccccccc} \texttt{A} & +1 \Rightarrow & \texttt{B} & +1 \Rightarrow & \texttt{C} & +1 \Rightarrow & \texttt{D} & +1 \Rightarrow & \texttt{E} & +1 \Rightarrow & \texttt{F} \\ \updownarrow & & \updownarrow & & \updownarrow & & \updownarrow & & \updownarrow & & \updownarrow \\ \texttt{C} & +2 \Rightarrow & \texttt{E} & +2 \Rightarrow & \texttt{A} & +2 \Rightarrow & \texttt{C} & +2 \Rightarrow & \texttt{E} & +2 \Rightarrow & \texttt{A} \end{array}$$

There is an insurmountable problem when we increment the bottom row by 2. Instead of getting the full alphabet, we only get three letters. If our encrypted message contains an `A`, we don't know if the original letter was a `C` or an `F`, as both encrypt to `A`. Similarly, we don't know how to decrypt a `C` (it could be either an `A` or a `D`) or an `E` (which could be either a `B` or an `E`). As we have six letters in our restricted alphabet, there are six choices for the second row's increment: 0, 1, 2, 3, 4, or 5. Of course, we can't increment by 0, as then we just get the same letter each time. Let's investigate the other five possibilities and see if we can detect a pattern of which ones work and which ones fail. For definiteness, we start with a `C` under the `A`.

	A	B	C	D	E	F	
+1	C	D	E	F	A	B	works
+2	C	E	A	C	E	A	fails
+3	C	F	C	F	C	F	fails
+4	C	A	E	C	A	E	fails
+5	C	B	A	F	E	B	works

Of the five possible shifts for the second row of our six letter alphabet, only shifting by 1 and 5 work. These are the only two shifts that lead to the second row being a rearrangement of the entire alphabet; all other shifts lead to just a subset of the alphabet.

So, what's special about 1 and 5 in a six letter alphabet? Unfortunately, there are a lot of explanations possible with a small data set. It could be because they're the largest and smallest elements. Maybe what matters is that neither number is a proper divisor of 6. Or perhaps it's due to the fact that, if we write out the numbers in English, "one" and "five" alternate between consonants and vowels while "two", "three" and "four" do not. We need more data. Thus, let's increase the size of our alphabet to ten letters and see what happens, again starting with a `C` under `A`. Try it before looking at the answer below.

	A	B	C	D	E	F	G	H	I	J	
+1	C	D	E	F	G	H	I	J	A	B	works
+2	C	E	G	I	A	C	E	G	I	A	fails
+3	C	F	I	B	E	H	A	D	G	J	works
+4	C	G	A	E	I	C	G	A	E	I	fails
+5	C	H	C	H	C	H	C	H	C	H	fails
+6	C	I	E	A	G	C	I	E	A	G	fails
+7	C	J	G	D	A	H	E	B	I	F	works
+8	C	A	I	G	E	C	A	I	G	E	fails
+9	C	B	A	J	I	H	G	F	E	D	works

Let's summarize what we've found. When we had six letters, the shifts that failed were 2, 3 and 4; with ten letters, the shifts that failed were 2, 4, 5, 6, and 8. We now have additional support for one of our explanations for the patterns: the shift fails if and only if it shares a divisor greater than 1 with the size of the alphabet. Note the shifts of size 1 and 5 are relatively prime to 6, while the shifts of 1, 3, 7, and 9 are relatively prime to 10.

What's happening is that we've gone from simply adding to adding and multiplying. Remember if we have an alphabet with m letters, then we can represent the letters by the numbers $0, 1, \ldots, m-1$. For the Caesar cipher on the full alphabet, we shifted `A` by a fixed amount and then wrapped the alphabet. If we send `A` to `C`, that corresponds to a shift of 2. We can represent this using functional notation by $f(0) = 2$. As `B` goes to `D`, we would have $f(1) = 3$. Noting that `C` goes to `D` gives us $f(2) = 4$, and we see that this Caesar cipher is just $f(n) = n + 2 \pmod{26}$.

Let's return to the six letter alphabet where we sent `A` to `C` and we increased by 2 each time in the bottom row. Since `A` goes to `C`, using function notation we have $f(0) = 2$. Next we found `B` went to `E`, so $f(1) = 4$, then `C` went to `A` giving $f(2) = 0$ (we don't write $f(2) = 6$ as we're doing all arithmetic modulo 6). Continuing, we find $f(3) = 2$, $f(4) = 4$ and $f(5) = 0$. We can express this more compactly by writing $f(n) = (2n + 2) \pmod{6}$. The 2 in front of the n is because we are increasing the bottom row by 2 each time, while the constant term 2 is from writing `C` under `A`.

Our analysis above suggests the following candidate to generalize the Caesar cipher. It's called the **affine cipher** since maps of the form $f(x) = ax + b$ are a special case of what are called affine transformations.

Whenever you introduce a new concept or term in mathematics, it's worth taking some time to think about the notation you use. Good notation can help you understand a problem by allowing you to quickly process the information. For the affine cipher, instead of denoting the function by f, we often write it as $f_{a,b}$. The advantage of this notation is that it highlights the rolls a and b play in the function, and allows us to easily distinguish different maps. Note that an affine cipher with $a = 1$ is just a Caesar cipher.

Affine cipher: Let $a, b \in \{0, 1, \ldots, 25\}$ such that a and 26 share no proper divisors (so the only integer dividing both is 1). Then $f_{a,b}(n) = an + b \pmod{26}$ is a valid encryption scheme and gives the affine cipher. We call the pair (a, b) the **key** for the affine cipher.

We give a proof that the above is a valid encryption scheme in §4.3. We will also show that if a shared a proper divisor with 26, then the above scheme would fail, encrypting different letters to the same letter. We discuss there how to find a simple function for decryption; however, for now we can easily decrypt by looking at what number is sent where. For example, if we have $f_{3,2}(n) = 3n + 2 \pmod{26}$, then $f_{3,2}(1) = 5$ (or `B` is sent to `F`), so if we receive a 5 in the ciphertext, then the corresponding character in the message text is a 1 (equivalently, if we receive an `F` then a `B` was sent).

The affine cipher introduces another level of security, but a determined attacker can crack it with only a bit more work than required for the Caesar cipher. How many possible keys are there? We have 26 choices for b. As a must be relatively prime to 26, there are 12 possibilities for a: 1, 3, 5, 7, 9, 11, 15, 17, 19, 21, 23 and 25. Thus there are only $12 \cdot 26 = 312$ possible affine ciphers (or 311 if we ignore the key that doesn't change the text). This is larger than the 25 possible Caesar ciphers, but still small enough that a deliberate brute force attack will succeed relatively quickly.

2.7. The Vigenère Cipher

We've seen that both the Caesar cipher and a generalization, the affine cipher, offer little security against a persistent attacker. There are only 26 keys to try for the Caesar cipher, and only 312 for the affine cipher. Of course, these are just two special cases of substitution alphabet ciphers. Remember there are 26! or approximately $4.03 \cdot 10^{26}$ possible substitution alphabet ciphers, and we've only explored 312. Perhaps a more complicated choice would provide adequate security?

For example, imagine our sender and receiver meet and agree upon a reordering of the alphabet. It's of course not always possible to have them meet and exchange a key, but this allows us to analyze the general substitution alphabet cipher in its best case. It seems like an attacker has a daunting task, as there are over 10^{26} possible keys to try. If he could try a trillion keys a second, it would still take over 12,000 years to try them all;

however, one of the most important lessons in cryptography is just because something seems hard, it doesn't mean it is hard.

It turns out all substitution alphabet ciphers have a grave weakness which can be exploited. They are all vulnerable to a frequency analysis. We describe this type of attack in detail in §4.5. Briefly, the problem is that we always encode each letter the same way. If we have a long ciphertext, we can look at the frequency distributions of the letters. Whatever letter is most common is probably the encryption of the most common letter in the English alphabet. We can then continue and look for the second most common letter, and so on, or we could look at pairs of letters.

This deficiency, however, suggests at least the shape of the solution and hints at the evolution of the subject. Practitioners needed a new way to encode messages such that letters are not always encrypted the same way. At the same time, they didn't want to completely throw away the Caesar and substitution alphabet ciphers. These are very easy to use: they're fast to encode and fast to decode. The question was whether or not the substitution alphabet cipher could be improved to make it impervious (or at the very least, more impervious) to frequency and other attacks.

The **Vigenère cipher**, invented in 1553 by Giovan Battista Bellaso and rediscovered by Blaise de Vigenère in 1586, addresses these issues by shifting letters at different places in the message by different amounts. This is an enormous advance over our previous methods. Since an `E` isn't always encrypted to the same letter, it seems a frequency attack might either fail, or at the very least be less effective. As you read below, remember that this system was designed and initially used almost 500 years ago. There were no powerful computers to do the laborious calculations, so what might seem trivial for computers today to attack might very well have been quite secure back then.

In all the ciphers we've studied, there's always been a key. It was one number for the Caesar cipher (the shift), a pair of numbers for the affine cipher (for the map $f_{a,b}(n) = an + b$), and a 26-tuple for the general substitution alphabet cipher. For the Vigenère cipher, the key is a word or a phrase. We call this the **keystream** of the cipher.

> **The Vigenère cipher:** To encrypt a message, we first choose a keyword or phrase, and write it repeatedly underneath the plaintext. We call the repeated keyword or phrase the **keystream**. We then "add" the plaintext and the keystream letter by letter, just as we did in a Caesar cipher. To decrypt, all we do is subtract rather than add.

Before doing an example, we quickly review how to add letters. Recall that we associate a number to each letter (from 0 for `A` to 25 for `Z`), and that addition is modulo 26. For example, `T` corresponds to the number 19 and `M` corresponds to the number 12. Thus `T` + `M` is `F` as $19 + 12 = 5 \pmod{26}$ (and 5 corresponds to `F`).

Imagine we want to encrypt the following (translated) passage from Kant with a Vigenère cipher with keyword `MORALS`.

> There is no possibility of thinking of anything at all in the world, or even out of it, which can be regarded as good without qualification, except a good will. Intelligence, wit, judgment, and whatever talents of the mind one might want to name are doubtless in many respects good and desirable, as are such qualities of temperament as courage, resolution, perseverance. But they can also become extremely bad and harmful if the will, which is to make use of these gifts of nature and which in its special constitution is called character, is not good.

We showed above that the first letter encrypts to an `F` (as `T` plus `M` is `F`). A similar calculation shows the next letter encrypts to a `V` as $7 + 14 = 21 \pmod{26}$ (which is `V`). If we transmit in blocks of five and run MORALS continuously under the text, we get the following.

```
THERE ISNOP OSSIB ILITY OFTHI NKING OFANY THING ATALL INTHE WORLD
MORAL SMORA LSMOR ALSMO RALSM ORALS MORAL SMORA LSMOR ALSMO RALSM
FVVRP AEBFP ZKEWS IWAFM FFEZU BBIYY ATRNJ LTWEG LLMZC IYLTS NOCDP

OREVE NOUTO FITWH ICHCA NBERE GARDE DASGO ODWIT HOUTQ UALIF ICATI
ORALS MORAL SMORA LSMOR ALSMO RALSM ORALS MORAL SMORA LSMOR ALSMO
CIEGW ZCLTZ XUHNH TUTQR NMWDS XACVQ RRSRG ARNIE ZAIKQ FSXWW INSFW

ONEXC EPTAG OODWI LLINT ELLIG ENCEW ITJUD GMENT ANDWH ATEVE RTALE
RALSM ORALS MORAL SMORA LSMOR ALSMO RALSM ORALS MORAL SMORA LSMOR
FNPPO SGTLY ACUWT DXWET PDXWX EYUQK ZTUMP UDEYL MBUWS SFSME CLMZV

NTSOF THEMI NDONE MIGHT WANTT ONAME AREDO UBTLE SSINM ANYRE SPECT
ALSMO RALSM ORALS MORAL SMORA LSMOR ALSMO RALSM ORALS MORAL SMORA
NEKAT KHPEU BUOYW YWXHE OMBKT ZFMAV ACWPC LBEDQ GJIYE MBPRP KBSTT

SGOOD ANDDE SIRAB LEASA RESUC HQUAL ITIES OFTEM PERAM ENTAS COURA
LSMOR ALSMO RALSM ORALS MORAL SMORA LSMOR ALSMO RALSM ORALS MORAL
DYACU AYVPS JICSN ZVADS DSJUN ZCIRL TLUSJ OQLQA GECSY SETLK OCLRL

GERES OLUTI ONPER SEVER ANCEB UTTHE YCANA LSOBE COMEE XTREM ELYBA
SMORA LSMOR ALSMO RALSM ORALS MORAL SMORA LSMOR ALSMO RALSM ORALS
YQFVS ZDGHZ OYHQF JEGWD OECPT GHKHP QOOEA WKAPV CZEQS OTCWY SCYMS

DANDH ARMFU LIFTH EWILL WHICH ISTOM AKEUS EOFTH ESEGI FTSOF NATUR
MORAL SMORA LSMOR ALSMO RALSM ORALS MORAL SMORA LSMOR ALSMO RALSM
POEDS SDAWU WARHY EHAXZ NHTUT WJTZE MYVUD WATKH PKQUZ FEKAT EAEMD

EANDW HICHI NITSS PECIA LCONS TITUT IONIS CALLE DCHAR ACTER ISNOT
ORALS MORAL SMORA LSMOR ALSMO RALSM ORALS MORAL SMORA LSMOR ALSMO
SRNOO TWTHT FUHJS AWOWR LNGZG KIEMF WFNTK OOCLP VOVRR LUFSI IDFAH

GOOD
RALS
XOZV
```

Decryption is as easy as encryption. All the recipient needs to do is repeatedly write the keyword under the ciphertext and subtract.

The Caesar cipher is the special case of the Vigenère cipher where the keyword has length 1. The power of the Vigenère cipher is the dramatic change in behavior from using a keyword of length 1 to a keyword of greater length. Let's look at just the first word of our plaintext, `THERE`, and its corresponding ciphertext, `FVVRP`. Note the first `E` encrypts to a `V` while the second `E` encrypts to a `P`. Thus a letter can be encrypted to different letters depending on where it is in the text. We can also look at what happens in the opposite direction. The first `V` in the ciphertext came from an `H`, while the second `V` came from an `E`. Thus the same letter can decrypt to different letters depending on where it's located.

The Vigenère cipher has many great features, and in fact was called the unbreakable cipher by many (though we'll see in sections §4.6 and §4.7 how to break it!) It's almost as easy to use as the Caesar cipher, but far more secure as there are significantly more possibilities. Attackers start off at a great disadvantage, as they don't even know the length of the keystream. Of course, there are some vulnerabilities. The sender and receiver have to somehow exchange this secret keystream, not only might this be hard, but if the keystream is compromised, then all subsequent messages can be deciphered. For example, the South used the Vigenère cipher in the Civil War, often with the keystream either `Manchester Bluff`, `Complete Victory`, or `Come Retribution`. Knowing that, it's easy to break a recently uncovered Southern message (see Exercise 2.11.45), though even without knowing the keystream, the message is easily broken today.

2.8. The Permutation Cipher

Substitution alphabet ciphers (including the Caesar and affine ciphers) work by relabeling the letters of the alphabet to disguise the original message. We can use frequency analysis (described in detail in Chapter 4) to figure out the plaintext by discovering how the alphabet was "relabeled". Frequency analysis is very powerful. All we need to do is count the percentage of time each letter occurs in the ciphertext and compare those to the distribution of letters, pairs of letters, and so on in the English language. Though it's a bit harder to implement, frequency analysis even works for breaking a Vigenère cipher.

In this section, we describe another method of encryption. While it's immune to frequency attacks, it too has issues that are exploitable by a clever eavesdropper.

The permutation cipher doesn't change the letters. Instead, it just moves them to different positions. This is a terrific advantage, as frequency analysis is now useless. The reason is that the frequency of letters in the plaintext and ciphertext are the same; all that changes is the ordering of the letters.

Before we can describe the cipher, we need to set some terminology.

Permutation: Let m be a positive integer, and consider the ordered set $\{1, 2, 3, \ldots, m\}$. There are $m!$ ways to reorder this set (with order mattering). Each of these orderings is called a **permutation**. We denote a permutation σ of m elements by writing the m numbers in increasing order on top, and then writing what each is sent to down below. For example,

$$\sigma = \begin{pmatrix} 1 & 2 & 3 & 4 & 5 \\ 4 & 2 & 3 & 5 & 1 \end{pmatrix}$$

is the permutation that sends the first element to the fourth place, the second and third elements are left alone, the fourth element is sent to the fifth position, and the fifth element is sent to the first. We let $\sigma(i)$ denote where the ith number is moved; for our example, $\sigma(1) = 4$, $\sigma(2) = 2$, and so on.

In this notation, the first row is the initial configuration while the second row tells us where things have been moved. For example,

$$\begin{pmatrix} 1 & 2 & 3 \\ 3 & 1 & 2 \end{pmatrix}$$

is the permutation which rotates three objects cyclically *to the right*. We started with three objects in the order (1 2 3), and they are permuted so that they end in the order (3 1 2). Each element has been moved to the right, and the last element has *wrapped around* to the first position. In the case of the plaintext block `VEN`, applying this permutation results in the ciphertext block `NVE`.

The permutation cipher: Choose a positive integer m and a permutation σ of the ordered set $\{1, 2, 3, \ldots, m\}$; σ is the key of the cipher. Break the message into blocks of length m. In each block, reorder the letters according to the permutation. Thus the first letter in a block is sent to the $\sigma(1)$ position, the second letter is sent to the $\sigma(2)$ position, and so on. To decrypt a message, apply the inverse permutation.

For example, consider the message `MEET AT TEN THIRTY`. We can break the message into blocks of three letters each: `MEE TAT TEN THI RTY`. Let's use the permutation $\begin{pmatrix} 1 & 2 & 3 \\ 2 & 3 & 1 \end{pmatrix}$; this *rotates* the letters in each block one place to the right (moving the right-most letter in each block to the first position). The message becomes `EME TTA NTE ITH YRT`. Finally, we group the letters into blocks of 5 to get the ciphertext to be transmitted: `EMETT ANTEI THYRT`. Remember we want to transmit messages in blocks to hide spacings and word lengths, as that information is a boon to an attacker. If we sent a ciphertext broken into blocks of length 3, that might tip an attacker as to the length of our permutation.

To decipher the message, the recipient breaks the message back into blocks of three, and reverses the permutation of the letters by rotating the letters in each block to the left (moving the left-most letter in each block to the last position).

In this example, encryption was done by rotation in blocks of 3, but the permutation cipher can work on blocks of arbitrary size with arbitrary permutations.

For another example, we can consider the permutation represented by

$$\begin{pmatrix} 1 & 2 & 3 & 4 & 5 \\ 5 & 3 & 4 & 2 & 1 \end{pmatrix}, \tag{2.1}$$

which acts on blocks of 5 objects. To use it to encipher the message `MEETA TTENT HIRTY` we simply *apply* the permutation to each of the blocks (since it's already grouped into blocks of the right size). The permutation in (2.1) tells us that the fifth letter is moved to the first position, the third letter to the second position, the fourth letter to the third position, the second letter to the fourth position, and finally the first letter goes to the fifth position. Applying this permutation to our message gives `AETEM TENTT YRTIH`.

To decipher the message, we need to find the permutation which *reverses* the permutation from (2.1). This is called the *inverse* of the permutation. This is the permutation that takes objects in the order (5 3 4 2 1) and puts them in the order (1 2 3 4 5). To find the permutation, we start by writing the given ordering (5 3 4 2 1) over the desired ordering (1 2 3 4 5).

$$\begin{pmatrix} 5 & 3 & 4 & 2 & 1 \\ 1 & 2 & 3 & 4 & 5 \end{pmatrix} \tag{2.2}$$

Now we rearrange columns so that the first row is in the standard increasing order.

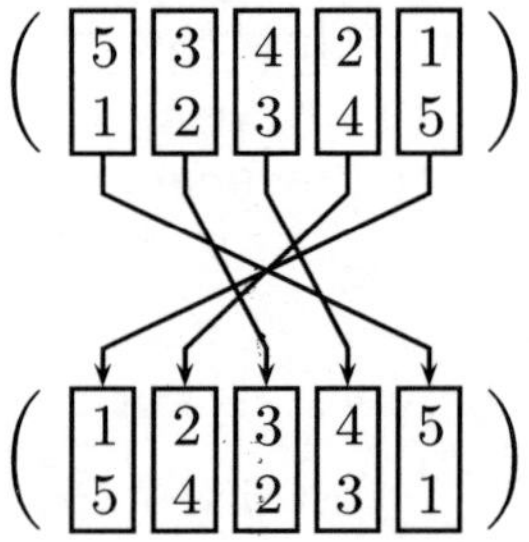

Thus the message `AETEM TENTT YRTIH` can be decrypted by applying the inverse permutation

$$\begin{pmatrix} 1 & 2 & 3 & 4 & 5 \\ 5 & 4 & 2 & 3 & 1 \end{pmatrix}$$

to each of the blocks of five.

There is one issue we haven't discussed yet which is essential for encoding: What if the message cannot be broken into blocks of the appropriate size? Imagine, for example, that we want to encrypt the message MEETA TTEN with the permutation

$$\begin{pmatrix} 1 & 2 & 3 & 4 \\ 3 & 4 & 2 & 1 \end{pmatrix}.$$

After grouping into blocks of 4 we have MEET ATTE N, and there is a leftover letter since the message length isn't a multiple of 4. Fortunately, there's a simple solution. In this case, we can **pad** the message by adding extra nonsense letters to the end: MEET ATTE NCTH. This encrypts to ETEM TETA THCN, or, after regrouping, ETEMT ETATH CN. When the recipient decrypts the message, all they have to do is discard any nonsense at the end that was added for padding, to get MEET AT TEN.

2.9. The Hill Cipher

Note: This section assumes familiarity with basic properties of matrix multiplications, and it may safely be omitted.

The **Hill cipher** was invented by Lester Hill in 1929. It's another attempt to address an inherent weakness of the substitution alphabet ciphers, which always encrypt the same letter the same way, no matter where that letter is in the message. We've discussed how frequency analysis can exploit this weakness. Our first attempt around this was the Vigenère cipher. At first glance, this appears to avoid this problem as the same letter is encrypted differently depending on where it is in the message; however, the improvement is not as great as we might have hoped. While the same letter can be encrypted differently, all that matters is the position of our letter in the message modulo the length of the keystream. For example, if the keyword has length 20, then any E in the third, 23rd, 43rd, 63rd, and so on positions in the message are encrypted the same, as we add the same keyword letter to each.

We see that in the Vigenère cipher the encryption of a letter depends weakly on its location; all that matters is where it is in the message. The Hill cipher adds another level by having the encryption depend not just on the location, but also on what letters are nearby. This novelty is a great feature of the Hill cipher. Even though it too is vulnerable to attack, it was an important advance in the growth of ciphers.

In order to describe the Hill cipher we need to talk about some standard facts about matrices, which are worth knowing. The reader comfortable with matrices can skim the next few paragraphs (the only nonstandard item is looking at matrices modulo 26, especially given what that means for invertibility). To keep the exposition simple, we concentrate on 2×2 matrices below. While this will limit which Hill ciphers we can study, the generalizations to larger ciphers will hopefully be clear.

Matrix operations: Consider 2×2 matrices $A = \begin{pmatrix} a & b \\ c & d \end{pmatrix}$ and $B = \begin{pmatrix} e & f \\ g & h \end{pmatrix}$, and a column vector $\overrightarrow{v} = \begin{pmatrix} x \\ y \end{pmatrix}$. We define a 2×1 matrix (which is a column vector with two entries) $A\overrightarrow{v}$ by

$$A\overrightarrow{v} = \begin{pmatrix} a & b \\ c & d \end{pmatrix}\begin{pmatrix} x \\ y \end{pmatrix} := \begin{pmatrix} ax + by \\ cx + dy \end{pmatrix},$$

and a 2×2 matrix AB by

$$AB = \begin{pmatrix} a & b \\ c & d \end{pmatrix}\begin{pmatrix} e & f \\ g & h \end{pmatrix} := \begin{pmatrix} ae + bg & af + bh \\ ce + dg & cf + dh \end{pmatrix}.$$

The special matrix $\begin{pmatrix} 1 & 0 \\ 0 & 1 \end{pmatrix}$ is called the identity matrix, and is denoted I.

Two matrices are **equivalent modulo 26** if their corresponding entries are equivalent modulo 26. We say the matrix B is a modulo 26 inverse to the matrix A if AB and BA are equivalent to I modulo 26. If A has an inverse, we often denote it by A^{-1} and say A is **invertible** modulo 26.

The **Hill cipher** tries to thwart frequency attacks by having a letter's encryption depend not just on its location in the text, but also on what letters are near it. The key is a square matrix modulo 26. Again, to keep the linear algebra to a minimum we only deal with 2×2 matrices as keys, but those who know determinants and inverses for general $n \times n$ matrices are encouraged to generalize the arguments below to larger keys.

The Hill cipher (2×2 case): Let A be an invertible 2×2 matrix modulo 26; this is the **key** for the Hill cipher. Break the message into blocks of length two. Write each block as a column vector, apply the matrix A, and reduce modulo 26. This gives the ciphertext. To decode, break the received message into blocks of length two, and apply the inverse matrix.

Let's encrypt the message `ET PHONE HOME`. We break this into blocks of length 5 for transmission purposes, which really obscures the message (`ETPHO NEHOM E`). We need to choose a 2×2 matrix as a key; let's use $\begin{pmatrix} 3 & 6 \\ 1 & 3 \end{pmatrix}$. We split the message into blocks of length 2, since this is the number of rows (and columns) in the matrix: `ET PH ON EH OM ET`. Note that we had to pad the last block since the message had an odd number of letters. Next, each block of two letters is treated as a column vector of numbers modulo 26. Since `E` corresponds to the number 4 and `T` corresponds to the number 19, the first block of our message corresponds to the matrix $\begin{pmatrix} 4 \\ 19 \end{pmatrix}$. After converting the rest of the blocks, we see that the message corresponds to the following list of column vectors:

$$\begin{pmatrix} 4 \\ 19 \end{pmatrix}, \begin{pmatrix} 15 \\ 7 \end{pmatrix}, \begin{pmatrix} 14 \\ 13 \end{pmatrix}, \begin{pmatrix} 4 \\ 7 \end{pmatrix}, \begin{pmatrix} 14 \\ 12 \end{pmatrix}, \begin{pmatrix} 4 \\ 19 \end{pmatrix}.$$

Up to now, all we have done is rewrite the message as numbers arranged in columns. The encryption step is to multiply each of these matrices by the key matrix modulo 26. For example, the first matrix becomes

$$\begin{pmatrix} 3 & 6 \\ 1 & 3 \end{pmatrix}\begin{pmatrix} 4 \\ 19 \end{pmatrix} = \begin{pmatrix} 3\cdot 4 + 6\cdot 19 \\ 1\cdot 4 + 3\cdot 19 \end{pmatrix} = \begin{pmatrix} 126 \\ 61 \end{pmatrix} = \begin{pmatrix} 22 \\ 9 \end{pmatrix} \pmod{26}.$$

After carrying out the rest of the multiplication, we get the column vectors

$$\begin{pmatrix} 22 \\ 9 \end{pmatrix}, \begin{pmatrix} 9 \\ 10 \end{pmatrix}, \begin{pmatrix} 16 \\ 1 \end{pmatrix}, \begin{pmatrix} 2 \\ 25 \end{pmatrix}, \begin{pmatrix} 10 \\ 24 \end{pmatrix}, \begin{pmatrix} 22 \\ 9 \end{pmatrix}.$$

To get the ciphertext, we simply convert back to letters, obtaining `WJ JK QB CZ KY WJ`, and then regroup. Thus the ciphertext is `WJJKQ BCZKYQ G`.

Decryption works by multiplying by the inverse of the encryption matrix; as not all matrices are invertible, we see that not all matrices can serve as keys. Not surprisingly, we chose our key matrix to be invertible modulo 26. Remembering the inverse of 3 modulo 26 is 9 (which is written $3^{-1} = 9 \pmod{26}$), the inverse modulo 26 of the encryption matrix is

$$\begin{pmatrix} 3 & 6 \\ 1 & 3 \end{pmatrix}^{-1} = (3)^{-1}\begin{pmatrix} 3 & -6 \\ -1 & 3 \end{pmatrix} = 9\begin{pmatrix} 3 & 20 \\ 25 & 3 \end{pmatrix} = \begin{pmatrix} 1 & 24 \\ 17 & 1 \end{pmatrix} \pmod{26}.$$

To decrypt the first block of ciphertext, `WJ`, we multiply it by the decryption matrix:

$$\begin{pmatrix} 1 & 24 \\ 17 & 1 \end{pmatrix}\begin{pmatrix} 22 \\ 9 \end{pmatrix} = \begin{pmatrix} 4 \\ 19 \end{pmatrix} \pmod{26}.$$

We've recovered the first plaintext vector, corresponding to the first two plaintext letters `ET`. The rest of the message is found similarly.

Note that decryption works because multiplying by the inverse matrix reverses the original multiplication. A more formal way of viewing this is that we start with a vector $\overrightarrow{v}$, we apply our key matrix and get $A\overrightarrow{v}$, and then apply the decrypt matrix A^{-1}, yielding $A^{-1}(A\overrightarrow{v})$. We haven't proved that matrix multiplication is associative, but it is, and this means we can regroup the parentheses. Thus our attempted decryption $A^{-1}(A\overrightarrow{v})$ equals $(A^{-1}A)\overrightarrow{v}$, but the last is just $\overrightarrow{v}$ since $A^{-1}A$ is the identity matrix modulo 26, and $I\overrightarrow{v} = \overrightarrow{v}$ for all vectors. We can see this by revisiting our decryption. In our example, we have

$$\begin{aligned} \begin{pmatrix} 1 & 24 \\ 17 & 1 \end{pmatrix}\begin{pmatrix} 22 \\ 9 \end{pmatrix} &= \begin{pmatrix} 1 & 24 \\ 17 & 1 \end{pmatrix}\begin{pmatrix} 3 & 6 \\ 1 & 3 \end{pmatrix}\begin{pmatrix} 4 \\ 19 \end{pmatrix} \\ &= \begin{pmatrix} 1 & 0 \\ 0 & 1 \end{pmatrix}\begin{pmatrix} 4 \\ 19 \end{pmatrix} = \begin{pmatrix} 4 \\ 19 \end{pmatrix} \pmod{26}. \end{aligned}$$

2.10. Summary

In this chapter we explored some of the history of cryptographic tools. For over a thousand years the Caesar cipher and its descendants were used. These substitution alphabet ciphers have many good features, most especially their ease of use. In the course of our studies we saw the power of modular arithmetic. We know how to add numbers, but what does it mean to add two letters? By converting letters to numbers, we can add and then we just convert back afterwards.

We also saw the dangers of using these simple encryption schemes. In particular, the Caesar cipher and the affine cipher are both extremely vulnerable to brute force assaults, while a general substitution alphabet cipher can be cracked by frequency analysis. This vulnerability is the greatest weakness of the substitution alphabet ciphers and is due to the fact that each letter is always encrypted the same way. The Vigenère cipher was a major advance and designed to thwart such attacks, as now the encryption of a letter depends not only on what letter it is but it's location in the text. Other attempted solutions are the permutation ciphers and the Hill ciphers. While all of these schemes are more secure, each is vulnerable to attack, and none may safely be used in an era of powerful and accessible computing.

Before describing these attacks in detail in Chapters 4 and 5, we first take a brief detour and explore the Enigma and Ultra in Chapter 3. The reason is that this serves as an excellent bridge from our historical tour and descriptions of methods to the later attacks. The Enigma is a natural outgrowth of these substitution alphabet ciphers, but with so many additional layers of complexity added that it'll take an entire chapter to describe it. We concentrate on how the Enigma worked, and only briefly touch upon how it was successfully cracked by the Allies.

2.11. Problems

• From §2.1: Ancient Cryptography

EXERCISE 2.11.1. *One of the biggest problems in cryptography is to make sure messages are correctly received. If parts of the message are omitted or altered, the meaning can drastically change (just imagine what could happen if you get the wrong firing coordinates for an attack). Show that the slave-barber method does quite well here; specifically, why is there no danger of the wrong message being received?*

EXERCISE 2.11.2. *Research and write about other ancient methods of encryption. In particular, what was known in Egypt? Research the Greek* scytale.

• From §2.2: Substitution Alphabet Ciphers

EXERCISE 2.11.3. *Here are two sentences found on the web. If you remove their punctuation, one of them is appropriate for this section; which one and why? How can the other be fixed and used?*

- *TV quiz jock, Mr. Ph.D., bags few lynx.*
- *Few black taxis drive up major roads on quiet hazy nights.*

EXERCISE 2.11.4. *One of the* 26! *substitution alphabet ciphers would be a very poor choice. Which one, and why?*

EXERCISE 2.11.5. *If a substitution alphabet cipher switches letters, prove that it must switch at least two letters. This means that there are at least two letters that are not sent to themselves. How many different substitution alphabet ciphers are there where* 24 *letters are sent to themselves and only two letters are switched? Fortunately the percentage of all substitution alphabet ciphers that only switch two letters is negligibly small—find this percentage.*

EXERCISE 2.11.6. *How many substitution alphabet ciphers are there that only switch the five vowels amongst themselves and leave the* 21 *consonants alone? How many are there that switch the* 21 *consonants amongst themselves and leave the five vowels alone? Finally, how many are there that switch the five vowels amongst themselves and the* 21 *consonants amongst themselves?*

EXERCISE 2.11.7. *How many substitution alphabet ciphers are there that do not switch the letter* `A`*?*

EXERCISE 2.11.8. *The definition of a substitution alphabet cipher is that no two letters are sent to the same letter. Show that this implies every letter appears once and only once as the image of a letter. For example, show that it is impossible to miss the letter* `Y`.

EXERCISE 2.11.9. *(Hard) How many substitution alphabet ciphers are there such that at least one letter is* not *switched? What about when exactly one letter is not switched? What percentage of all substitution alphabet ciphers have exactly one letter that is not switched?*

• From §2.3: The Caesar Cipher

EXERCISE 2.11.10. *Encrypt the message* `MATH` *with the Caesar cipher with* 4 *as the key.*

EXERCISE 2.11.11. *Encrypt the message* `CRYPTO` *with the Caesar cipher with* 6 *as the key.*

EXERCISE 2.11.12. *The message* `QIIX PEXIV` *was encrypted using the Caesar cipher with* 4 *as the key. Decrypt the message.*

EXERCISE 2.11.13. *The message* `SKKZ NKXK` *was encrypted using a Caesar cipher. Decrypt the message.*

EXERCISE 2.11.14. *There are 26! distinct substitution alphabet ciphers. What fraction of these are also Caesar ciphers?*

EXERCISE 2.11.15. *Consider the following twist to the Caesar cipher: instead of writing the alphabet in the second row and shifting it, now we write the alphabet backwards and shift. For example, we first write the alphabet backwards*

Z Y X W V U T S R Q P O N M L K J I H G F E D C B A

and then shift by 3, which gives

A	B	C	D	E	F	G	H	I	J	K	L	M	N	O	P	Q	R	S	T	U	V	W	X	Y	Z
↕	↕	↕	↕	↕	↕	↕	↕	↕	↕	↕	↕	↕	↕	↕	↕	↕	↕	↕	↕	↕	↕	↕	↕	↕	↕
C	B	A	Z	Y	X	W	V	U	T	S	R	Q	P	O	N	M	L	K	J	I	H	G	F	E	D

Notice that exactly *two letters are left unchanged, the* B *and the* O. *Prove or disprove: no matter what shift is used for this method, there will be either zero, one, or two letters left unchanged, and each possibility can happen.*

EXERCISE 2.11.16. *After reading §2.3, one of your friends (a classicist who loves everything Roman) decides to use the Caesar cipher to communicate with you. You're worried, however, as there are only 26 possible shifts, and thus an attacker can decipher a message in at most 26 attempts. You tell your friend. To increase security, he decides that he will choose 100 different shifts, and encrypt the message with the first shift, then encrypt the encrypted message with the second shift, and so on until all 100 shifts are used. As there are $26^{100} \approx 3 \cdot 10^{141}$ ways to choose 100 shifts when order matters, your friend is convinced that the resulting message is safely encrypted and there is no danger of an attacker deciphering it in any reasonable amount of time. Is your friend correct? Why?*

EXERCISE 2.11.17. *In the Caesar cipher every letter is encrypted the same way; we add the same number to each. To increase security, your friend suggests that instead we choose a shift for the first letter, and then each subsequent letter is encrypted by adding the result of the previous letter encryption. The advantage of this method is that we now have variable shifts. For example, if we had the word* CAN *and a shift of three, then* C *encrypts to* F, *which is 6. We now add 6 to* A *and get* G, *which is 7. Finally, adding 7 to* N *gives* U. *Thus our ciphertext is* FGU. *The encryption process is still manageable, though not as simple as before; however, have we gained any security? Why or why not?*

The last few problems are examples of recursive functions.

EXERCISE 2.11.18. *Let $f(n)$ be the number of ways to place 2×1 dominoes on a $2 \times n$ board. (a) Show that $f(1) = 1$ and $f(2) = 2$. (b) Prove the following recursive formula: $f(n) = f(n-1) + f(n-2)$. (c) Deduce that $f(n)$ is the nth Fibonacci number, where the Fibonacci numbers are $F_1 = 1$, $F_2 = 2$, $F_3 = 3$, $F_4 = 5$, and in general $F_n = F_{n-1} + F_{n-2}$.*

EXERCISE 2.11.19. *Let $f(n)$ be the number of ways to match $2n$ people into n pairs of 2, where all that matters is which people are paired together; in other words, it doesn't matter what number pair someone is in, nor whether or not they're the first or second person in the pair. (a) Prove $f(2n) = (2n-1)f(2n-2)$. (b) Show $f(2) = 1$ and thus deduce that $f(n) = (2n-1)!!$, where the double factorial of an odd number means the product of all odd numbers down to 1.*

- **From §2.4: Modular Arithmetic**

EXERCISE 2.11.20. *Is 15 congruent to 171 modulo 26? Is 11 congruent to 57 modulo 13?*

EXERCISE 2.11.21. *Is 1776 congruent to 1861 modulo 26? Is 1701 congruent to 35 modulo 26?*

EXERCISE 2.11.22. *Find the reductions of 15 and 19 modulo 7, and 11 and 18 modulo 3.*

EXERCISE 2.11.23. *Find the reductions of 11 and 5 modulo 3, and 29 and 19 modulo 26.*

EXERCISE 2.11.24. *Show an integer is divisible by 3 if and only if the sum of its digits is divisible by 3; similarly, show an integer is divisible by 9 if and only if the sum of its digits is divisible by 9. For example, the number 1234 is not divisible by 3 as $4+3+2+1 = 10$ is not divisible by 3, but 1233 is, as 3 divides $3+3+2+1 = 9$.*

EXERCISE 2.11.25. *Show a number is divisible by 11 if and only if the alternating sum of its digits is divisible by 11. Thus 16962 is divisible by 11 as its alternating sum is $2-6+9-6+1 = 0$, which is divisible by 11.*

EXERCISE 2.11.26. *Is there an integer x such that x both congruent to 2 modulo 4 and also to 3 modulo 7? If yes, is there just one such x, or more than one?*

EXERCISE 2.11.27. *Is there an integer x such that x is both congruent to 2 modulo 30 and also to 3 modulo 72? If yes, is there just one such x, or more than one?*

EXERCISE 2.11.28. *The previous two problems involved searching for solutions to simultaneous congruences. Imagine you're given numbers a_1, a_2, n_1, and n_2, and you want to find an x such that $x = a_1 \pmod{n_1}$ and $x = a_2 \pmod{n_2}$. Come up with as efficient an algorithm as possible to find such solutions, or show they do not exist.*

- **From §2.5: Number Theory Notation**

EXERCISE 2.11.29. *Find the smallest prime greater than 144.*

EXERCISE 2.11.30. *Find the greatest common divisor of 91 and 144.*

EXERCISE 2.11.31. *Write* 1800 *as a product of prime powers.*

EXERCISE 2.11.32. *If* $x = 2^6 3^9 7^4$ *and* $y = 2^4 5^6 7^2 11^3$, *find the greatest common divisor of* x *and* y.

• From §2.6: The Affine Cipher

EXERCISE 2.11.33. *Show that we can represent the encryption from the six letter alphabet with* A *going to* C *and shifting each letter in the bottom row by* 3 *by* $f_{3,2}(n) = 3n + 2 \pmod 6$. *If instead we shifted each letter in the bottom by* 4 *show that* $f_{4,2}(n) = 4n+2 \pmod 6$, *while a shift of* 5 *would give* $f_{5,2}(n) = 5n + 2 \pmod 6$.

EXERCISE 2.11.34. *Show that we can represent the encryption from the ten letter alphabet with* A *going to* C *and a shift of* 7 *in the bottom row by* $f_{7,2}(n) = 7n + 2 \pmod{10}$. *More generally, show that we can represent the encryption from the ten letter alphabet with* A *going to* C *and a shift of* k *in the bottom row by* $f_{k,2}(n) = kn + 2 \pmod{10}$.

EXERCISE 2.11.35. *Consider the standard* 26 *letter alphabet. Prove that* $f_{11,2}(n) = 11n+2 \pmod{26}$ *is a valid affine cipher, while* $f_{12,2}(n) = 12n+2$ *is not.*

EXERCISE 2.11.36. *Consider an alphabet with* 13 *letters. How many valid affine ciphers are there? What if the alphabet has* 13^2 *letters? What if it has* 91 *letters?*

EXERCISE 2.11.37. *Consider an alphabet with* p *letters for a prime* p. *How many valid affine ciphers are there? What if the alphabet has* p^2 *letters?*

EXERCISE 2.11.38. *Let* p *and* q *be two distinct primes, and consider an alphabet with* pq *letters. How many valid affine ciphers are there?*

EXERCISE 2.11.39. *Your friend from Exercise* 2.11.16 *is back. This time, he decides to increase the security of the encryption by choosing* 100 *affine ciphers and applying them, one at a time, to the message. He is convinced that this has greatly increased the security. Is he right?*

• From §2.7: The Vigenère Cipher

EXERCISE 2.11.40. *Encrypt* `FOLLO WTHEY ELLOW BRICK ROAD` *with the keyword* `OZ`.

EXERCISE 2.11.41. *Decrypt* `LOSVW AZBSH DHQID ARSLG EL`, *encrypted with the Vigenère cipher using* `SHOES` *as a key.*

EXERCISE 2.11.42. *How many keywords are there of length* 5*? How many keywords are there of length at most* 5*? Find a formula for the number of keywords of length at most* n.

EXERCISE 2.11.43. *In general, the longer the keyword, the greater the security; however, there is a cost. A longer keyword requires agents to remember more. One possibility is to use a palindrome. (A palindrome is a word or sentence that is read the same backwards and forwards. Famous examples include* "`Madam, I'm Adam`" *and* "`A man, a plan, a canal: Panama`".) *If a palindrome keyword of length* 1701 *is used, how many (and which!) letters much an agent remember?*

EXERCISE 2.11.44. *There are* 26! *substitution alphabet ciphers. Find the smallest* n *so that the number of keywords of length at most* n *exceeds* 26!. *We may interpret this value as telling us how big the keywords need to be before we have more options than a generic substitution alphabet cipher. Find the smallest* m *so that there are at least* 26! *keywords of length exactly* m.

EXERCISE 2.11.45. *Decrypt the message* `SEANWIEUIIUZHDTGCNPLBHXGKO ZBJQBFEQTXZBWJJOYTKFHRTPZWKPVURYSQVOUPZXGGOEPHCKUASFKIPWPLVOJI ZHMNNVAEUDXYFDURJBOVPASFMLVFYYRDELVPLWAGJFSXGJFXSBCUHAPMCMPHIJ MVBTASETOVBOCAJDSVQU`. *This message was sent almost* 150 *years ago to General Pemberton, the Confederate commander of Vicksburg. The message was never delivered, probably because by the time the courier arrived the Union flag was flying over the city.* Hint: It's likely that the message began with some variant of "Dear General Pemberton". Parts of the rest of the hint are encrypted with a Vigenère cipher with keyword `SOUTH` so we don't accidentally ruin your fun in case you want to try your hand at decryption. The message starts with `YSHEWWAVXYLCH`. The text was encrypted with one of the standard Confederate keywords, `EOHVOWGNXYTZOYM` (the others were `UCGISWHYOPUHIKF` and `UCGXYWHLBIMHCHU`).

- **From §2.8: The Permutation Cipher**

EXERCISE 2.11.46. *The message* `XIMTI LLAPU` *was encrypted with the permutation cipher with key* $\begin{pmatrix} 1 & 2 & 3 & 4 & 5 \\ 3 & 2 & 1 & 5 & 4 \end{pmatrix}$. *Decrypt it.*

EXERCISE 2.11.47. *The message* `IESEZ EHTAD Y` *was encrypted with the permutation cipher with key* $\begin{pmatrix} 1 & 2 & 3 & 4 & 5 \\ 3 & 2 & 1 & 5 & 4 \end{pmatrix}$. *Decrypt it.*

EXERCISE 2.11.48. *Give an encryption for the message* `PERMU TATIO N` *using the permutation*

$$\begin{pmatrix} 1 & 2 & 3 & 4 \\ 3 & 4 & 2 & 1 \end{pmatrix}.$$

EXERCISE 2.11.49. *Give an encryption for the message* `ONEIF BYLAN D` *using the permutation*

$$\begin{pmatrix} 1 & 2 & 3 & 4 \\ 3 & 4 & 2 & 1 \end{pmatrix}.$$

EXERCISE 2.11.50. *Imagine we have a message of length* 121, *and we want to encrypt it with a permutation cipher of length* 10. *As our length is not a multiple of* 10, *we need to add nine garbage characters at the end of the message. What's dangerous about choosing* `ZZZZZZZZZ`?

EXERCISE 2.11.51. *We saw that while there are over* $4 \cdot 10^{26}$ *distinct substitution alphabet ciphers, many of them (such as the Caesar and affine ciphers) are trivial to break by brute force. Similarly, some permutation ciphers are easily broken. We say a permutation is a* ***transposition*** *if it exchanges two numbers and leaves the rest alone. Discuss why transpositions are poor keys for a permutation cipher. What percentage of permutations are transpositions?*

EXERCISE 2.11.52. *Let* n *be an even number. Consider the special permutations where we can break the* n *numbers into* $n/2$ *pairs such that the effect of the permutation is just to switch each element of a pair. For example, if* $n = 8$ *we could take as pairs* $(1,5)$, $(2,8)$, $(3,4)$, *and* $(6,7)$ *and get the special permutation* $\begin{pmatrix} 1 & 2 & 3 & 4 & 5 & 6 & 7 & 8 \\ 5 & 8 & 4 & 3 & 1 & 7 & 6 & 2 \end{pmatrix}$. *If* $n = 26$, *what percentage of the permutations are special? More generally, find an expression for the number of special permutations of* $2m$ *objects.*

EXERCISE 2.11.53. *List all the permutations of three elements, and list all the permutations of four elements. It's important to have a good, methodical way of enumerating the permutations so one doesn't miss anything.* Hint: You can use the enumeration of the permutations of three elements to "build" the permutations of four elements.

EXERCISE 2.11.54. *We define the product* $\sigma_2\sigma_1$ *of two permutations* σ_1 *and* σ_2 *as the result of first applying* σ_1 *and then applying* σ_2. *If* $\sigma_1 = \sigma_2 = \sigma$, *we write* σ^2 *for the product.*

- *Prove the product of two permutations is a permutation.*
- *Show* σ^2 *is the identity permutation (i.e., it sends each number back to itself) if* σ *is any permutation of two elements. Is* σ^3 *the identity whenever* σ *is a permutation of three elements? What about* σ^4 *when* σ *is a permutation of four elements?*

EXERCISE 2.11.55. *Our friend from Exercise* 2.11.16 *is back. Worried about the security of the permutation cipher, he decides to choose* 100 *permutations of six elements, and apply them one after the other. How many permutations of six elements are there, and how many ways can he choose* 100 *permutations of six elements (order matters; this means it matters which is his first choice, which is his second, and so on). If he argues as he did in Exercise* 2.11.16, *he will believe he has gained enormously in terms of security. How hard does he think it will be for someone to guess his choice of* 100 *permutations of six elements? Do we need to figure out his choices to crack the code? How many possibilities are there to check?*

- **From §2.9: The Hill Cipher**

EXERCISE 2.11.56. *Carry out the following matrix multiplications modulo* 26 :

(a) $\begin{pmatrix} 3 & 2 \\ 0 & 15 \end{pmatrix} \begin{pmatrix} 13 & 2 \\ 8 & 22 \end{pmatrix}$.

(b) $\begin{pmatrix} 5 & 11 \\ 2 & 3 \end{pmatrix} \begin{pmatrix} 22 & 8 \\ 4 & 19 \end{pmatrix}$.

EXERCISE 2.11.57. *Carry out the following matrix multiplications modulo* 26 :

(a) $\begin{pmatrix} 11 & 3 \\ 2 & 5 \end{pmatrix} \begin{pmatrix} 19 & 2 \\ 3 & 2 \end{pmatrix}$.

(b) $\begin{pmatrix} 0 & 5 \\ 25 & 15 \end{pmatrix} \begin{pmatrix} 13 & 3 \\ 4 & 4 \end{pmatrix}$.

EXERCISE 2.11.58. *In general, matrix multiplication is not commutative; this means that AB is not always equal to BA. Find two 2×2 matrices A and B such that $AB \neq BA$.*

EXERCISE 2.11.59. *Check that the formula below gives the inverse of a 2×2 matrix by carrying out the matrix multiplication (you should get the identity matrix):*

$$\frac{1}{ad - bc} \begin{pmatrix} d & -b \\ -c & a \end{pmatrix} \begin{pmatrix} a & b \\ c & d \end{pmatrix}.$$

EXERCISE 2.11.60. *How must the inverse formula from Exercise* 2.11.59 *change if we only desire an inverse modulo* 26? Hint: The hardest part is figuring out what to do with the factor $1/(ad - bc)$.

EXERCISE 2.11.61. *Find the inverses of the following matrices (or indicate "no inverse" when there is none). When you find an inverse, check it by multiplying by the original matrix to get the identity.*

(a) $\begin{pmatrix} 11 & 3 \\ 2 & 5 \end{pmatrix}^{-1}$.

(b) $\begin{pmatrix} 19 & 2 \\ 3 & 2 \end{pmatrix}^{-1}$.

(c) $\begin{pmatrix} 0 & 5 \\ 25 & 15 \end{pmatrix}^{-1}$.

(d) $\begin{pmatrix} 13 & 3 \\ 4 & 4 \end{pmatrix}^{-1}$.

EXERCISE 2.11.62. *How many matrices modulo* 26 *are invertible?*

EXERCISE 2.11.63. *How many matrices modulo* 6 *are invertible? Modulo* 10? *Modulo* 14? *Modulo* 22?

EXERCISE 2.11.64. *Notice all of the numbers from the previous two exercises are of the form $2p$ for a prime p. Is there a pattern for how many such matrices are invertible? If yes, try and prove it.*

EXERCISE 2.11.65. *How many matrices modulo* 4 *are invertible? Modulo* 9? *Modulo* 25?

EXERCISE 2.11.66. *Notice all of the numbers from the previous exercise are of the form p^2 for a prime p. Is there a pattern for how many such matrices are invertible? If yes, try and prove it.*

EXERCISE 2.11.67. *Recall that B is the inverse of A if $AB = BA = I$ (here we are looking at true inverses, and not inverses modulo* 26). *Prove that if A has an inverse, then the inverse is unique; in other words, if C also satisfies $CA = AC = I$, then $C = B$. If instead we looked at inverses modulo* 26, *must the inverse be unique?*

EXERCISE 2.11.68. *Encrypt the message* `MATRI CES` *using the Hill cipher with key* $\begin{pmatrix} 7 & 3 \\ 5 & 2 \end{pmatrix}$.

EXERCISE 2.11.69. *The message* `IWUTP NIWGZ` *was encrypted with the Hill cipher using the matrix* $\begin{pmatrix} 5 & 17 \\ 4 & 15 \end{pmatrix}$ *as the key. Find the corresponding decryption matrix and decrypt the message.*

EXERCISE 2.11.70. *While we can generalize the* 2×2 *Hill cipher to larger matrices, it's worth examining what happens if we go in the opposite direction. Specifically, what would a* 1×1 *Hill cipher correspond to?*

Chapter 3

Enigma and Ultra

In Chapter 2 we discussed the classical ciphers; we'll see in Chapters 4 and 5 how to easily break them. This chapter is about the Enigma, one of the most famous cryptographic systems ever. It's similar to the ciphers we've discussed in that it encrypts letters to letters, but in a significantly complicated way. The Germans used it during World War II, and believed it provided them with perfect security in their communications. We'll start with a quick review of its history, then move on to some of the mathematics needed to study it. The analysis illustrates a common theme in cryptography: frequently, it only takes elementary mathematics to state or describe a problem, but solving it is another story entirely! Essentially all we need in this chapter is elementary combinatorics, at the level of counting how many ways there are to choose a fixed number of people from a larger group (with and without the order of choice mattering).

As the Enigma was in fact broken, it's natural to ask what went wrong, or, from the point of view of the Allies, what went right. The German cryptographers weren't fools. Using these counting arguments, we'll see why they thought there was no chance of the Allies reading their messages, and then see how little mistakes in implementation gave the Allies the needed opening into its secret.

3.1. Setting the Stage

The movie *Patton* is about, not surprisingly, the life of General George S. Patton. In an early scene he arrives in Africa shortly after *Operation Torch*, the Allied landings in Africa in November 1942. The U.S. forces have just been routed by the Germans, and Patton takes charge. Shortly thereafter, his units engage and soundly defeat elements of German Field Marshall Rommel's army. In one of the most memorable moments of the movie, Patton surveys the victory and exclaims: "*Rommel, you magnificent bastard, I READ YOUR BOOK!*"

Rommel *had* written a book on tank strategies, *Infanterie Greift An* (Infantry Attacks), which was published a few years before WWII in 1937; however, it is rare to have such insights into your foe. As you would expect, there is a huge advantage if your enemy is kind enough to describe their methods, preferences, targets, and strategy to you. Thus, most of the time, each side goes to great lengths to keep these secret; however, different units must be able to communicate quickly and securely with each other. In the midst of a battle, commanders need to redeploy their forces, either to attack or to reinforce weakened positions. They can't wait hours for a message to be encrypted, transmitted, and decoded—these must be done almost instantly or the information is useless. A vulnerability may exist for only a few minutes, and you have to act fast. At the same time, you don't want the enemy to decode your orders and change their plans.

For centuries, cryptographers had been working on this problem, trying to devise a way to let select people exchange information quickly without others unraveling the message. With the Enigma, the Germans believed they had found a practical and essentially impenetrable solution.

The **Enigma** was a machine used by the Germans in World War II and before, to encrypt and decrypt messages. As we'll see when we read about all the different wirings and rotors and parts (see Figure 3.1 for an illustration), it's a very complicated contraption with an incredibly large number of possibilities that need to be tried to successfully decrypt a message. Because the number of possibilities was so large (over 10^{100} as we see below, significantly more than the approximately $4.03 \cdot 10^{26}$ substitution alphabet ciphers discussed in Chapter 2), the Germans were very confident that they had a secure means to communicate, and they sent many important messages via the Enigma. Fortunately for the Allies, the Germans made several mistakes in using the Enigma, and these errors were exploitable and led to a significant decrease in security. The Allied efforts to decrypt the German Enigma traffic are known as **Ultra**.

How valuable were these decryptions? Estimates range from shortening World War II by two years to preventing a German victory. The Allies were able to use the decoded messages for a variety of purposes, ranging from determining Germany's long-term strategic goals to dealing with more immediate threats, such as the location of German submarines hunting Allied convoys.

The Germans placed absolute confidence in the security of the Enigma. As one German cryptographer stated (see [**32**]), "From a mathematical standpoint we cannot speak of a theoretically absolute solvability of a cryptogram, but... the solvability is so far removed from practical possibility that the cipher system of the machine, when the distribution of keys is correctly handled, must be regarded as virtually incapable of solution." All branches, all levels of the military in the Third Reich used the Enigma freely. In fact, a large part of Hitler's idea of *blitzkrieg* was based on the Enigma: as General Erhart Milch stated "the real secret is speed—speed of attack

FIGURE 3.1. A German Enigma machine. (Image from Wikipedia Commons.)

through speed of communication", which the mechanized Enigma provided German commanders, as well as total security. Or so they believed.

Breaking the Enigma was a daunting task. The number of possible combinations used to produce a message was enormous. Given that the Germans used a new one every day, trying to guess that combination was prohibitively difficult. Fortunately, due to espionage, mathematics, and some amazing inspiration, the Allies were able to decode German messages daily. As it turns out, many of the components of the security of the Enigma could be divorced from each other and dealt with separately, which reduced the difficulty of the task significantly.

It's impossible to do the Enigma and Ultra justice in a short chapter. Our goal here is to give a brief overview of how the Enigma worked, and how the Allies beat it, and to talk just a little about the importance of codes in war. For more on the subject, see the excellent treatises [**26, 32, 45, 54, 56**].

Before we turn to the mathematics of the subject, it's worth looking at the Enigma one last time (Figure 3.1). One of its great advantages is the ease of use. There are two keyboards on the top: one is like a typewriter with keys to press, the other has bulbs underneath the letters that light up. After setting everything up to encrypt a message, all the operator has to do is type; each time he presses a key, what that letter encodes to lights up. To decrypt a message, again after choosing the correct settings, the person on the other end just types in the encrypted message. Each time she types a letter, the letter that lights up is what that letter decrypts to. In other words, encryption and decryption is done at the speed of the typist! There is no difficult math problem to be solved on either end by the operators; the machine takes care of everything. This is a very desirable feature for battlefield situations.

3.2. Some Counting

There are several reasons for studying the Enigma early in a cryptography course. It's one of the most important examples ever, and its successful decryption changed the fate of the world. It's a natural outgrowth of the classical ciphers discussed in Chapter 2. Mathematically, it's extremely accessible. All we need is some elementary combinatorics. We begin by quickly reviewing these through some basic problems, isolating the functions that will appear in later sections when we turn to analyzing the Enigma.

All our arguments below are based on a very important principle, the **multiplicativity of combinations**.

> **The multiplicativity of counting:** If we have several decisions to make, the total number of possible decisions is the product of the number of options for each decision.

Sometimes we are fortunate, and the various choices are independent of each other; sometimes they are not, and the counting is more involved. A standard example involves someone getting dressed and having to choose an outfit.

Question #1: *Assume Alan has four different shirts and three different pairs of pants. How many distinct outfits (of one shirt and one pair of pants) can he create?*
Answer: Alan must make two decisions. He has to chose a shirt and he has to choose a pair of pants. Note that the choice of shirt has no effect on the choice of a pair of pants. Similarly, the choice of pants has no impact on our choice of shirt. There are four ways to choose a shirt, one for each shirt in the closet. Similarly he has three choices for a pair of pants. Thus there are $4 \cdot 3$ ways to choose a shirt and a pair of pants.

To see this, let's label our shirts $\{\text{shirt}_1, \text{shirt}_2, \text{shirt}_3, \text{shirt}_4\}$ and the pants $\{\text{pants}_1, \text{pants}_2, \text{pants}_3\}$. The possible combinations are

$$\begin{array}{ccc} (\text{shirt}_1, \text{pants}_1), & (\text{shirt}_1, \text{pants}_2), & (\text{shirt}_1, \text{pants}_3), \\ (\text{shirt}_2, \text{pants}_1), & (\text{shirt}_2, \text{pants}_2), & (\text{shirt}_2, \text{pants}_3), \\ (\text{shirt}_3, \text{pants}_1), & (\text{shirt}_3, \text{pants}_2), & (\text{shirt}_3, \text{pants}_3), \\ (\text{shirt}_4, \text{pants}_1), & (\text{shirt}_4, \text{pants}_2), & (\text{shirt}_4, \text{pants}_3). \end{array}$$

When doing problems like this, it is important to enumerate the possibilities in a consistent way so no options are missed. Here we went through all ensembles by first considering everything where the shirt was our first shirt. After listing all those cases, we then went through all the outfits where we chose our second shirt. Next we considered sets where we had our third shirt, and finally we looked at choosing the fourth shirt. In the next problem we show another approach to exhausting all the options by drawing a tree of all the possibilities.

What makes this problem simpler than later ones is that choosing a shirt has no impact on choosing pants, and vice versa. Our next example requires us to make multiple choices, but each choice comes from the same candidate list. Thus, once we choose an object it is no longer available for future decisions. The answer is again obtained by multiplying possibilities, but now these numbers have some mutual dependence.

Let's start with a basic example. We have essentially already done this when we counted the number of substitution alphabet ciphers in Chapter 2, but it's worth revisiting as we need a generalization.

Question #2: *How many ways can you arrange three people in a line?*
Answer: Suppose we want to seat our friends Alan, Bernhard, and Charles. In this problem the order of the three people matter. Thus Alan, Charles, and Bernhard is different than Charles, Bernhard, and Alan. A good way to approach this is to fill the three available slots one at a time. How many choices do we have for the first position? Before anyone is assigned, we have three people: Alan, Bernhard, and Charles. We have to choose one of them for the first position. We're now left with two people, and two open spots. We don't need to know *which* two are left, only that there is *one* person fewer. We thus have two ways to choose which person is second. For the last spot, we only have one choice left, and we must put the remaining person in that final place. The total number of arrangements is six, coming from multiplying the three ways to choose the first person, the two ways to choose the second, and one way to choose the last.

A good way to view this is to look at this as a tree (see Figure 3.2). We start with all positions open. We have three choices for the first person, which creates three branches. At the next stage, we have two remaining people to choose from. Each of our initial branches then branches again, but this time into just two possibilities, one for each remaining person. Note that while the people involved in the split differ from branch to branch, the

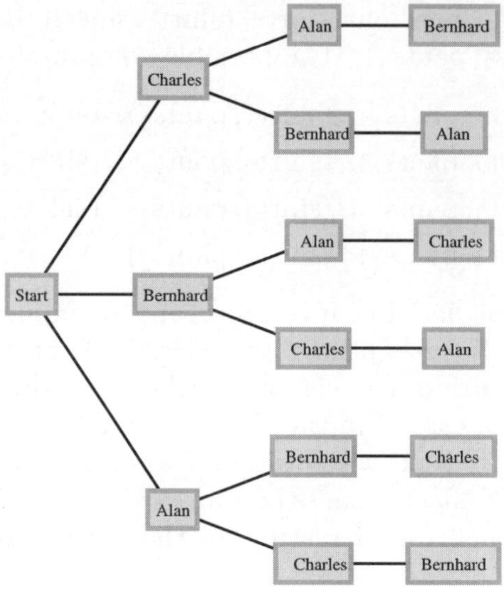

FIGURE 3.2. Tree diagram indicating the number of ways to put three people in a line (when order matters). Note we end up with $6 = 3 \cdot 2 \cdot 1$ possible arrangements corresponding to the 6 leaves of the tree.

number of people involved is always two. Finally, each branch is continued once more as we assign the third person. As only one person is left to assign, the choice is forced upon us.

Arguing along the same lines, we see that if we had four people to arrange in a line (with order mattering), the number of possibilities would be $4 \cdot 3 \cdot 2 \cdot 1 = 24$. These descending products occur so frequently in the subject that we have a special name and notation for them. This is the **factorial function** from Chapter 2, and we write 3! for $3 \cdot 2 \cdot 1$ and 4! for $4 \cdot 3 \cdot 2 \cdot 1$. Notice that this function grows *very* rapidly; 10! is ten times larger than 9!, which is nine times larger than 8!, and so on. For each positive integer n, we have $n! = n \cdot (n-1)!$.

As the above problem suggests, there is a combinatorial interpretation of the factorial function: $n!$ is the number of ways of arranging n people in a line, when order matters. It turns out to be useful to define 0!. Though it might seem surprising, the right definition is to take 0! to be 1. We may interpret this as follows: There is only one way to do nothing!

We now generalize the previous question.

Question #3: *How many ways can you line up three of five people?*
Answer: Again, this is a problem where order matters. If the five people are Alan, Bernhard, Charles, Danie, and Ephelia, then choosing Charles, Danie, and Alan is counted as different than choosing Danie, Alan, and Charles. The solution to this problem is similar to the previous. We have already solved how to choose five people from five: the answer is 5!, or $5 \cdot 4 \cdot 3 \cdot 2 \cdot 1$. The

difference now is that we're not choosing all five people. We're only choosing three, and thus we stop after the third choice. Our answer is therefore $5 \cdot 4 \cdot 3$, or 60. Notice how quickly the number of possibilities grow; drawing a tree with all the possibilities is no longer practical!

There is another way to write down the answer. Instead of writing $5 \cdot 4 \cdot 3$, we could instead write $5!/2!$. The reason is

$$\frac{5!}{2!} = \frac{5 \cdot 4 \cdot 3 \cdot 2 \cdot 1}{2 \cdot 1} = 5 \cdot 4 \cdot 3.$$

What we're doing is multiplying by 1 in a clever way, and multiplying by 1 won't change the answer. Here we wrote 1 as $2!/2!$. The reason this is a good idea is that the product $5 \cdot 4 \cdot 3$ looks a lot like a factorial, but it isn't. It's missing the $2 \cdot 1$. By including that factor, we obtain a factorial. The problem, of course, is that we can't just multiply part of an expression; thus if we multiply by $2 \cdot 1$ we must also divide by it. If instead we had to choose four people from nine with order mattering, the answer is $9 \cdot 8 \cdot 7 \cdot 6$. To make this into $9!$ we need the factor $5!$. We thus multiply and divide by this, and find the number of ways to choose four people from nine (when order matters) is $9!/5!$.

More generally, if we want to take k people from n people with order mattering, it's just $n \cdot (n-1) \cdots (n-(k-1))$. Why is this the product? Clearly, the first term is n, then $n-1$, then $n-2$. What is the last term? We need to have k terms in all. The first term, n, is really $n-0$. The second term is $n-1$, and notice that we are subtracting one less than the number term it is. Thus to find the kth term, we must subtract off one less than k, or $k-1$. Thus the last term is $n-(k-1)$. We've just shown that the number of ways to choose k people from n people (with order mattering) is $n(n-1) \cdots (n-(k-1))$. This wants to be a factorial, but it doesn't go all the way down to 1. To remedy this, let's argue as before and multiply by $(n-k)!/(n-k)!$. This won't change the answer as we're just multiplying by 1, but now we have all the numbers from n down to 1 being multiplied together, and we see that the answer is just $n!/(n-k)!$.

Let's summarize what we have done so far.

> The number of ways to order n people (where order matters) is $n! = n(n-1) \cdots 2 \cdot 1$, while the number of ways to choose and order k people from n is $n!/(n-k)!$ (assuming k is an integer at most n). It is convenient to set $0!$ equal to 1. Some books write nPk for $n!/(n-k)!$.

So far, we've looked at how to arrange people (or more generally anything) when order matters. What happens when you just need to pick some people and the order in which you pick them doesn't matter? We can view this as choosing a subset of a group to serve on a committee, and all committee members are equal; conversely, what we did earlier (arranging people with order mattering) would be assigning people to a group but designating one the president, one the vice president, one the treasurer, and so on.

For example, imagine we want to pick four people from a group of four people. How many different ways could we do this? Since there are only four people, and we have to pick four, there's no freedom. We *have* to pick all four, and thus there's only one way to pick four people from four people when order does not matter. Equivalently, we can say there is only one group of four people that can be made from four people. Let's consider a more involved case.

Question #4: *How many ways are there to pick four people from a group of nine people, when it doesn't matter which order the people are selected?*
Answer: Now things are more interesting, and it isn't immediately clear what the answer should be. Fortunately, there's a nice way to view this problem. Let's first consider the easier problem of how many ways there are to choose four people from nine. We looked at this earlier, and saw that the answer is $9 \cdot 8 \cdot 7 \cdot 6$ or $9!/5!$. This is the number of ways of choosing four people from 9, but order matters. The clever idea is to realize that this is precisely $4!$ more than the number of ways to choose four people from nine when order does not matter. Why? For each group of four people chosen, there are $4!$ ways to order them. Each of these $4!$ ways has been counted in the $9!/5!$ ways to choose four people from nine with order mattering. Thus, if we take $9!/5!$ and divide by $4!$, we remove the extra counting, we remove the ordering we don't care about and obtain the answer. The reason this works is that *all* groups of four people have the same number of arrangements.

There is nothing special with choosing four people from nine. If instead we wanted to take k people from n, not caring about the order, we would argue similarly. We first note that there are $n!/(n-k)!$ ways to choose k people from n when order matters. For each group of k people there are $k!$ ways to assign an ordering; thus each group of k people is counted $k!$ times among the $n!/(n-k)!$ orderings. If we divide by $k!$, we remove this counting, and we find that the number of ways to choose k people from n people, not caring about the order in which they are chosen, is simply $\frac{n!}{k!(n-k)!}$.

It was worth defining the factorial function and giving it special notation because of how often it appears in combinatorial problems. Similarly, it is worth introducing some notation for $\frac{n!}{k!(n-k)!}$ so we don't have to write this out every single time.

Binomial coefficients: Let n and k be nonnegative integers, with $k \leq n$. We set

$$\binom{n}{k} := \frac{n!}{k!(n-k)!},$$

and read this as n *chooses* k; remember we set $0!$ to be 1. This is called a **binomial coefficient**. The top refers to the number of options we have, and the bottom to how many we are choosing. It is the number of ways to choose k objects from n when order does not matter. Alternative notations are $C(n,k)$ or nCk.

Binomial coefficients have many useful properties. Doing some simple algebra, we see that $\binom{n}{k} = \binom{n}{n-k}$, because

$$\binom{n}{n-k} = \frac{n!}{(n-k)!(n-(n-k))!} = \frac{n!}{(n-k)!k!} = \frac{n!}{k!(n-k)!} = \binom{n}{k}.$$

While the above argument proves the claim, it's not particularly enlightening. There's a much better argument. Note $\binom{n}{k}$ is the number of ways to choose k people from n when order doesn't matter. We can look at this another way. If we choose k people from n, we can view this as *excluding* $n-k$ people from n. Thus $\binom{n}{k}$ must be the same as $\binom{n}{n-k}$. This proof illustrates a central idea in combinatorics: *if you can tell an appropriate story, you can give an interpretation to quantities and see the relationships.*

When encountering new ideas, it's usually a good idea to test simpler cases where you have some intuition. Let's imagine the case where we have three people, and we want to choose a group of three and it doesn't matter what order we choose. Clearly there's only one way to do this (as we have to take everyone), which agrees with our formula since $\binom{3}{3} = \frac{3!}{3!0!} = 1$ (remember $0! = 1$, and in fact now we see why this is a good convention). What if we want to choose three people from four, where order doesn't matter? Our formula tells us there are $\binom{4}{3} = \frac{4!}{3!1!} = 4$ distinct possibilities. To conserve space, let's give the four people the boring names A, B, C, and D. There are 24 ways to choose three of four. There are six ways to order A, B, and C if they are the three people chosen: {A,B,C}, {A,C,B}, {B,A,C}, {B,C,A}, {C,A,B}, {C,B,A}. These six different orderings have the same people, and thus only contribute one possibility. Similarly there are six ways to order the people A, B, and D, another six to order A, C, and D, and finally six more for B, C, and D. Note $6 = 3!$, and thus the number of unordered possibilities is $4!/3! = 24/6 = 4$, as each unordered possibility of three people can be ordered $3!$ ways.

To sum up, we have shown

Combinatorics and binomial coefficients: Let n and k be integers, with $0 \le k \le n$.

- The number of ways to choose k objects from n objects *with order mattering* is $n(n-1)\cdots(n-(k-1))$, which may be written as $n!/(n-k)!$.
- The number of ways to choose k objects from n objects *with order not mattering* is $\frac{n!}{k!(n-k)!}$, which is denoted by the **binomial coefficient** $\binom{n}{k}$.

Armed with our building block functions of factorials and binomial coefficients, we can now analyze the Enigma's complexity and see why the Germans were unconcerned about Allied attempts to crack their codes.

3.3. Enigma's Security

If the Enigma were used properly, it would have been extremely secure. While today computers are everywhere, during World War II the situation was completely different. People often had to examine all the tedious combinations by hand; in fact, the need to quickly exhaust numerous possibilities was a huge impetus in the birth of computers. As we'll soon see from analyzing the Enigma's workings, the Germans were justified in believing that the Enigma would provide secure communications. Fortunately for the Allies, however, the Germans had several practices that greatly weakened the actual security. Before describing their errors, let's take a look at what exactly made the Enigma seem so impregnable. This section relies heavily on *The Cryptographic Mathematics of Enigma* by Dr. A. Ray Miller [**32**]. It's a short pamphlet from the NSA and available online, and we highly recommended it to anyone interested in additional reading.

3.3.1. The Plugboard

There are five things that contributed to the cryptographic strength of the Enigma. The first was a **plugboard** with 26 holes for 13 connector cables, each hole associated with a letter (see Figure 3.3). The operator would plug in these cables as designated in the code book, which in effect would switch the outputs of the connected letters. For instance, if the code book indicated a cable should be plugged into the holes for `E` and `F`, the outputs for those two letters would be switched. In that case, the word `effect` would become `feefct`. If instead `O` and `Y` were connected and `P` and `N` were connected, then `effect` would appear unchanged, but `pony` would become `nypo`. The more pairings we have, the more garbled the words become.

The operator could use anywhere between zero and thirteen cables. Not all letter connections are possible. You can't connect two letters to the same letter, or a letter to itself. Let's see how these choices affect the security of a message. Specifically, let's determine how many different letter swaps are available. We first consider the simplest possibility: don't use any cables (and thus no message is changed in typing). There's only one way to do nothing. Another way of looking at this is that there is only one assignment of cables connecting pairs of letters that results in no swaps: use no cables!

Now let's look at a slightly more difficult example. Let's use just one cable. In this case, we're going to switch two letters (like we did for `E` and `F` before). How many different ways could we do this? Well, we have 26 letters, and we want to *choose* two of them to switch. This is the definition of the binomial coefficient $\binom{26}{2}$ from last section, as this represents the number of ways to choose two objects from 26 when order doesn't matter. One of the hardest parts of combinatorial problems like these is keeping track of when order matters and when order doesn't matter. At the end of the day, we can't tell if we first plugged in one end of the cable to `E` and then to `F`, or the other way around. It doesn't matter; all that matters is that `E` and `F` are

FIGURE 3.3. The Enigma's plugboard, with O and Y connected and P and N connected. (Image from Wikipedia Commons.)

now connected. Thus, for these cable problems, it is always choosing with order immaterial. As $\binom{n}{k} = \frac{n!}{k!(n-k)!}$, we see $\binom{26}{2} = \frac{26!}{2!(24)!} = \frac{26\cdot 25}{2\cdot 1}$, or 325.

We've now figured out how many ways to use zero cables (just one way, which is $\binom{26}{0}$) or one cable ($\binom{26}{2}$ or 325 ways). What about two cables? A natural thought is that it should be the number of ways of choosing four letters from 26, with order not mattering, or $\binom{26}{4}$. While this looks like the next natural term in the pattern, it isn't quite right. The reason is that we need to do more than just choosing four letters; after we choose the four letters, we must pair them into two groups of two. For example, if we choose the four letters A, E, L, and Y, then there are three different ways to pair them into two groups of two: we could have the pairings A-E and L-Y, or A-L and E-Y, or A-Y and E-L. Note that the pairing A-E and L-Y is the same as the pairing E-A and L-Y or L-Y and A-E, and so on, as all that matters is which two are paired and not the order. Because of this, we see it is going to be a lot more work to determine the number of ways to use exactly two cables.

There are several ways we can arrange this calculation. One nice way is to first choose the four letters to be paired with the two cables; there are $\binom{26}{4}$ ways to do this. We then have to split the four letters into two pairs of two; we don't care which pair is listed first or which letter is listed first in each pair. How many ways are there to do this? Similar to our analysis on the number of ways to choose k objects from n objects when order doesn't matter, it's easier to first impose an order and then remove it. What we do here is put labels on the pairs (*first pair, second pair*), and then remove the labels. If we have four letters, there are $\binom{4}{2}$ ways to choose two letters to be the first pair. This leaves us with two remaining letters, which must be chosen to be the second pair; there are $\binom{2}{2}$ ways to choose the two remaining letters for the second pair. Multiplying, we see that there are $\binom{4}{2}\binom{2}{2}$ ways to pair the four letters into two pairs of two, with the pairs being labeled *first pair, second pair.* We now remove the labels. How many ways are there to assign the two labels to the two pairs? There are 2! ways—we have two choices for the label for the first pair, and then one remaining label for the second. Thus, we have over-counted each desired pairing of four letters into two pairs (not caring about which is the first and which is the second pair) by a factor of two. The answer is not $\binom{4}{2}\binom{2}{2}$, but instead $\binom{4}{2}\binom{2}{2}/2!$. Combining this with the number of ways to choose four letters from 26, we see there are $\binom{26}{4}\binom{4}{2}\binom{2}{2}/2!$ ways to use two cables. Multiplying this out gives 44,850, which is much larger than the 325 possibilities with just one cable!

To highlight the method, let's consider three cables. There are several ways to look at this problem. We'll first solve it by modifying our previous argument. We have three cables; as each cable connects two letters, we must choose six letters out of 26. There are $\binom{26}{6}$ ways to do this. Now that this has been done, we have to pair the six letters into three groups of two. It doesn't matter which group is called the first pair or the second or the third or, within a pair, which letter is chosen first. A good way to count the possibilities is to look at how many ways we can pair the six letters into three pairs with the pairs labeled, and then remove the labels to reduce the over-counting. There are $\binom{6}{2}$ ways to choose two letters from the six letters for the first pair, $\binom{4}{2}$ ways to choose the two letters from the remaining four letters for the second pair, and then $\binom{2}{2}$ ways to choose two letters from the remaining two letters for the third pair. There are 3! ways to assign labels to the three pairs; thus we have over-counted by a factor of 3!, which we must remove. Putting this all together, the number of ways to use three cables is $\binom{26}{6}\binom{6}{2}\binom{4}{2}\binom{2}{2}/3!$.

If we multiply out the last factor, $\binom{6}{2}\binom{4}{2}\binom{2}{2}/3!$, we notice something interesting. It's

$$\begin{aligned}\binom{6}{2}\binom{4}{2}\binom{2}{2}\frac{1}{3!} &= \frac{6!}{2!4!}\ \frac{4!}{2!2!}\ \frac{2!}{2!0!}\ \frac{1}{3!}\\ &= \frac{6\cdot 5}{2}\ \frac{4\cdot 3}{2}\ \frac{2\cdot 1}{2}\ \frac{1}{3\cdot 2\cdot 1}\\ &= \frac{6\cdot 5\cdot 4\cdot 3\cdot 2\cdot 1}{6\cdot 4\cdot 2}\\ &= 5\cdot 3\cdot 1.\end{aligned}$$

This result suggests a useful generalization of the factorial notation.

The double factorial: We write $n!!$ to mean the product of every other term, down to 2 if n is even and 1 if n is odd. We call this the **double factorial**. Thus $5!! = 5 \cdot 3 \cdot 1$ while $6!! = 6 \cdot 4 \cdot 2$.

While we could have set $n!!$ to be the factorial of $n!$, that expression just doesn't turn up that often in problems, whereas the product of every other term frequently arises. We can define anything we want—the question is what's useful to define. Here, it turns out that every other product is useful, and worth giving a special name. Thus $5!! = 5 \cdot 3 \cdot 1$, $6!! = 6 \cdot 4 \cdot 2$, $7!! = 7 \cdot 5 \cdot 3 \cdot 1 = 7 \cdot 5!!$ and so on.

We just showed that the number of ways to pair six letters into groups of two, where the pairs are unlabeled and it doesn't matter if a letter is listed first or second in a pair, is $5!!$. One can similarly show that if we had to pair eight letters into groups of two, then the answer would be $7!!$, and in general if we had $2p$ letters that had to be paired into groups of two, then the answer would be $(2p-1)!!$.

One way to see the last claim is to proceed by **induction**. We provide a quick introduction to proofs by induction in the appendix to this chapter, §3.7. We know the claim is true when we have six letters. Imagine now we have eight letters. The first letter has to be paired with one of the remaining seven letters; there are seven ways to do this. We now have six letters remaining that must be paired in groups of two. We know there are $5!!$ ways to do this, and thus the number of ways to pair eight letters in groups of two is just $7 \cdot 5!!$, which is $7!!$. We now turn to ten letters. Consider the first letter. It must be paired with one of the remaining nine letters; there are nine ways to do this. We're now left with eight letters that must be matched in pairs; however, we've just shown that the number of ways to do this is $7!!$. Thus the number of ways to pair ten letters in groups of two is just $9 \cdot 7!!$ or $9!!$. Arguing along these lines proves the claim in general.

Based on our calculations above, we see that if we have p cables connecting $2p$ letters, then the number of possibilities is $\binom{26}{2p}(2p-1)!!$; the $\binom{26}{2p}$ comes from the number of ways to choose $2p$ letters from 26, and the $(2p-1)!!$ comes from the number of ways to match them in pairs.

To get the total number of combinations for any number of cables used, we just add the results for having p cables, where p ranges from 0 to 13. We can just add them because using one cable results in a different output than when we don't use any cables (`effect` versus `feefct`), that is, one setting was independent from all other settings. When we add these together (using $(-1)!! = 1$ to allow us to write things in a common language), we get

$$\binom{26}{0}(-1)!! + \binom{26}{2}1!! + \binom{26}{4}3!! + \binom{26}{6}5!! + \cdots + \binom{26}{26}25!!$$
$$= 532985208200576 \approx 5.32 \cdot 10^{14}. \tag{3.1}$$

This is an *enormous* number of possibilities. To put it in perspective, imagine the Allies could check a million combinations a second (this is far more than was possible back in the 1940s). How long would it take to go through the approximately $5 \cdot 10^{14}$ possible cable connections? There are about 31,536,000 seconds in a year (60 seconds in a minute, 60 minutes in an hour, 24 hours in a day and 365 days in a year). In one year, checking at a rate of a million wirings per second, the Allies would be able to examine 31,536,000,000,000 or about $3.15 \cdot 10^{13}$, less than a tenth of the possibilities! To make matters worse, the wiring chosen changes—after all this work the Allies would only know the wiring for one particular day, and would have to start all over again for the next day.

To sum up, even assuming the ability to check a million wirings a second (which was well beyond what the Allies could do) only gives us a 10% chance of finding one day's wiring *in an entire year's time!* We can begin to see why the Germans were confident in the security of Enigma, especially as this is just the first of many pieces. We'll now discuss the rotors, whose complication dwarfs the contribution of the cables.

3.3.2. The rotors and reflector

The **rotors** were the second component contributing to the strength of the Enigma (see Figure 3.4). These rotors took 26 inputs and wired them to 26 outputs. The number 26 is no accident, as the effect of the rotor was to provide another exchange between letters. This means the Germans could have constructed 26! different rotors. Why is this? Let's look at how this wiring was done. Each rotor has two sets of the alphabet, and the rule is we connect each letter in the first alphabet with a letter in the second alphabet, never connecting two letters to the same letter (this means we can't send A and B on the first alphabet to C on the second, but we could send A to A).

Let's consider the letter A. We can connect it to any of the 26 letters in the second alphabet. Now that we've connected A somewhere, what about the wire for B? As before, we can wire it to any one of the remaining outputs. There are only 25 available now because the wire from the letter A is connected to one of them already. The wire from C can connect to any of the 24 remaining outputs, and so on, so we see that there are 26! different choices. Note this problem is the same as our earlier exercise of arranging n people in a line. We may regard the letters in the first alphabet as positions 1, 2, 3, ..., 26, and the second alphabet is then 26 people we need to order on the line, with order counting.

As the standard Enigma machines had three rotors, the number of possibilities for the three rotors is $26!^3$ (26! ways to choose each rotor), which is approximately $6.559 \cdot 10^{79}$. We thought $5.32 \cdot 10^{14}$ was bad; checking rotors at a rate of a *trillion* possibilities per second means we could investigate about $3.2 \cdot 10^{25}$ triples of rotors in a year. As the universe has only existed for a few billion years (a billion is 10^9), we see that if we check a trillion combinations per second for the entire life of the universe, then we're not even at 10^{40}, and we need to get up to 10^{79}. Actually, the Allies had one piece of information

FIGURE 3.4. The stack of rotors inside an Enigma machine, consisting of three rotors and *Umkehrwalze-B* (the reflector). (Image from Wikipedia Commons from user Matt Crypto.)

on the German rotors: they knew the Germans didn't reuse a rotor in the machine, and thus the three rotors had to be distinct. Therefore, instead of 26^3 possible triples of rotors, in fact it was $26! \cdot (26! - 1) \cdot (26! - 2)$. The difference between these two is negligible, and the first 26 digits of the two numbers are the same. (The difference between these two numbers is about $4.87 \cdot 10^{53}$. We're not saying that this large number is negligible; what's negligible is the savings. We've saved much less than 10^{-20} percent!)

The next variable was how the rotors were initially positioned. Operators could rotate the rotors to any initial position they wanted. Since there were 26 different positions, and three rotors, the number of possible initial positions is $26 \cdot 26 \cdot 26$, or $26^3 = 17,576$.

The fourth component to consider is the **reflector**. The reflector was really just like a plugboard with 13 cables. Each letter was connected to another letter via a wire, essentially switching the letters, just like the plugboard. The difference was that the reflector never changed, and switched all 26 letters. From our analysis earlier, we see that the number of possible reflector arrangements is just

$$\binom{26}{26}(26-1)!! \;=\; 7,905,853,580,625 \;\approx\; 7.9 \cdot 10^{12}.$$

The last contribution to the Enigma's complexity were **notches** on the rotors. These notches controlled when the next rotor would move forward by one letter. The rotor positioned on the right would rotate every time a key was pressed, the second rotor would turn every 26 key strokes, and the last rotor would rotate once for every 26 rotations of the second rotor, or once every $26 \cdot 26 = 676$ keystrokes. The notches determined when the second and third rotor first turned. If the first notch were in a particular position, the second rotor would turn for the first time after, say, three keystrokes. If it were in a different position, the second rotor would not rotate until, say, 19 keys had been pressed. The notch on the first rotor could be in one of 26 positions, and similarly for the second rotor. Since the reflector didn't rotate, the notch on the third rotor did nothing. Therefore, the number of possibilities from this component of the Enigma is just $26^2 = 676$. The notches provided an enormous amount of security. Without the notches, each letter is always encrypted the same way. This means the messages would be vulnerable to a frequency analysis. We'll discuss this attack in greater detail in Chapter 4, but the gist is easy to describe: if you have a very long message in English, the most common letter is almost surely an E. Using letter frequencies (or, even better, frequencies for pairs and triples of letters), we can make very good guesses at how the letters are being switched. The advantage of the notches is that at first an E is encrypted as a Y, and then maybe as an N, then perhaps as an R, and so on. By constantly changing how each letter is encrypted, it's impossible to attack the code with a frequency analysis.

To sum up, the number of possibilities introduced by the rotors, notches and reflector is

$$26!(26!-1)(26!-2) \cdot (26)^3 \cdot \binom{26}{26}(26-1)!! \cdot 26^2 \approx 6.16 \cdot 10^{99}.$$

In the previous section we looked at the plugboard. Its contribution of about 10^{14} looked impressive, but that complexity is dwarfed by these other features.

3.3.3. Possible Enigma States

Now comes the fun part—we'll see why the Germans had such faith in the Enigma. Since each component could be changed without affecting any other part, to get the total number of Enigma configurations we just need to multiply our previous numbers together. The result is simply astonishing. In its full glory, we get

$$\begin{aligned}&3,283,883,513,796,974,198,700,882,069,882,752,878,379,955,\\&\quad 261,095,623,685,444,055,315,226,006,433,615,627,409,666,\\&\quad 933,182,371,154,802,769,920,000,000,000,\end{aligned}$$

or approximately $3 \cdot 10^{114}$.

This number is so much larger than what we've seen before that it takes a while to view this in a meaningful way. Currently we believe the universe is about 14 billion years old, or approximately $3.2 \cdot 10^{22}$ seconds. Using guestimates from the number of stars in the universe, one gets that there should be about 10^{80} atoms in the universe (give or take a few orders of magnitude). Imagine each such atom were a supercomputer devoted entirely to checking Enigma possibilities and capable of looking at a billion (i.e., 10^9) setups per second. If these supercomputers had been running nonstop since the creation of the universe, by today they would only have checked a little less than 10^{112} possibilities. In other words, in such a situation of unbelievably fantastic computer power at our disposal (and for such an incredibly long period of time), we would have less than a 1% chance of having found the correct wiring for a given day! It's no wonder that the Germans felt the Enigma provided more than enough of a challenge to the Allies, and were not worried about their messages being read.

3.4. Cracking the Enigma

In the last section we showed that the Germans had more than 10^{114} possible ways to configure their Enigma machines. Against that sort of number, how could the Allies possibly have broken the Enigma? Fortunately for the Allies, the Germans made several mistakes in its use. For a very good read and more detail on some of the ways the Enigma was weak, we recommend *The Code Book* by Simon Singh [**45**], from which much of the information in this chapter is taken. Another good source is the NSA pamphlet *Solving the Enigma—History of the Cryptanalytic Bombe* by Jennifer Wilcox [**54**], available online.

Before looking at some of the German mistakes in implementation, let's look at some things that will bring that terrifying number of 10^{114} down a little right off the bat. For one thing, suppose we can solve every challenge the Enigma throws at us except how the plugboard is arranged. Then what we have is essentially a monoalphabetic cipher, i.e., a substitution alphabet cipher. In other words, it just switches letters and always switches them the same way. We will see how to use frequency analysis to solve this puzzle in Chapter 4.

Moreover, if not all the letters are switched, it might be possible to just guess what the words are. For instance, take the phrase "*sieben und zwanzigste Bataillon*", which is German for twenty-seventh battalion. If you were reading a message from the 27th Battalion, you might guess that "*aiebenundzwsnzigatebstsillon*" really means "*siebenundzwanzigstebataillon*", and that `S` and `A` have been switched. Continuing in this manner, you could, perhaps, find all the plugboard settings. So, assuming we can handle everything else, we've reduced the number of settings we have to check by about 10^{14}. Unfortunately this still leaves us the problem of solving the rest, and the number of combinations there is of the order 10^{100}, but we've made some progress.

We now turn to the rotors. We need to choose three different rotors. There are 26! possibilities for each rotor, which led to approximately $6.559 \cdot$

10^{79} triples of distinct rotors. This is a huge part of the remaining 10^{100} possibilities, and if we can make good progress here, we're well on our way to deciphering the Enigma. It's important to remember the distinctions and differences between theory and practice. While in theory a German soldier could have any set of rotors on him, a little thought shows that there is no way that he can carry all $26! \approx 4 \cdot 10^{26}$ rotors with him in the field. Reasons for this range from the obvious (it would weigh too much, there's no way German factories could churn out that many rotors) to the more subtle (if he had even 10^{10} rotors on him, it would take a long time to find the right ones to use, and the Enigma is supposed to provide not only secure but also rapid communication). So, naturally, in practice there would only be a few available rotors. While this looks promising, unfortunately we still don't know the exact layout of the rotors. How can we solve this problem?

Instead of trying to magically guess the right layout, the Allies used espionage. Even before the war started, the French were able to obtain documents that gave the layout of all of the rotors in use at the time, and during the war, the Allies occasionally captured Enigma machines. Thus, instead of having to test $6.559 \cdot 10^{79}$ different rotor wirings, they could test five (or in the German navy's case ten, as the navy correctly felt that more security was needed). This is an enormous savings, and without this precious information no further progress could have been made.

Next, let's look at the notches. Remember the notches determine when the rotors advance by one letter, and that only the first two notches affect the code. Therefore, you can imagine that if you could solve every other part of the arrangement, you could just decode until the first notch advanced, which would cause the decoded message to suddenly become gibberish again. Whenever that happened, you would know that another notch had advanced, and in this way, you could easily determine the initial settings of the notches.

All we have left are the rotor settings. If there were some way to separate finding the rotor settings from all the other problems, we could solve the entire cryptogram. As it turns out, there are two ways in which this was done. The first was developed by Marion Rejewski and the Polish cryptanalysts before the war, and the second was developed by Alan Turing and the British once the war had begun.

Once the Germans switched over to using the Enigma in 1926, most European intelligence agencies quickly gave up decoding German messages. There was one nation, however, that could not afford to relax: Poland. They realized that being between Germany and the Soviet Union was a dangerous place to be, especially since they now controlled formerly German areas. As a result, they needed all the intelligence they could get from communications intercepts.

When the French got a contact in the German signals corps, the Poles requested all the information that the French had on the Enigma. They also did something that had never been done before: they hired mathematicians as cryptanalysts. In the past, cryptanalysts were generally linguists, or people who were good with words, but now, with the advent of the mathematical complexity put forth by the Enigma, a different sort of mind was required. These two moves proved to be a winning combination. They got the wiring information from espionage, and the ability to put that information to good use from the mathematicians.

The Germans distributed what was called the *day key* in a different code book every month. The day key specified the settings for the Enigma for that day: which rotors went where, which letters to switch, and so on. The day key, however, was used only to transmit three letters at the beginning of the message, which indicated what the receiver should set his rotors to for the rest of the message. Except, that's not exactly right. Rather than transmit just those three letters, for much of the war the letters were transmitted twice to avoid operator error. This was an enormous mistake, and allowed Rejewski to break the Enigma.

The following example is taken from Simon Singh's *The Code Book* [**45**]; for more details, see his chapter "Cracking the Enigma". Suppose you intercept a message that starts with `PEFNWZ`. You know that `P` and `N` are the same letter, but are enciphered differently because of the way the Enigma works. Similarly, `E` and `W`, and `F` and `Z` are the same letters, respectively. After receiving many messages and using this information over and over, you might tabulate the relationships in the following way. You know that the first and fourth letter are the same. If you intercepted four messages that started with `LOKRGM`, `MVTXZE`, `JKTMPE`, and `DVYPZX`, you would know that `L` and `R`, `M` and `X`, `J` and `M`, and `D` and `P` are all respectively the same letter. If you did this enough times, you might come up with a table like this:

A	B	C	D	E	F	G	H	I	J	K	L	M
F	Q	H	P	L	W	O	G	B	M	V	R	X

Let's look at this closely. What is `A` linked to? `F`, as we can see by looking in the second row below `A`. What is `F` related to? Looking at the second row just below `F`, we see `W`. And below `W` we see `A`. Well that's interesting. We've found a chain of letters that loops back on itself: `A`→`F`→`W`→`A`. Clearly, if the day key were set differently, we would have a different chain, maybe longer, maybe shorter, and probably different letters. However, and this is really important, the length of the chains is determined purely by the rotor settings! The plugboard might change what letter goes where, but the length of the chain would stay the same regardless of the plugboard setting. Using this knowledge allowed Rejewski to *fingerprint* a specific day key. He then set about using the Cipher Bureau's Enigma replica to catalogue *every single possible rotor setting* and the link numbers associated with it. As you can imagine, this was an enormous and tedious task. However, since the number of rotor settings is only about 105,000, it *was* doable if the need was great enough, and the need was definitely great. It took the Cipher

Bureau more than a year to do it (while this is a long time, it is *magnitudes* less than the life of the universe!), but after that, all they had to do was build the table of relationships, find the size of the chains it generated, and look it up in the catalogue. As we've already established, once this was accomplished, the other parts of the Enigma code (the plugboard, notches, and the like) were possible to break.

Later in the war, however, the Germans realized their mistake, and stopped transmitting the three letters for the message key twice. Once this happened, it was no longer possible to build the table of relationships, and the catalog was next to useless. Enter Alan Turing and Bletchley Park (where Ultra was headquartered). Turing and his associates had been concerned that the Germans would do just that, and had developed an ingenious plan to get around it. This new method required what is known as a *crib*. A **crib** is a word that you know will occur in the plaintext of the message. For instance, one of the most common cribs the British made use of was wetter, the German word for weather. When intercepting messages from a weather station, the word wetter is likely to occur. This time, though, instead of using letters encrypted the same way to find chains, Turing wanted to find loops occurring between plaintext letters and ciphertext letters. Suppose, for example, that you knew the word `wetter` was encrypted to `ETJWPX`. Let's view this as a table of relationships:

W	E	T	T	E	R
E	T	J	W	P	X

What does `W` go to? It goes to `E`, and `E` goes to `T`. What does `T` go to? Well, one `T` goes to `W`, forming a chain. If, somehow, you could find rotor settings that enciphered `W`, `E`, `T`, and `R` in such a way as to create this pattern, you would have a likely candidate for the rotor settings in the message key. Once you had the message key, deducing the day key was a simple matter. The British quickly built machines to do this, and were soon up and running.

3.5. Codes in World War II

While the point of this chapter is to describe the Enigma and Ultra, we would be remiss if we didn't mention some of the consequences of deciphering German messages, as well as some of the other cryptographic issues and challenges faced during the war. Numerous books have been written about each of these incidents; our purpose here is to highlight some of these issues and to excite you to read more.

(1) *Battle of the Atlantic.* Throughout the entire war, there was a constant struggle between the Allies, desperate to get food, supplies, and troops from North America to assist the Allied cause around Europe, and the Axis, desperate to sink these loaded ships. The Germans relied heavily on submarine warfare to disrupt and

sink Allied shipping. Cracking the Enigma provided a decisive advantage to the Allies, who could not only adjust convoy routes to avoid enemy subs but could also use this intelligence to seek out and sink these ships. There are numerous other examples where decrypted intelligence played a key role in the distribution of Allied forces, ranging from aerial defense in the Battle of Britain (July 10–October 31, 1940), to Operation Vengeance on April 18, 1943 (the successful attack on Japanese Admiral Yamamoto's plane, leading to his death).

(2) *Coventry: November 14, 1940.* Coventry was an important industrial city for the British, manufacturing many military items. On November 14, 1940, the Germans launched a devastating raid on the city. Some sources say the British knew about the impending attack, but only through an Ultra decrypt, which is why no additional defensive measures or warnings happened. One of the greatest fears of the British was that the Germans would realize their Enigma had been compromised. Thus, Allied forces could never act on Ultra decrypts unless there was a convincing story that would ascribe their actions to something other than having broken the Enigma. While the truth about Coventry may be unknown, this was not an academic problem, and it occurred in various forms throughout the war. For example, Allied forces launched many aerial reconnaissance missions in the Mediterranean to find the Axis ships. The Allies already knew the location of these ships from Ultra decrypts, but it was imperative that the Germans and Italians see the Allied planes, and thus believe that this was how their positions were discovered (see, for example, the Battle of Cape Matapan, March 27–29, 1941). Many pilots were shot down and died to preserve the Ultra secret.

(3) *Battle of Midway: June 4–7, 1942.* After a horrible defeat at Pearl Harbor on December 7, 1941, American forces were on the defensive in the Pacific and lost numerous engagements to the Japanese. It's hard to do justice to this battle in a paragraph; it was one of the turning points in the war, a massively decisive victory for the Americans. One aspect of the battle is worth mentioning: Japanese codes were partially broken, and the Allies knew an attack was coming. Unfortunately, the target was given a code name, so even though the message was decrypted, all battle orders listed the target as `AF`. Based on the code name and other items, many believed that `AF` was the Japanese designation for Midway. If this were true, the Americans could deploy their forces appropriately and surprise the Japanese; however, if `AF` happened to refer to another target.... There was thus a pressing need to quickly determine whether or not `AF` was Midway. The solution was ingenious. Commander Joseph

J. Rochefort and his team at Station Hypo hit on an elegant solution. They had Midway send a message that their water distillation plant was damaged and request fresh water. Shortly afterwards, decrypted messages stated that `AF` was in need of water.

(4) *Operation Fortitude: March–June 1944.* In war, there are secrets, and then there are ***SECRETS***. We've already discussed Ultra, which fell into this category—if the Germans had realized their codes were compromised, they could have easily upgraded their security with disastrous effects. Another super-secret was the Manhattan Project, the development of the atomic bomb which ended the war with Japan in the Pacific. One more worth mentioning is D-Day. This was the Allied invasion of Europe. The fact that the Allies were planning to invade Europe in force wasn't a secret—the Axis knew such discussions were in place from the time America entered the war. What was unclear was *when* the invasion would be and *where*. Both of these were closely guarded secrets on a need-to-know basis. Though the Allies eventually decided upon landings in Normandy, France, there were other candidate sites. One was Pas de Calais, also in France. The advantage of a landing here is that this was the closest point between British and French soil. Operation Fortitude was part of the Allied effort of disinformation to convince the Germans that the main attack would be in Calais. General Patton was given command of a fictitious army unit with simulated radio traffic, fake equipment, and false information leaked to known German agents. The deception succeeded brilliantly; even after the Normandy landings, Hitler held Panzer tank units in reserve. He was convinced these landings were a feint, and numerous units that could have attacked the Allied beachheads in Normandy sat idle waiting for an attack on Calais that never came. There is a great lesson to be learned here: one can wreak havoc on the enemy through careful misinformation, leading to a disastrous allocation of resources.

There is a common theme above: it's not enough to crack the enemy's code; one has to decide what to do with the information, which ranges from acting on the message to deciding it was a ploy designed to mislead. These decisions are some of the hardest optimization problems ever faced, with trade-offs between what is best in the short term having to be weighed against the long-term advantage of decrypting future messages.

3.6. Summary

The Enigma was a complex machine, among the most advanced of its time. It had lots of great features: it could quickly encode and decode messages, it was extremely portable and easy to use, and most importantly, it appeared perfectly secure. However, due to espionage, mistakes in how it was used, and the brilliant ingenuity of some very talented people, it was broken.

We've made some simplifications in the few comments we made about breaking the Enigma. Numerous people from many nations labored for years in these attempts; there is no way to do justice to their efforts or give credit to all in just a few pages. Our goal instead is to give a flavor for both the problem and the mathematics behind it. For more details on how it was broken and for more technical readings, we recommend [**32, 45, 54**].

3.7. Appendix: Proofs by Induction

In this section we give a very brief introduction to **proofs by induction**. This is one of the most powerful proof techniques. For a detailed exposition of this and other methods of proof, see Appendix A of [**33**], though almost any first book in proofs covers induction.

The general framework is for each positive integer n we have some statement $P(n)$. Perhaps $P(n)$ is: *The sum of the first n positive integers is $n(n+1)/2$.* Or maybe it is: *For each positive integer n, $11^{n+1} + 12^{2n-1}$ is a multiple of* 133. The difficulty is that we want to show $P(n)$ is true *for all* n. Often it's easy to verify $P(n)$ holds for a *specific* n.

Unfortunately, we can never prove it holds for all n by checking finitely many n. At best we can gain some confidence, but the pattern might stop at some point. For example, note $\frac{16}{64} = \frac{1}{4}$ and $\frac{19}{95} = \frac{1}{5}$. A young mathematician notices this, and conjectures that this is true because whenever you divide a pair of two digit numbers, if you have the same number on the diagonal, you just cancel that to get the answer. Thus we remove the two 6's in the first pair to get $1/4$, and the two 9's in the second pair to get $1/5$. Our friend checks this on another number, $\frac{49}{98}$. Canceling the 9's gives $\frac{4}{8}$, which reduces to $\frac{1}{2}$ and this *is* $\frac{49}{98}$! Of course, this is not how we divide numbers; it's just a coincidence that it works for so many. If we try $\frac{12}{24}$, then the canceling method says it should equal $\frac{1}{4}$, but it is actually $\frac{1}{2}$. This example illustrates the need to methodically check all cases as well as the danger of being misled by a small number of examples.

Proofs by induction: Let $P(n)$ be a statement for each positive integer n. To show $P(n)$ holds for all n it suffices to show two items.

(1) **Base case:** Show $P(1)$ is true.
(2) **Inductive step:** Show that whenever $P(n)$ is true, then $P(n+1)$ is true.

Why is this enough? We know $P(n)$ true implies $P(n+1)$ is true. If we take $n = 1$, we obtain that $P(1)$ is true implies $P(2)$ is true; however, by the base case we know $P(1)$ is true. Thus we deduce $P(2)$ is true. We now take $n = 2$ and find that $P(2)$ is true implies $P(3)$ is true. Since we've just shown $P(2)$ is true we now see that $P(3)$ is true. The process continues.

Many lecturers liken this to dominoes falling on each other, with each case implying the next.

The following is the standard example of a proof by induction. Let $P(n)$ be the claim that the sum of the first n integers equals $n(n+1)/2$; mathematically, this is the same as $1+2+\cdots+n = n(n+1)/2$. The base case is easily handled, as $1 = 1(1+1)/2$. We now turn to the inductive step. We assume $P(n)$ is true, and now we must show that $P(n+1)$ is true. In other words, we must show $1+2+\cdots+n+(n+1) = (n+1)(n+1+1)/2$.

Let's analyze the left-hand side, $1+2+\cdots+n+(n+1)$. As we are assuming $P(n)$ is true, we should look for its statement in our quantity. A little inspection finds it: it's just the first n terms. We thus group the sum of the first $n+1$ integers as

$$(1+2+\cdots+n)+(n+1).$$

By induction, the sum of the first n integers is $n(n+1)/2$; this is simply the statement $P(n)$, which we get to assume is true when doing the inductive step. We substitute this in and do some algebra:

$$\begin{aligned} 1+2+\cdots+n+(n+1) &= (1+2+\cdots+n)+(n+1) \\ &= \frac{n(n+1)}{2}+(n+1) \\ &= \frac{n(n+1)}{2}+\frac{2(n+1)}{2} = \frac{(n+1)(n+2)}{2} \\ &= \frac{(n+1)(n+1+1)}{2}, \end{aligned}$$

which proves the inductive step.

It can take a while to master inductive arguments, but the benefit is great as it allows you to prove statements for all n simultaneously. If you want more practice, here are some fun statements to prove.

(1) Prove $1+3+5+\cdots+(2n-1) = n^2$ (the sum of the first n odd numbers is n^2).
(2) Prove $1^2+2^2+\cdots+n^2 = n(n+1)(2n+1)/6$.
(3) For the bold: $1^3+2^3+\cdots+n^3$ equals an^3+bn^2+cn+d for some constants a, b, c, and d. Find these constants and prove the equality. This problem is significantly harder than the previous ones as you are not told what the sum equals.
(4) The **Fibonacci numbers** are defined recursively as follows: $F_0 = 0, F_1 = 1$, and $F_{n+2} = F_{n+1} + F_n$. Prove

$$F_n = \frac{1}{\sqrt{5}}\left(\frac{1+\sqrt{5}}{2}\right) - \frac{1}{\sqrt{5}}\left(\frac{1-\sqrt{5}}{2}\right)^n;$$

this is **Binet's formula**.
(5) Prove $11^{n+1}+12^{2n-1}$ is a multiple of 133.

3.8. Problems

• From §3.2: Some Combinatorics

EXERCISE 3.8.1. *How many ways are there to choose three people from ten people, when order does not matter? What if order does matter?*

EXERCISE 3.8.2. *How many ways are there to choose* at most *three people from ten people, when order does not matter? What if order does matter?*

EXERCISE 3.8.3. *At a party there are six men and six women. How many ways are there to order them in a line such that men are in six odd positions (i.e., first, third, fifth, and so on) and women are in the even positions (second, fourth, sixth, and so on)?*

EXERCISE 3.8.4. *Five men and eight women go to a dance. How many ways are there to choose three men and three women? How many ways are there to choose three men and three women so that each of the chosen men dances with* exactly *one of the chosen women?*

EXERCISE 3.8.5. *Imagine we have a circular table with three seats; on each seat is a label, so it matters who sits in each seat. How many ways are there for three people to sit in the three seats? What about four people in four seats, and five people in five seats?*

EXERCISE 3.8.6. *Imagine the following harder version of Exercise* 3.8.5. *We still have a circular table, but now there are no longer labels on each seat. What matters is not where you sit, but the relative ordering of people. In other words, who is on your left and who is on your right. Thus an ordering of Alice, Bob, and Charlie is really Alice, Bob, Charlie, Alice, Bob, Charlie, Alice, Bob, and so on as we keep going around the table, and is the same as Bob, Charlie, Alice, Bob, Charlie, Alice, Bob, Charlie and so on. How many ways are there now for three people to sit at such a table with three seats? For four people at a table with four seats? For five people at a table with five seats? Do you see a pattern.*

EXERCISE 3.8.7. *Redo Exercise* 3.8.6 *for n people at such a table with n seats.*

EXERCISE 3.8.8. *Redo Exercise* 3.8.6 *for m people at such a table with n seats, where $m > n$.*

EXERCISE 3.8.9. *The following example is typical of many lotteries: We're given the set of numbers $\{1, 2, \dots, 36\}$ and we must choose six of them for our ticket. All that matters is which six numbers we choose; it doesn't matter the order we choose them. How many ways are there to do this?*

EXERCISE 3.8.10. *Consider the lottery described in Exercise* 3.8.9. *If all sets of six numbers are equally likely to be the winning ticket, what is the probability your ticket wins? What is the probability you have exactly five of the six numbers correct? What is the probability you have at least five of the six numbers correct?*

EXERCISE 3.8.11. *Evaluate* $\sum_{k=0}^{n} \binom{n}{k}$ *for* $n = 1, 2, 3,$ *and* 4. *Try and guess the pattern.*

EXERCISE 3.8.12. *Evaluate* $\sum_{k=0}^{n} \binom{n}{k}$ *for* n *a positive integer.*

EXERCISE 3.8.13. *Evaluate* $\sum_{k=0}^{n}(-1)^n \binom{n}{k}$ *for* $n = 1, 2, 3,$ *and* 4. *Try and guess the pattern.*

EXERCISE 3.8.14. *Evaluate* $\sum_{k=0}^{n}(-1)^n \binom{n}{k}$ *for* n *a positive integer.*

EXERCISE 3.8.15. *Show* $\sum_{k=0}^{n} \binom{n}{k}\binom{n}{n-k}$ *equals* $\binom{2n}{n}$. Hint: Tell a story. Imagine there are $2n$ people, with n men and n women.

EXERCISE 3.8.16. *The following is a standard riddle, though the numbers vary depending on who tells it. Imagine there are seven generals and a safe with many locks. You assign the generals keys in such a way that* ***every*** *set of four generals has enough keys between them to open* ***all*** *the locks; however,* ***no*** *set of three generals is able to open all the locks. Show that one solution involves 35 locks.* Hint: 35 is a binomial coefficient!

The next few problems involve a deck of cards. A standard deck has 52 cards: 13 spades, 13 hearts, 13 diamonds, and 13 clubs. Each suit starts with an ace and then has a 2, 3, ..., 10, ending with a jack, a queen, and a king. Thus there are four aces, four 2's and so on. The ordering of the numbers is A, 2, 3, 4, 5, 6, 7, 8, 9, 10, J, Q, K, A; yes, an ace is considered both high and low.

EXERCISE 3.8.17. *How many ways are there to choose five cards from 52? Here the* order *in which you get the cards is immaterial; all that matters is* which *cards are in your hand.*

EXERCISE 3.8.18. *How many ways are there to choose five cards from 52 so that you have all the aces? How many ways are there to choose five cards from 52 so that you have four cards with the same number.*

EXERCISE 3.8.19. *How many ways are there to choose five cards from 52 so that all the cards are in the same suit?*

EXERCISE 3.8.20. *How many ways are there to choose five cards from 52 so that you have two cards with the same number?* Hint: This problem is more tricky than a lot of people think.

EXERCISE 3.8.21. *How many ways are there to choose five cards from 52 so that the numbers of the five cards are in increasing order, with each card exactly one higher than the previous? Thus* $2, 3, 4, 5, 6$ *would work, but not* $Q, K, A, 2, 3$ *(we are not allowed to wrap around, but we* are *allowed to use an ace as high or low).*

EXERCISE 3.8.22. *Look up the standard poker hands (without wild cards). Calculate the probabilities of each hand. If the math is done correctly, the higher ranked hand should have* fewer *ways of being realized.*

EXERCISE 3.8.23. *Building on the previous hand, look up what it means for a card to be a* wild card *in poker. Does the ordering change when we allow wild cards?*

The next few problems are harder, but can be solved with binomial coefficients and elementary combinations. It's important to note that elementary does *not* mean easy; it just means we don't need any advanced mathematics.

EXERCISE 3.8.24. This problem is hard, but can be solved with binomial coefficients *if* you look at it the right way. *Imagine we have ten identical cookies and five (distinct) people: Alice, Bob, Charlie, Danie, and Eve. How many different ways can we divide the cookies among the people? Since the cookies are identical, all that matters is how many cookies a person receives, not which cookies.*

EXERCISE 3.8.25. *Building on the previous problem, imagine now that instead of distributing the ten cookies to the five people that we allow ourselves the option of* not *giving out all of the cookies. Find a simple formula for the number of possibilities.*

EXERCISE 3.8.26. *A harder version of the lottery allows repeated numbers, though order does not matter. Imagine we have to choose six numbers from* $\{1, 2, \ldots, 36\}$*, but now the same number can be used multiple times. If order mattered it would be simple: there would be* 36^6 *possibilities (we would have* 36 *options for the first number,* 36 *for the second, and so on); however, we're interested in the case where order does not matter. Thus choosing the numbers in the order* $3, 4, 3, 5, 21, 3$ *is the same as choosing them in the order* $21, 5, 4, 3, 3, 3$*. How many possible tickets are there now?*

- **From §3.3: Enigma's Security**

EXERCISE 3.8.27. *Calculate the number of combinations when* 11 *cables are used, and when* 13 *cables are used. Why does it make sense for the number with* 11 *cables to be larger than for the number with* 13 *cables?*

EXERCISE 3.8.28. *Which value of* p *do you think gives the greatest contribution to the sum (equation* (3.1)*)? Why? Compute all the values and see if you're right.*

EXERCISE 3.8.29. *In the Enigma we used* p *cables in the plugboard, with* $1 \le p \le 13$*. Note that all a cable can do is connect two letters and switch them. Imagine now we have a supercable, which works as follows. We have two sets of pairs of letters, say* $\mathcal{L}_1, \mathcal{L}_2$ *and* $\mathcal{L}_3, \mathcal{L}_4$*. The supercable works so that if we type* $\mathcal{L}_1$ *we get* $\mathcal{L}_3$*, if we type* $\mathcal{L}_2$ *we get* $\mathcal{L}_4$*, but now when we* *type* $\mathcal{L}_3$ *we get* $\mathcal{L}_2$ *(and* not $\mathcal{L}_1$*, which is what we would have with a regular cable), and finally when we type* $\mathcal{L}_4$ *we get* $\mathcal{L}_1$*. How many combinations can we get if we have one supercable? What if we have two, or three, or four?*

EXERCISE 3.8.30. *Is there any way to build the supercable discussed in the previous exercise from regular cables?*

EXERCISE 3.8.31. *Based on the definition of the factorial and the double factorial, what do you think the triple factorial should be? Why?*

EXERCISE 3.8.32. *Compute* $(2n)!!$ *for* $n = 1, 2, 3,$ *and* 4*. The data suggests there is a simple function* $f(n)$ *such that* $(2n)!! = f(n) \cdot n!$*. What do you think* f *is?*

EXERCISE 3.8.33. *Building on the previous problem, find a simple function* f *such that* $(2n)!! = f(n) \cdot n!$*.*

EXERCISE 3.8.34. *Consider the function*

$$g(m, n) = (2m)!!/((2n)!!(2m - 2n)!!)$$

for $m \geq n$ *even numbers. Find a simple formula for* $g(m, n)$*.* Hint: Try a few special values of m, n to build intuition.

EXERCISE 3.8.35. *In the rotors for the Enigma, it's possible to connect a letter to a copy of itself; for example, we could send* `A` *to* `B`*,* `B` *to* `C`*,* `C` *to* `A`*, and then* `D` *to* `D`*,* `E` *to* `E`*, and so on. What if we added a restriction that no letter could go to a copy of itself? It's now a lot harder to find the number of such rotors. If you know about the Principle of Inclusion/Exclusion, that's a great way to attack this. Instead, let's consider "simpler" alphabets. Find the number of such rotors if there are only two letters in our alphabet. What if there are three letters? What about four letters? (As the number of letters in the alphabet grows, the ratio of number of these restricted rotors to the number of rotors converges to* $1 - 1/e$*.)*

EXERCISE 3.8.36. *Consider all the pieces in the Enigma, from the plugboard to the reflector. Which step adds the most number of combinations?*

EXERCISE 3.8.37. *Keeping all else the same, how many combinations would there be if the Germans used four rotors? If they used five rotors (the German navy did use five rotors)?*

EXERCISE 3.8.38. *In the previous exercise we explored the effect of adding additional rotors. What would happen if instead the Germans added additional plugboards?*

• **From §3.4: Cracking the Enigma**

EXERCISE 3.8.39. *Read* Solving the Enigma—History of the Cryptanalytic Bombe *by Jennifer Wilcox* [**54**]*, and write a short report on it.*

EXERCISE 3.8.40. *Marion Rejewski and Alan Turing played key roles in cracking the Enigma. Write a short biography about one of them.*

EXERCISE 3.8.41. *For a while, the Germans erred by transmitting the initial three letter grouping twice; discuss what would have happened if instead they transmitted it three times, or four. How would knowing this affect the decryption attacks and the Enigma's security?*

EXERCISE 3.8.42. *Research the British capture of the German U-boat U-110, and write a short report.*

- **From §3.5: Codes in World War II**

EXERCISE 3.8.43. *Research one of the following four topics mentioned in the text:* Battle of the Atlantic, Coventry, Battle of Midway, *and* Operation Fortitude. *Discuss the role cryptography played in your choice.*

EXERCISE 3.8.44. *Write a short report about the Japanese code* Purple, *and the Allied efforts to break it.*

EXERCISE 3.8.45. *Write a short report about the Navajo code, and the key players in convincing the U.S. military to try it.*

Chapter 4

Classical Cryptography: Attacks I

In Chapters 2 and 3 we explored substitution alphabet ciphers and some of their descendants. While these are easy to use, they're also easy to break. They're especially vulnerable to frequency analysis, which is the main theme of this chapter.

The effectiveness of frequency analysis was a driving force in the push for better encryption systems. We saw several possibilities in Chapter 2. The great advance of systems such as the Vigenère, Permutation, and Hill ciphers is that the same letter is not always encrypted the same way. Not surprisingly, this complication increases the security. While the last two are impervious to frequency analysis, we'll see in Chapter 5 that these extensions also have issues.

Even though the substitution alphabet and Vigenère ciphers are vulnerable to frequency analysis, our time spent studying them is still time well spent. These have the advantage of allowing for fast encryption, and in many situations speed is extremely important. It turns out that these fast methods can be combined with more secure, but slower, methods to lead to secure, efficient encryption. We'll discuss this in greater detail after we've developed secure systems such as RSA (see Chapter 8).

4.1. Breaking the Caesar Cipher

The goal of any cipher is to allow two people to communicate securely. For our substitution alphabet ciphers, if Alice and Bob both know the secret key, then they can easily encrypt and decrypt messages. The security of a system is a function of how hard it is for a third person to decrypt messages without knowing the key. In this section we explore the security of the Caesar cipher, and we see (not unsurprisingly!) that it's a very insecure method! Thousands of years ago, when few people knew how to read and even fewer knew about the method, it sufficed. Today, of course, a straight Caesar cipher can't be used.

There's really no need to introduce the method of **frequency analysis** to attack the Caesar cipher. There are only 25 possible keys (or 26 if we allow for the possibility that our sender didn't encrypt the message!), and even an impatient attacker will quickly crack the code. We choose to use frequency analysis here for two reasons. The most important is that we need frequency analysis for later ciphers, and it's worth seeing it first on an easier example. The second is that frequency analysis suggests in what order we should try the 25 possibilities. In other words, we get an intelligent ordering of shifts to try.

Consider the intercepted message

`T QZFYO ESP MLR`

which was encrypted with a Caesar cipher. Even without knowing the key, we have a lot of information. For example, we know that the message begins with a one-letter word; it's precisely for this reason that we often remove spaces when sending! Assuming the message is in English, the letter `T` must have been encrypted either from the letter `A` or from the letter `I`.

From Table 2.1 on p. 25 we see that `T` corresponds to the number 19 and `A` to the number 0, which means that for `A` to be encrypted to `T`, the key would have to be 19. Based on this guess, we can try decrypting the message as if it was encrypted with 19 as the key.

	T	Q	Z	F	Y	O	E	S	P	M	L	R
	19	16	25	5	24	14	4	18	15	12	11	17
subtract 19:	0	-3	6	-4	5	-5						
MOD 26:	0	23	6	22	5	21						
	A	X	G	W	F	V						

Since the beginning is gibberish, there's no need to decode the rest of the message; it is unlikely that 19 is the key. Let's try the other possibility, that `T` was the encryption of `I`. Since `T` corresponds to 19 and `I` corresponds to 8, this means the encryption key is $19 - 8 = 11$. Trying that, we find the following.

	T	Q	Z	F	Y	O	E	S	P	M	L	R
	19	16	25	5	24	14	4	18	15	12	11	17
subtract 11:	8	5	14	-	13	3	-	7	4	1	0	6
				6			7					
MOD 26:	8	5	14	20	13	3	19	7	4	1	0	6
	I	F	O	U	N	D	T	H	E	B	A	G

We've broken the message. Notice that *if we can guess just one letter correctly, we can break any message encrypted with a Caesar cipher.*

It's clear that the spacing of a message provides valuable information for deciphering it. For this reason, encoded messages are usually written without their original spacing. For example, if we wanted to send the message `WHEN WILL YOU RETURN` using the Caesar cipher with 10 as a key, we first break the message into groups of five letters, ignoring the original spacing: `WHENW ILLYO URETU RN`. (We don't have to break it into groups of five letters, of course; all that matters is that we agree upon some length.)

If we encrypted this message with 16 as a key, we get `LWTCL XAAND JGTIJ GC`. If someone without the key intercepts the message, they would have to break it without knowing the lengths of any words. In particular, our "trick" of looking at one-letter words is no longer available. The intended recipient, using the key, can easily recover the message `WHENW ILLYO URETU RN`, and understand it even without the correct spacing. Of course, some phrases have a very different meaning depending on how they are punctuated, but we ignore such issues for now.

Even without the word spacing intact, it's easy to break the cipher. Imagine we have intercepted the following message, encrypted using the Caesar cipher with an unknown key:

`THTWW CPEFC YLQEP CESCP POLJD`

The letters which appear most frequently in this message are `C` (four times) and `P` (4 times). The most common letter in the English language is `E`, so it is likely that `E` might have been encrypted to either `C` or `P`. `E` corresponds to the number 4, and `C` corresponds to the number 2, so for `E` to be encrypted to `C` the key would have to be 24, since $4 + 24 = 28$, and 28 MOD $26 = 2$. Decrypting with a key of 24 gives `VJVYY ERGHE` ..., which is nonsense. Since this didn't work, we guess instead that `E` was encrypted to `P`; in this case, the key would have been $15 - 4 = 11$. Decrypting with 11 as the key gives

`IWILL RETUR NAFTE RTHRE EDAYS`

and so the message is "I will return after three days". This technique for breaking codes is called **frequency analysis**, since it uses the ordinary frequency of letters in the English language to figure out how a message was encrypted. Table 4.1 shows the frequencies of letters in Project Gutenberg's collection of public-domain English-language books.

Notice that the letter `C` in the ciphertext above corresponded to the letter `R` in the correctly decoded plaintext; even though `C` was just as common as `P` (which turned out to be `E`) and the letter `R` is only the ninth most common letter in English. This example serves as a warning about the dangers in applying frequency analysis to short messages. If the message is short, there can be a significant variation from the limited letter frequencies, and thus a lot of trial and error might be required.

It appears that in Caesar's time, his cipher was never broken, although there is a reference by the writer Aulus Gellius to a "rather ingeniously written treatise by the grammarian Probus" concerning Caesar's cryptographic techniques. The earliest surviving account of a work describing how to break the cipher is "A Manuscript on Deciphering Cryptographic Messages", written in the ninth century by the Arab philosopher, scientist, and mathematician al-Kindi. This is the first known description of frequency analysis.

TABLE 4.1. Frequencies of letters in English text, from 9,481 English works from Project Gutenberg; see `http://www.cryptograms.org/letter-frequencies.php`.

1	e	12.58%	14	m	2.56%
2	t	9.09%	15	f	2.35%
3	a	8.00%	16	w	2.22%
4	o	7.59%	17	g	1.98%
5	i	6.92%	18	y	1.90%
6	n	6.90%	19	p	1.80%
7	s	6.34%	20	b	1.54%
8	h	6.24%	21	v	0.98%
9	r	5.96%	22	k	0.74%
10	d	4.32%	23	x	0.18%
11	l	4.06%	24	j	0.15%
12	u	2.84%	25	q	0.12%
13	c	2.58%	26	z	0.08%

Of course, as we've discussed there is another way to break the Caesar cipher—**the method of brute force!** One of the great themes in cryptography is that usually it's not enough to be able to do something; you have to be able to do it quickly. In most cases there are just too many possibilities to check to make a brute force exhaustion of all the possibilities practical. The Caesar cipher is one of the few exceptions. There are only 26 possible Caesar ciphers. If we know a message was encrypted with a Caesar cipher, all we need to do is figure out the key. The key is an integer between 0 and 25, and almost surely not 0 (if it were 0, there would be no encryption). Thus we just shift by 1, 2, ..., and 25 and see which number works. We chose to discuss guessing letters in detail because, even though it isn't needed here, it will become very important in cracking the improvements of the Caesar cipher. Also, a quick frequency analysis approach suggests a more efficient *order* in which to try the 25 possible keys.

4.2. Function Preliminaries

Before analyzing the affine cipher, it's worthwhile to introduce some notation and definitions. Though the notation and formalism can seem overwhelming at first, the effort is well worth it. The purpose of notation is to highlight what is going on. Thus instead of using f to denote a certain function later we'll write $f_{a,b}$, as the subscripts a and b will remind us of key properties of the function. As you continue in mathematics, the formulas become longer and more involved; good notation helps you see what matters at a glance.

As an attacker, our goal is to undo the sender's encryption. It's useful to visualize the sender as applying some function to the text, and thus the attacker's job is to find an inverse function to return to the original text. We formally define this and related concepts, and then follow with some examples.

Below we consider **functions** between sets S, T, and U. A function associates a unique output to each input, though it can associate the same output to different inputs. If f maps elements of S to T, we denote this by writing $f : S \to T$; this means the function f takes elements of S as input and outputs elements of T.

For example, let S and T be the set of real numbers, and let $f(x) = x^2$. This is a function; we associate an output to each input, and we don't associate multiple outputs to any inputs. It's fine that $f(1) = f(-1)$; there's no problem if different inputs go to the same output. If, however, $f(1) = 2$ and 3, *then* we would have a real problem, as it wouldn't be clear what $f(1)$ equals.

Let's do another example, one which is more useful for cryptography. Let S be the set of letters in the alphabet, $\{\text{A}, \text{B}, \dots, \text{Z}\}$, T the set of integers $\{\dots, -2, -1, 0, 1, 2, \dots\}$, and f takes an element of S as input and outputs its numerical value. Thus $f(A) = 0$, $f(B) = 1$, $\dots$, $f(Z) = 25$. Notice that each letter is sent to one and only one number.

Below we define some of the basic operations on functions. These will be very helpful when we start analyzing the affine cipher and other topics in the course.

Composition, inverse function, modular inverses.

- **Composition:** If f is a function from S to T, and g is a function from T to U, then we can define a function from S to U. We denote this function by $g \circ f$. We define it as follows: given any s in S, we set $(g \circ f)(s) = g(f(s))$. This means we first send s to $f(s)$, and then apply g to $f(s)$. We call $g \circ f$ the *composition*, and say g composed with f. Note we first apply f to our input, and then we apply g.
- **Inverse function:** If f is a function from S to T, then a function g from T to S is its *inverse function* if $(g \circ f)(s) = s$ for all s in S. Informally, applying g undoes the effect of applying f, returning us to our starting value. We often write f^{-1} for the inverse function to f.
- **Modular inverse:** Let x be an integer. If there exists a y such that yx is equivalent to 1 modulo n, then we say y is a *modular inverse* to x. Not all numbers are invertible (n is never invertible modulo n), and inverses are not unique (if y is an inverse to x, so too is $y+n$, $y+2n$, $y-n$ and so on). If x is invertible modulo n, we write x^{-1} for the *unique* inverse in $\{0, 1, \dots, n-1\}$.

For example, if $f(x) = x^2 + 3$ and $g(x) = \sqrt{x-2}$, then

$$(g \circ f)(x) \;=\; g(f(x)) \;=\; g(x^2+3) \;=\; \sqrt{(x^2+3)-2} \;=\; \sqrt{x^2+1}.$$

This is because $g(x) = \sqrt{x-2}$ should be read as if we apply g to an input of x, then we get an output of $\sqrt{x-2}$. In this example, we applied g to an input of x^2+3, so the output is $\sqrt{(x^2+3)-2}$. Notice that in general $g \circ f$ is not the same as $f \circ g$. For these functions, $(g \circ f)(0) = g(f(0)) = g(3) = \sqrt{3-1} = \sqrt{2}$; however, $(f \circ g)(0) = f(g(0)) = f(\sqrt{0-3})$, and we run into trouble as we cannot take the square-root of a negative number.

While it is hard to write down a nice formula for an inverse function in general, for many examples this can be done without too much difficulty. Imagine that $f(x) = 3x + 7$, and we want to find an inverse function g. If g is the inverse of f, then $(g \circ f)(x) = x$. This means $g(f(x)) = x$, or $g(3x+7) = x$. Let us make the inspired guess that g is a linear function, just like f. If that is true, then $g(x) = ax + b$ for some real numbers a and b. So, g takes an input and returns b more than a times that input. For us, $g(3x+7) = x$ becomes $a(3x+7) + b = x$. Grouping like terms together, we find $(3a-1)x = -(7a+b)$. As this must hold for all x, we need $3a-1$ and $7a+b$ to be zero, implying $a = 1/3$ and $b = -7a = -7/3$. Thus our inverse function is $g(x) = \frac{1}{3}x - \frac{7}{3}$. As a check,

$$(g \circ f)(11) \;=\; g(f(11)) \;=\; g(3 \cdot 11 + 7) \;=\; g(40) \;=\; \frac{1}{3}40 - \frac{7}{3} \;=\; 11.$$

Later we'll write $f_{3,7}(x)$ instead of $f(x)$ to emphasize the value of the coefficients, and we'll be similarly explicit with g.

Finally, let's look at some inverses modulo 26. The inverse of 17 modulo 26 is 23, as $23 \cdot 17 = 391$, which is equivalent to 1 modulo 26. Similarly the inverse of 3 modulo 26 is 9, as $9 \cdot 3 = 27$, which is equivalent to 1 modulo 26. However, 13 is not invertible modulo 26. If it were, there would be some y such that $13y$ is equivalent to 1 modulo 26. That would entail $13y = 1 + 26m$ for some integer m. Rearranging, we find $13y - 26m = 1$; the left-hand side is always divisible by 13 but the right-hand side never is. This is a contradiction, and thus 13 is not invertible modulo 26. The failure to have an inverse is due to the fact that 13 and 26 have a common divisor.

4.3. Modular Arithmetic and the Affine Cipher

Our goal in this section is to discuss the mathematics needed to prove that the affine cipher (introduced in §2.6) is a workable means for encryption, and that there is a unique decryption. Understanding how the cipher works is the first step towards cracking it, which we do next in §4.4. To prove every claim requires more mathematics than we've covered up to now, specifically, we need the Euclidean algorithm. This is discussed in detail in Chapter 8. Fortunately, the Euclidean algorithm is only needed to show the validity of the affine cipher for arbitrary sized alphabets; for our 26-letter alphabet, we can verify all the needed claims by brute force. An interested reader should return to this section after reading about the Euclidean algorithm and try to generalize these results.

TABLE 4.2. Multiplication table modulo 26.

	A	B	C	D	E	F	G	H	I	J	K	L	M	N	O	P	Q	R	S	T	U	V	W	X	Y	Z
	0	**1**	**2**	**3**	**4**	**5**	**6**	**7**	**8**	**9**	**10**	**11**	**12**	**13**	**14**	**15**	**16**	**17**	**18**	**19**	**20**	**21**	**22**	**23**	**24**	**25**
0	0	0	0	0	0	0	0	0	0	0	0	0	0	0	0	0	0	0	0	0	0	0	0	0	0	0
1	0	1	2	3	4	5	6	7	8	9	10	11	12	13	14	15	16	17	18	19	20	21	22	23	24	25
2	0	2	4	6	8	10	12	14	16	18	20	22	24	0	2	4	6	8	10	12	14	16	18	20	22	24
3	0	3	6	9	12	15	18	21	24	1	4	7	10	13	16	19	22	25	2	5	8	11	14	17	20	23
4	0	4	8	12	16	20	24	2	6	10	14	18	22	0	4	8	12	16	20	24	2	6	10	14	18	22
5	0	5	10	15	20	25	4	9	14	19	24	3	8	13	18	23	2	7	12	17	22	1	6	11	16	21
6	0	6	12	18	24	4	10	16	22	2	8	14	20	0	6	12	18	24	4	10	16	22	2	8	14	20
7	0	7	14	21	2	9	16	23	4	11	18	25	6	13	20	1	8	15	22	3	10	17	24	5	12	19
8	0	8	16	24	6	14	22	4	12	20	2	10	18	0	8	16	24	6	14	22	4	12	20	2	10	18
9	0	9	18	1	10	19	2	11	20	3	12	21	4	13	22	5	14	23	6	15	24	7	16	25	8	17
10	0	10	20	4	14	24	8	18	2	12	22	6	16	0	10	20	4	14	24	8	18	2	12	22	6	16
11	0	11	22	7	18	3	14	25	10	21	6	17	2	13	24	9	20	5	16	1	12	23	8	19	4	15
12	0	12	24	10	22	8	20	6	18	4	16	2	14	0	12	24	10	22	8	20	6	18	4	16	2	14
13	0	13	0	13	0	13	0	13	0	13	0	13	0	13	0	13	0	13	0	13	0	13	0	13	0	13
14	0	14	2	16	4	18	6	20	8	22	10	24	12	0	14	2	16	4	18	6	20	8	22	10	24	12
15	0	15	4	19	8	23	12	1	16	5	20	9	24	13	2	17	6	21	10	25	14	3	18	7	22	11
16	0	16	6	22	12	2	18	8	24	14	4	20	10	0	16	6	22	12	2	18	8	24	14	4	20	10
17	0	17	8	25	16	7	24	15	6	23	14	5	22	13	4	21	12	3	20	11	2	19	10	1	18	9
18	0	18	10	2	20	12	4	22	14	6	24	16	8	0	18	10	2	20	12	4	22	14	6	24	16	8
19	0	19	12	5	24	17	10	3	22	15	8	1	20	13	6	25	18	11	4	23	16	9	2	21	14	7
20	0	20	14	8	2	22	16	10	4	24	18	12	6	0	20	14	8	2	22	16	10	4	24	18	12	6
21	0	21	16	11	6	1	22	17	12	7	2	23	18	13	8	3	24	19	14	9	4	25	20	15	10	5
22	0	22	18	14	10	6	2	24	20	16	12	8	4	0	22	18	14	10	6	2	24	20	16	12	8	4
23	0	23	20	17	14	11	8	5	2	25	22	19	16	13	10	7	4	1	24	21	18	15	12	9	6	3
24	0	24	22	20	18	16	14	12	10	8	6	4	2	0	24	22	20	18	16	14	12	10	8	6	4	2
25	0	25	24	23	22	21	20	19	18	17	16	15	14	13	12	11	10	9	8	7	6	5	4	3	2	1

We need to multiply numbers modulo 26. We record the results in Table 4.2. Each alphabet has its own multiplication table; fortunately, it is a one-time cost to compute.

Before proving our claims for all valid keys (a, b), let's do an example with the key $(3, 10)$. We'll take `MEETA TTEN` as our message. First, we encrypt the message (Table 4.2 is a big help when carrying out the multiplications). Letting x represent a character of plaintext and y the corresponding character of ciphertext, the encryption is simply $y = 3x + 10 \pmod{26}$.

	M	E	E	T	A	T	T	E	N
x	12	4	4	19	0	19	19	4	13
$y = 3x + 10 \pmod{26}$	20	22	22	15	10	15	15	22	23
	U	W	W	P	K	P	P	W	X

Assume now we receive `UWWP KP PWX` and we know the message was encrypted according to the congruence $y = 3x + 10 \pmod{26}$. One way to decrypt this is of course to just write down the lookup table (what does `A` go to, what does `B` go to, and so on) and use that. Instead, we try to find a nice inverse function to undo the encryption. While it isn't immediately clear that such an inverse function should exist, it's not unreasonable to search for it.

When decrypting the message, we know y and are trying to find x. Let's solve this congruence for x. First we subtract 10 from both sides of the congruence $y = 3x + 10 \pmod{26}$ and get

$$y - 10 \;=\; 3x \pmod{26}.$$

Since -10 is equivalent to 16 modulo 26, we may instead look at

$$y + 16 \;=\; 3x \pmod{26}.$$

Finally, we want to multiply both sides by the multiplicative inverse of 3 modulo 26. Consulting our table, we see that's 9 (as 9 times 3 is equivalent to 1 modulo 26), and find

$$9(y + 16) \;=\; 9 \cdot 3x \pmod{26},$$

which simplifies to

$$9y + 14 \;=\; x \pmod{26}.$$

We've found the decryption congruence for the key (3,10), and can use it to decrypt the message:

	U	W	W	P	K	P	P	W	X
y	20	22	22	15	10	15	15	22	23
$x = 9y + 14 \pmod{26}$	12	4	4	19	0	19	19	4	13
	M	E	E	T	A	T	T	E	N

There are two very important lessons worth extracting from this example:

- We had to multiply by the inverse of 3 to find the decryption congruence. This brings up an important point: for the same reason that the multiplication cipher requires a key which is relatively prime to 26, **the number a in a key (a, b) used for the affine cipher must be relatively prime to 26**. If not, then we cannot find an inverse, and there will be no suitable decryption congruence.
- Decrypting a message encoded with the key $(3, 10)$ is the same as encrypting a message with the key $(9, 14)$; however, writing the second key as $(9, 14)$ hides a lot of what is going on. If we write 3^{-1} for the multiplicative inverse of 3 modulo 26, then we can write $(9, 14)$ as $(3^{-1} \bmod 26, -10 \cdot 3^{-1} \bmod 26)$. This example suggests that there should be an easy way to decrypt a message, using a similar function to our encryption.

These results hold in general and are not peculiar to encrypting with the key $(3, 10)$. For example, imagine we now take $(5, 12)$ as our key. Thus encrypting sends x to $y = 5x + 12 \pmod{26}$. To decrypt, we solve for x in terms of y. We subtract 12 and get $y - 12 = 5x \pmod{26}$, and then multiply by the inverse of 5 modulo 26. The inverse is 21, since $5 \cdot 21 = 105$, which is 1 modulo 26. Thus $x = 21(y - 12) \pmod{26}$, or after simplifying $x = 21y + 8 \pmod{26}$. Our decryption key is $(21, 8)$, and notice that $21 = 5^{-1} \pmod{26}$ and $8 = -12 \cdot 5^{-1} \pmod{26}$.

We now return to our two claims about the affine cipher. The first is that $f_{a,b}(n) = an+b \pmod{26}$ is a valid encryption method if a and 26 have no proper common divisors, while the second is that it fails if some integer greater than 1 divides both a and 26. Remember we are writing $f_{a,b}$ and not f to emphasize the roll of the two parameters, a and b. The algebra gets a little long below; while the expressions would be shorter without the subscripts, the point of having them is to remind us what each function is. The purpose of notation is not to write things as concisely as possible, but to represent concepts so that we can easily follow the arguments.

We start with the second claim, which is easier. Imagine a and 26 have a common divisor. Let's denote it by d, with $2 \le d \le 25$ (since a is at most 25). One way to show that this system is invalid is to show that two different numbers encrypt to the same number. The easiest number to study is 0, which encrypts to b. Fortunately another number encrypts to b. We can write $a = md$ and $26 = kd$, and then note that

$$f_{a,b}(k) \;=\; a\cdot k+b \;=\; md\cdot k+b \;=\; m\cdot kd+b \;=\; 26m+b \;=\; b \pmod{26}.$$

We've just found two distinct integers that are encrypted to the same letter, which means decryption is impossible. .

We now know that it is *necessary* that a and 26 have no common factors; the difficulty is in showing that this is sufficient. It would take us too far afield right now to develop the theory of modular multiplication in full generality to attack this and related problems, but as promised we do this later since this is a key ingredient in RSA (discussed in Chapter 8). For now, we'll be content with saying that a modest brute force computation shows that if a is relatively prime to 26, then $f_{a,b}(n) = an + b \pmod{26}$ does list the 26 letters of the alphabet in a new order.

Of course, it's not enough to be able to encrypt a message; we need to be able to decrypt it as well. Thus, we need to find an inverse function $g_{a,b}$ for the function $f_{a,b}$. This means $g_{a,b}(f_{a,b}(n)) = n$. For the original Caesar cipher, $a = 1$ and our encryption function was $f_{1,b}(n) = n + b \pmod{26}$, corresponding to a shift of b. We decrypted by shifting back by b, so our decryption function was $g_{1,b}(m) = m - b \pmod{26}$. Note

$$g_{1,b}(f_{1,b}(n)) \;=\; g_{1,b}(n+b \bmod 26) \;=\; n+b-b \bmod 26 \;=\; n.$$

We used the fact that $(x+y \bmod 26)+z \bmod 26 = x+y+z \bmod 26$ (which you are asked to prove in Exercise 4.9.17); in other words, you can remove the multiples of 26 whenever you like.

Since the decryption and encryption functions for the Caesar cipher were of the same form, it is reasonable to see if the same is true for the affine cipher. Further evidence supporting this guess is provided by our examples with keys $(3,10)$ and $(5,12)$. Let's look for a $g_{a,b}$ of the same shape as $f_{a,b}$. We try $g_{a,b}(m) = cm + d$, where c and d will be functions of a and b.

We have

$$g_{a,b}(f_{a,b}(n)) \;=\; g_{a,b}(an + b \bmod 26) \;=\; c(an+b) + d \bmod 26,$$

which becomes

$$g_{a,b}(f_{a,b}(n)) \;=\; (ac)n + (bc+d) \bmod 26.$$

We leave it to the reader to justify the interchange of multiplication by c and removing copies of 26. We want the above to be n, so $d = -bc \pmod{26}$, while c has to be chosen so that $ac \pmod{26} = 1$. The first condition is easily satisfied, but the second requires some explanation. We write a^{-1} for the multiplicative inverse of a modulo 26 (which only exists if a and 26 have no proper common divisor). Thus $g_{a,b} = f_{a^{-1},-ba^{-1}}$, which is of the same form as the encryption function. In other words, given an encryption key (a,b), the decryption key is $(a^{-1}, -b \cdot a^{-1})$ (where the inverses and multiplication are done modulo 26).

We check this against our earlier example where $(a,b) = (3,10)$ was the encryption key. As $3^{-1} = 9 \pmod{26}$ and $-10 \cdot 3^{-1} = -90 = 14 \pmod{26}$, the result is consistent with the example. For our second example we chose the key $(5,12)$. Since $5^{-1} = 21 \pmod{26}$ and $-12 \cdot 5^{-1} = 8 \pmod{26}$, this example is also consistent with our theoretical result.

Table 4.2 is the multiplication table modulo 26. It's not elegant, but it is a one-time calculation and it does show that if a and 26 don't share a proper divisor, then there is always a c so that $ac = 1 \pmod{26}$. For example, if $a = 5$, then $c = 21$, while if $a = 11$, then $c = 19$.

Armed with this table, we can always find multiplicative inverses for a modulo 26 whenever a and 26 don't have a common factor. Whenever we write $ac = 1 \pmod{26}$ what we mean is that there is a c so that $ac = 1 + 26m$ for some integer m. If a and 26 have a common nontrivial divisor greater than 1, then a is not invertible modulo 26; however, if a and 26 are relatively prime then a is invertible modulo 26. This can be seen by a direct examination of the various possibilities for a, and thus there are 12 invertible a's modulo 26: 1, 3, 5, 7, 9, 11, 15, 17, 19, 21, 23, and 25.

Isolating what we've done so far, we have valid encryption and decryption functions for the affine cipher so long as a has no common factor with 26.

In §2.5 we introduced the notation **relatively prime** to denote no common factors; that is to say two integers x and y are **relatively prime** if the only positive integer dividing both is 1 (i.e., they have no proper common divisor, or equivalently their greatest common divisor is 1).

Encryption and decryption for the affine cipher: Let $a, b \in \{0, 1, \dots, 25\}$ with a and 26 relatively prime. Then $f_{a,b}(n) = an + b \pmod{26}$ is a valid encryption scheme and is decrypted by applying the function $g_{a,b} = f_{a^{-1},-ba^{-1}}$.

For example, we encrypt the message MEETA TTEN with the affine cipher with $a = 3$ and $b = 0$. Thus $f_{3,0}(n) = 3n \pmod{26}$, so to encrypt we multiply each number by 3 and then reduce modulo 26.

	M	E	E	T	A	T	T	E	N
	12	4	4	19	0	19	19	4	13
times 3	10	12	12	5	0	5	5	12	13
	K	M	M	F	A	F	F	M	N

To decrypt, we need the multiplicative inverse of 3 modulo 26. Looking at Table 4.2, we see that $3 \cdot 9 = 1 \pmod{26}$, so the decryption function is $f_{3^{-1},-0\cdot 3^{-1}} = f_{9,0}$, or more simply it is just multiplication by 9.

	K	M	M	F	A	F	F	M	N
	10	12	12	5	0	5	5	12	13
times 9	12	4	4	19	0	19	19	4	13
	M	E	E	T	A	T	T	E	N

Here the multiplication and MOD steps are shown as a single step; so, for example, K decrypts to M because $9 \cdot 10 = 90 = 12 \pmod{26}$. Reducing 90 (mod 26) can be done quickest with division: $26\overline{)90}$ gives a remainder of 12.

As before, using the multiplication table (Table 4.2), we see that 1, 3, 5, 7, 9, 11, 15, 17, 19, 21, 23, and 25 have multiplicative inverses modulo 26. Thus there are 12 choices for a in the affine cipher, and 26 choices for b, leading to $12 \cdot 26 = 312$ ciphers. Before the advent of computers, it would be a bit tedious for someone to go through all 312 possibilities (or 311 if we remove the trivial encoding which does nothing). That said, it's a one-time cost to enumerate these possible ciphers and create a decryption book. So, while the affine cipher is more secure than the Caesar cipher, it still isn't too secure.

In the next section we discuss attacks on affine ciphers. While these are not needed since brute force can succeed in a reasonable amount of time, it's worth discussing these attacks as they *are* needed for more sophisticated encryptions.

4.4. Breaking the Affine Cipher

If an eavesdropper's only approach to breaking an encryption system is to try all possible keys, the affine cipher is already doing much better than the Caesar cipher; we now have 312 keys instead of 26 (or 311 versus 25). However, just like the Caesar cipher, it is possible to break the affine cipher *without* having to try all the keys.

Assume we have intercepted the following message, encrypted with the affine cipher.

```
MCCLL IMIPP ISKLN UHCGI MCKBI XCUMT IPLKX
LRIGW MCXLA MWALV CCDGJ KXYCR
```

We again try to use frequency analysis to break the message. Counting shows that the most common letters in the message are `C`, `I`, and `L`, which occur 9, 7, and 7 times, respectively. Since "E" is the most common letter in English text, it is natural for us to guess that the ciphertext letter `C` was encrypted from the plaintext letter `E`.

We work backwards to break the message. The message was encrypted using the formula

$$y = ax + b \pmod{26}, \tag{4.1}$$

where the pair (a, b) is the affine cipher key and a is relatively prime to 26. Our guess that `E` encrypts to `C` means that for the plaintext $x = 4$ (an `E`) we get the ciphertext $y = 2$ (a `C`). Plugging these values into equation (4.1), we find

$$2 = 4a + b \pmod{26}. \tag{4.2}$$

Can we solve this congruence to figure out the key (a and b) and decrypt the message? Absolutely not! We have only one congruence, but two unknowns. When solving a system of equalities we need as many equations as unknowns; it's the same here. We thus need to find another congruence. Note it's reasonable that we need two congruences, as the key for the affine cipher is two numbers. The Caesar cipher's key was just one number, which is why knowing one letter sufficed there.

We can make another guess based on frequency analysis. Looking at Table 4.1, we see that `T` is the second most common letter in the English language. It's therefore natural to guess that `T` in the plaintext was encrypted to either `I` or `L` (the most common letters in the ciphertext after `C`). We guess that `T` was encrypted to `I`, which implies that $y = 8$ for $x = 19$. Plugging this into equation (4.1) gives

$$8 = 19a + b \pmod{26}. \tag{4.3}$$

We solve the system of congruences

$$\begin{cases} 2 = 4a + b \pmod{26}, \\ 8 = 19a + b \pmod{26} \end{cases} \tag{4.4}$$

for a and b. As there are two congruences and two unknowns, there's cause for optimism. One way to solve a system of congruences or equations is by subtracting multiples of one equation from the other one to eliminate a variable. In this case, subtracting the second congruence from the first one eliminates b, and gives

$$-6 = -15a \pmod{26},$$

which is equivalent to

$$20 = 11a \pmod{26}.$$

We solve for a by multiplying both sides by the multiplicative inverse of 11 (mod 26), which Table 4.2 tells us is 19. We find

$$19 \cdot 20 = 19 \cdot 11a \pmod{26},$$

and so

$$16 = a \pmod{26}. \tag{4.5}$$

Unfortunately, this can't be right. Recall that a must always be relatively prime to 26 for the affine cipher to work; thus one of the guesses must be wrong. Let's still guess that `E` encrypts to `C`, but now let's guess that `T` is encrypted to `L`. Now our system of congruences is

$$\begin{cases} 2 = 4a + b \pmod{26}, \\ 11 = 19a + b \pmod{26}. \end{cases} \tag{4.6}$$

After similar algebra as above, we get $a = 11$ and $b = 10$.

This analysis suggests that the key is $(11, 10)$. It's possible (especially since the message was rather short) that we were unlucky with frequency analysis, so we don't know that this is actually the key until we try decrypting the message.

To decrypt the message, we need to find the decryption congruence. If the encryption congruence is

$$y = 11x + 10 \pmod{26},$$

then solving this congruence for x gives the decryption congruence

$$x = 19y + 18 \pmod{26}.$$

We could have used our result from page 90. There we found an explicit, closed form expression for the decryption function given the encryption key. This also gives $(19, 18)$. Using this to decrypt the message gives

	M	C	C	L	L	I	M	I	P	P	...
y	12	2	2	11	11	8	12	8	15	15	...
$x = 19y + 18 \pmod{26}$	12	4	4	19	19	14	12	14	17	17	...
	M	E	E	T	T	O	M	O	R	R	...

As the beginning looks like English words, it's very likely this guess was correct. Continuing, we find the whole message decrypts to `MEET TOMORROW AT FIVE COME ALONE IMPORTANT DOCUMENTS MUST BE EXCHANGED` (in the interest of readability, we restored the spaces).

When solving systems of congruences, the number of solutions can sometimes be greater than 1 (although still small). Consider, for example, the situation where we have intercepted the message `B FNPKK D CDI`, encrypted with an affine cipher with unknown key. The original word spacing is still intact, thus it seems natural to guess, for example, that `B` corresponds to the plaintext letter `I`, and `D` corresponds to the plaintext letter `A`. These guesses lead to the system

$$\begin{cases} 1 = 8a + b \pmod{26}, \\ 3 = 0a + b \pmod{26}, \end{cases} \tag{4.7}$$

which, upon subtracting, give the congruence

$$24 = 8a \pmod{26}. \tag{4.8}$$

Unlike in the previous example, however, the coefficient of a here does not have an inverse modulo 26. In fact, Table 4.2 shows that $8 \cdot 3 = 24 \pmod{26}$ and $8 \cdot 16 = 24 \pmod{26}$ are both true congruences, thus we need to consider both $a = 3$ and $a = 16$ as possible solutions. Fortunately, we can immediately rule out the solution $a = 16$, since a must be relatively prime to 26 for the affine cipher to work. Plugging $a = 3$ back into one of the original congruences gives $b = 3$, and at this point, the decryption formula can be found and used as in the previous example. Using the formula from page 90, the decryption key is just $(a^{-1}, -ba^{-1})$ modulo 26, or $(3^{-1}, -3 \cdot 3^{-1}) = (9, 25)$.

Note that breaking the affine cipher was more of a nuisance than breaking Caesar's cipher. In addition to having to solve a system of congruences, we had to make *two* correct guesses from frequency analysis to come up with the correct key. Nevertheless, it is a weakness that discovering the correct decryption of two letters is enough to break the whole cipher. The substitution alphabet cipher, covered in the next section, requires substantially more guesswork to break.

4.5. The Substitution Alphabet Cipher

The Caesar and affine ciphers always use the same rules for encoding a letter, regardless of its position in the message. If an E in one part of the plaintext is encrypted to the letter O, then *all* E's in the plaintext are encrypted to the letter O. These ciphers are just special cases of the substitution alphabet cipher, which works by specifying an arbitrary substitution for letters in the alphabet. The advantage of these two ciphers is their simplicity; rather than having to specify an ordering of all 26 letters, it suffices to give one or two pieces of information. Of course, this simplicity means there are very few such ciphers (only 26 Caesar ciphers, and just 312 affine ciphers). There are $26! \approx 4.03 \cdot 10^{26}$ substitution alphabet ciphers!

For example, consider the following substitution alphabet cipher:

A	B	C	D	E	F	G	H	I	J	K	L	M	N	O	P	Q	R	S	T	U	V	W	X	Y	Z
C	T	U	A	F	H	L	J	M	Z	Q	B	S	O	N	X	V	W	D	Y	P	I	E	R	K	G

The message

WHENW ILLYO URETU RN

encrypts to

EJFOE MBBKN PWFYP WO

Here decryption works by reading the substitution table in reverse.

A key for the substitution alphabet cipher consists of a table like the one given above. As there is no longer any underlying structure, we have to specify the ordering of all 26 letters.

The number of possible keys is so great, in fact, that it is practically impossible to break the cipher just by guessing keys. This is an enormous improvement over the Caesar and affine ciphers; in those cases, there are so few keys that simple brute force suffices. With the substitution alphabet cipher, the number of possible keys is so great that, even using a modern desktop computer, this could take on the order of billions of years to try all the keys.

Do not be misled by this large number. Just because it would take billions of years to try all the keys doesn't mean that there isn't a better approach. We don't need to try all the keys; we just need to find the key that works. While we used frequency analysis to attack the Caesar and affine keys before, there it was more for convenience than need. There were so few keys that enumerating all the possibilities was a valid approach. Here frequency analysis is truly needed, as we neither can, nor want to, try over 10^{26} possible keys! As long as the message is long enough, we can crack it using frequency analysis. This technique is so easy to use and so powerful that many newspapers have messages encrypted by a substitution alphabet cipher as a puzzle for their readers' enjoyment.

Consider, for example, the following ciphertext.

```
GAYRI NGQKI CYHHY HCBLC IBOIZ VBYZI ELPQY BBYHC KVTIZ QYQBI ZLHBT
IKGHU GHELP TGOYH CHLBT YHCBL ELLHR ILZBN YRIQT ITGEJ IIJIE YHBLB
TIKLL UTIZQ YQBIZ NGQZI GEYHC KSBYB TGEHL JYRBS ZIQLZ RLHOI ZQGBY
LHQYH YBGHE NTGBY QBTIS QILPG KLLUB TLSCT BGAYR INYBT LSBJY RBSZI
QLZRL HOIZQ GBYLH
```

In order to apply frequency analysis, we need the letter counts. The most common letters are B, I, L, Y, and H, occurring 26, 25, 24, 23, and 20 times, respectively. From the letter frequencies in Table 4.1, it's reasonable to assume that the plaintext letters T and E correspond to some of these most common letters.

If we assume that E was encrypted to B and T was encrypted to I, we can make the following substitutions:

```
GAYRI NGQKI CYHHY HCBLC IBOIZ VBYZI ELPQY BBYHC KVTIZ QYQBI ZLHBT
    t     t         e   te t   e  t       ee       t     et    e
IKGHU GHELP TGOYH CHLBT YHCBL ELLHR ILZBN YRIQT ITGEJ IIJIE YHBLB
t                    e     e        t  e    t   t     tt t    e e
TIKLL UTIZQ YQBIZ NGQZI GEYHC KSBYB TGEHL JYRBS ZIQLZ RLHOI ZQGBY
 t      t     et      t         e e          e   t        t    e
LHQYH YBGHE NTGBY QBTIS QILPG KLLUB TLSCT BGAYR INYBT LSBJY RBSZI
       e       e   e t   t        e       e     e  e    e    e  e
QLZRL HOIZQ GBYLH
        t    e
```

There is something strange about this substitution, however: nowhere does the pattern T_E appear, which would mean that the word "the" never appears in the passage. While this is possible, it's more likely that the substitution should be the other way around (which gives us four candidates for "the"). Switching T and E gives

```
GAYRI NGQKI CYHHY HCBLC IBOIZ VBYZI ELPQY BBYHC KVTIZ QYQBI ZLHBT
    e     e         t   et e   t  e       tt       e     te    t
IKGHU GHELP TGOYH CHLBT YHCBL ELLHR ILZBN YRIQT ITGEJ IIJIE YHBLB
e                    t     t        e  t    e   e     ee e    t t
TIKLL UTIZQ YQBIZ NGQZI GEYHC KSBYB TGEHL JYRBS ZIQLZ RLHOI ZQGBY
 e      e     te      e         t t          t   e        e    t
LHQYH YBGHE NTGBY QBTIS QILPG KLLUB TLSCT BGAYR INYBT LSBJY RBSZI
       t       t   t e   e        t       t     e  t    t    t  e
QLZRL HOIZQ GBYLH
        e    t
```

There are now four instances of the pattern T_E: in the fifth block on the first line (ciphertext BOI), straddling the last block of the first line and the first block of the second line (ciphertext BTI), straddling the last block of the second line and the first block of the third line (ciphertext BTI), and in the fourth block of the fourth line (ciphertext BTI). Based on these occurrences, it seems reasonable to assume that T in the ciphertext corresponds to H in the plaintext and that the first instance BOI was just a coincidence (otherwise our long phrase would only have one "the"). Filling in this substitution, we get the following.

```
GAYRI NGQKI CYHHY HCBLC IBOIZ VBYZI ELPQY BBYHC KVTIZ QYQBI ZLHBT
    e     e         t   et e   t  e       tt      he     te    th
IKGHU GHELP TGOYH CHLBT YHCBL ELLHR ILZBN YRIQT ITGEJ IIJIE YHBLB
e           h        th    t        e  t    e h eh    ee e    t t
TIKLL UTIZQ YQBIZ NGQZI GEYHC KSBYB TGEHL JYRBS ZIQLZ RLHOI ZQGBY
he     he     te      e         t t h        t   e        e    t
LHQYH YBGHE NTGBY QBTIS QILPG KLLUB TLSCT BGAYR INYBT LSBJY RBSZI
       t     h t   the   e        t h   h t     e  th   t    t  e
QLZRL HOIZQ GBYLH
        e    t
```

As one would expect, our substitutions have produced several instances of TH in the plaintext, suggesting that we're on the right track. Continuing now with frequency analysis, the most common ciphertext letters for which we haven't assigned a substitution are L, Y, and H. Referring to Table 4.1, the most common English letters after *e* and *t* are *a*, *o*, and *i*. Notice, however, that the pattern LL occurs three times in the ciphertext: of the letters *a*, *o*, and *i*, only *o* appears commonly as a double letter in English, so it's natural to guess that L was substituted for O:

```
GAYRI NGQKI CYHHY HCBLC IBOIZ VBYZI ELPQY BBYHC KVTIZ QYQBI ZLHBT
    e     e         to  et e   t  e  o    tt      he     te  o th
IKGHU GHELP TGOYH CHLBT YHCBL ELLHR ILZBN YRIQT ITGEJ IIJIE YHBLB
e       o   h       oth    to  oo   eo t    e h eh    ee e    tot
TIKLL UTIZQ YQBIZ NGQZI GEYHC KSBYB TGEHL JYRBS ZIQLZ RLHOI ZQGBY
he oo  he     te      e         t t h   o    t   e o   o  e    t
LHQYH YBGHE NTGBY QBTIS QILPG KLLUB TLSCT BGAYR INYBT LSBJY RBSZI
o      t     h t   the   eo    oo t ho  h t     e  th o t    t  e
QLZRL HOIZQ GBYLH
 o  o   e    t o
```

We can also try frequency analysis on blocks of letters. For example, the three letter block YHC occurs five times in the ciphertext, more than any other triple. The most common English "trigrams" are *the*, *and*, and *ing*. Since our guesses so far rule out the *the*, it's natural to make the substitutions Y→A, H→N, and C→D:

```
GAYRI NGQKI CYHHY HCBLC IBOIZ VBYZI ELPQY BBYHC KVTIZ QYQBI ZLHBT
  a e     e danna ndtod et e   ta e  o  a ttand   he   a te  onth
IKGHU GHELP TGOYH CHLBT YHCBL ELLHR ILZBN YRIQT ITGEJ IIJIE YHBLB
e  n   n o  h  an dnoth andto  oon  eo t  a e h eh    ee e  antot
TIKLL UTIZQ YQBIZ NGQZI GEYHC KSBYB TGEHL JYRBS ZIQLZ RLHOI ZQGBY
he oo  he   a te      e   and   tat h  no    t    e o   o  e     t
LHQYH YBGHE NTGBY QBTIS QILPG KLLUB TLSCT BGAYR INYBT LSBJY RBSZI
on an at n   h ta  the   eo    oo t ho  h t     e  th o t    t  e
QLZRL HOIZQ GBYLH
 o  o n e    tyon
```

Unfortunately, there are indications that this last set of substitutions may be incorrect. For example, in the first line we now have have the blocks EDANNANDTODET and NOTHANDTO in the plaintext, on the first and second lines, respectively. Both of these blocks would seem more reasonable if A and D were replaced with I and G, respectively, suggesting that perhaps the ciphertext triple YHC corresponded to the trigram ING and not AND. Making these changes gives us

```
GAYRI NGQKI CYHHY HCBLC IBOIZ VBYZI ELPQY BBYHC KVTIZ QYQBI ZLHBT
  i e     e inni  ngtog et e   ti e  o  i tting   he   i te  onth
IKGHU GHELP TGOYH CHLBT YHCBL ELLHR ILZBN YRIQT ITGEJ IIJIE YHBLB
e  n   n o  h  in gnoth ingto  oon  eo t  i e h eh    ee e  intot
TIKLL UTIZQ YQBIZ NGQZI GEYHC KSBYB TGEHL JYRBS ZIQLZ RLHOI ZQGBY
he oo  he   i te      e   ing   tit h  no    t    e o   o  e     t
LHQYH YBGHE NTGBY QBTIS QILPG KLLUB TLSCT BGAYR INYBT LSBJY RBSZI
on in it n   h ti  the   eo    oo t ho  h t     e  th o t    t  e
QLZRL HOIZQ GBYLH
 o  o n e    tyon
```

Note the troublesome blocks are now EGINNINGTOGET and NOTHINGTO. At this point, we are basically playing hangman, trying to guess more substitutions by what is needed to make the revealed phrases words. For example, _EGINNINGTOGET seems like it could be BEGINNINGTOGET, suggesting the substitution K→B, while NOTHINGTO_O could be NOTHINGTODO, suggesting the substitution E→D. Making these substitutions yields

```
GAYRI NGQKI CYHHY HCBLC IBOIZ VBYZI ELPQY BBYHC KVTIZ QYQBI ZLHBT
  i e    be ginni ngtog et e   ti e do  i tting b he   i te  onth
IKGHU GHELP TGOYH CHLBT YHCBL ELLHR ILZBN YRIQT ITGEJ IIJIE YHBLB
eb n   ndo  h  in gnoth ingto doon  eo t  i e h eh d  ee ed intot
TIKLL UTIZQ YQBIZ NGQZI GEYHC KSBYB TGEHL JYRBS ZIQLZ RLHOI ZQGBY
heboo  he   i te      e  ding b tit h  no    t i  e o   o  e     t
LHQYH YBGHE NTGBY QBTIS QILPG KLLUB TLSCT BGAYR INYBT LSBJY RBSZI
on in it nd  h ti  the   eo   boo t ho  h t     e  th o t    t  e
QLZRL HOIZQ GBYLH
 o  o n e    tyon
```

Spanning the end of the second line and beginning of the third, the plaintext block INTOTHEBOO__ suggests the substitution U→K. In the third line, we have the plaintext INGB__TIT. The ING almost certainly represents the end of a word. It seems clear that the blank must be a vowel, and U seems the most likely candidate. The substitutions U→K and S→U give

```
GAYRI NGQKI CYHHY HCBLC IBOIZ VBYZI ELPQY BBYHC KVTIZ QYQBI ZLHBT
  i e    be ginni ngtog et e   ti e do  i tting b he   i te  onth
IKGHU GHELP TGOYH CHLBT YHCBL ELLHR ILZBN YRIQT ITGEJ IIJIE YHBLB
eb nk  ndo  h  in gnoth ingto doon  eo t  i e h eh d  ee ed intot
TIKLL UTIZQ YQBIZ NGQZI GEYHC KSBYB TGEHL JYRBS ZIQLZ RLHOI ZQGBY
heboo khe   i te      e  ding butit h  no    tu  e o   o e      t
LHQYH YBGHE NTGBY QBTIS QILPG KLLUB TLSCT BGAYR INYBT LSBJY RBSZI
on in it nd  h ti  theu  eo   bookt ho  h t     e  th o t    t  e
QLZRL HOIZQ GBYLH
 o  o n e    tyon
```

You should realize that we could choose to concentrate on many different parts of the message. It's somewhat similar to looking at a quadratic with integer coefficients and "seeing" the factorization. There's a lot of personal choice in the order of these steps, and what guesses you use to complete words. Moving on, another possibility presents itself on the first line, as TI_EDO__ITTING becomes TIREDOFSITTING under the substitutions Z→R, P→F, and Q→S. On the second line, ON_EO_T_I_E becomes ONCEORTWICE under the substitutions R→C, Z→R, and N→W. These five substitutions bring us to

```
GAYRI NGQKI CYHHY HCBLC IBOIZ VBYZI ELPQY BBYHC KVTIZ QYQBI ZLHBT
  ice   sbe ginni ngtog et er  tire dofsi tting b her siste ronth
IKGHU GHELP TGOYH CHLBT YHCBL ELLHR ILZBN YRIQT ITGEJ IIJIE YHBLB
eb nk  ndof h  in gnoth ingto doonc eortw icesh eh d  ee ed intot
TIKLL UTIZQ YQBIZ NGQZI GEYHC KSBYB TGEHL JYRBS ZIQLZ RLHOI ZQGBY
heboo khers ister   sre  ding butit h  no   rtu resor co  e rs t
LHQYH YBGHE NTGBY QBTIS QILPG KLLUB TLSCT BGAYR INYBT LSBJY RBSZI
onsin it nd  h ti stheu seof  bookt ho  h t   c e  th o t   ct re
QLZRL HOIZQ GBYLH
sorco n ers  tyon
```

At this point, it's not too hard to figure out the rest. This is quite common—after a while a tipping point is reached, and the final letters fall quickly. The plaintext is:

```
ALICE WASBE GINNI NGTOG ETVER YTIRE DOFSI TTING BYHER SISTE RONTH
EBANK ANDOF HAVIN GNOTH INGTO DOONC EORTW ICESH EHADP EEPED INTOT
HEBOO KHERS ISTER WASRE ADING BUTIT HADNO PICTU RESOR CONVE RSATI
ONSIN ITAND WHATI STHEU SEOFA BOOKT HOUGH TALIC EWITH OUTPI CTURE
SORCO NVERS ATION
```

There is no doubt that applying frequency analysis to the substitution alphabet cipher in this way can be tedious. Unlike the Caesar and affine ciphers, it's not enough to figure out just one or two of the substitutions; each one must be determined separately. That said, just the fact that it is possible at all, with a cipher that has such a large number of possible keys, indicates the power of frequency analysis. Messages are not random jumbles of letters, and frequency analysis allows the cryptographer to take advantage of that fact to break codes.

In fact, for many years mathematicians were not the primary people tasked with breaking codes. That honor and responsibility fell to linguists. Looking back on how we attacked the general substitution alphabet cipher, it's clear why!

4.6. Frequency Analysis and the Vigenère Cipher

The Vigenère cipher is designed to be resistant to frequency analysis. Consider the following graph of the frequencies of letters in the original plaintext from our Kant quote.

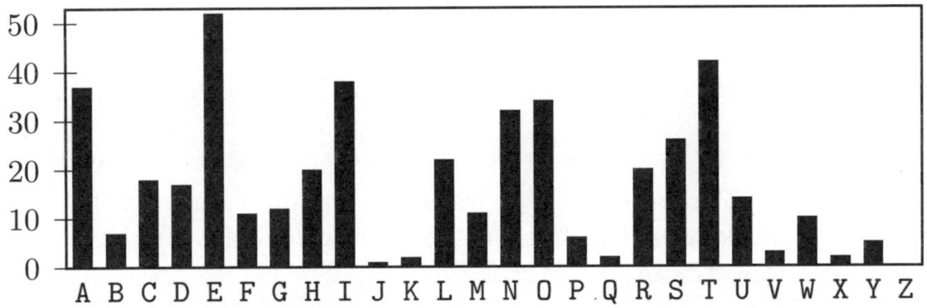

We can see that the plaintext looks like a typical English text: e's and t's are common, while letters like j's, x's, and z's are not.

If we had encrypted the message with the Caesar cipher (or even with any substitution alphabet cipher), then the same frequencies would occur in the distribution, but for different letters. For example, here's the distribution of the message after encryption by the Caesar cipher with a key of 5.

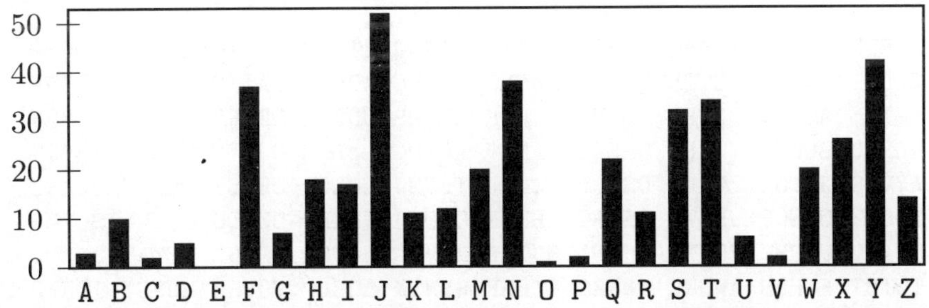

Frequency analysis is possible because we can guess that common letters in the ciphertext correspond to common English letters (e's, t's, and so on). In contrast, the following graph shows the distribution of letters in the ciphertext found by encrypting the plaintext with the Vigenère cipher (with `MORALS` as the key).

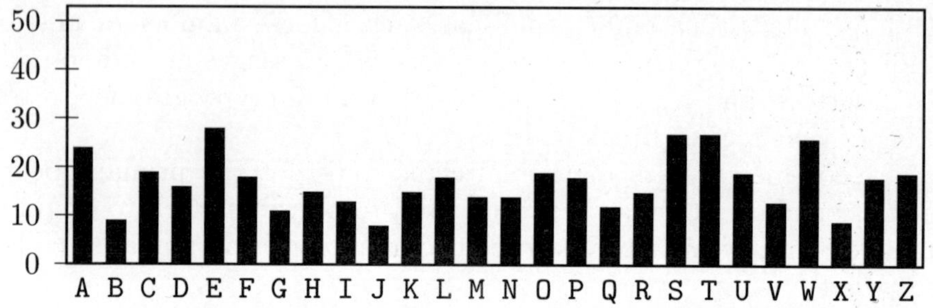

This distribution is very different than the frequency distribution of the plaintext; it's been "smoothed out" by the encryption. This makes the resulting distribution significantly less useful to an attacker. For example, given a common letter like W in the ciphertext, there doesn't appear to be a simple way to decide whether it's common because one of its corresponding plaintext letters was "very common", or because several of them were "fairly common". Further, we're seeing the result from a keyword of length five; the smoothing effect is typically more pronounced the longer the keyword. There's no clear way for an eavesdropper to use this frequency distribution to make guesses about the encryption key. This feature of the Vigenère cipher convinced many that it might be impossible to break. For a time, the cipher was actually known as "the unbreakable cipher", and no practical attack existed until 300 years after its introduction. It's worth dwelling on this for a moment. Three hundred years is a *long* time. So, even though it may be easy to crack now, the Vigenère cipher provided a significant level of security for a long time.

As standard frequency analysis is out, we need to be more clever. We'll assume that, somehow, our eavesdropper knows the length of the keyword. This is a major assumption, but it turns out to be a somewhat reasonable one (as we'll see when we discuss the Kasiski test in §4.7). For our Kant quote, the keyword has length 6. The eavesdropper now breaks the message into six groups where each group consist of letters that were shifted by the same amount in the encryption process. For example, for the current ciphertext

```
FVVRP AEBFP ZKEWS IWAFM FFEZU BBIYY ATRNJ LTWEG LLMZC IYLTS NOCDP
CIEGW ZCLTZ XUHNH TUTQR NMWDS XACVQ RRSRG ARNIE ZAIKQ FSXWW INSFW
FNPPO SGTLY ACUWT DXWET PDXWX EYUQK ZTUMP UDEYL MBUWS SFSME CLMZV
NEKAT KHPEU BUOYW YWXHE OMBKT ZFMAV ACWPC LBEDQ GJIYE MBPRP KBSTT
DYACU AYVPS JICSN ZVADS DSJUN ZCIRL TLUSJ OQLQA GECSY SETLK OCLRL
YQFVS ZDGHZ OYHQF JEGWD OECPT GHKHP QOOEA WKAPV CZEQS OTCWY SCYMS
POEDS SDAWU WARHY EHAXZ NHTUT WJTZE MYVUD WATKH PKQUZ FEKAT EAEMD
SRNOO TWTHT FUHJS AWOWR LNGZG KIEMF WFNTK OOCLP VOVRR LUFSI IDFAH
XOZV
```

the first group of letters is F, E, E, F, etc. (every sixth letter starting with the first one). The second group is V, B, W, M, etc. (every sixth letter starting with the second), and so on. The letters in each of these groups were encrypted the same way, since the keyword lines up the same way for each of them (all letters in the first group were encrypted by adding the letter `M`, all letters in the second group were encrypted by adding the letter `O`, and so on). The important thing to notice is that this means that frequency analysis should work on each group of letters when the groups are considered *separately*! Even though all the letters of the message were not encrypted in a consistent way, resulting in an unhelpful frequency distribution, each of the groups of letters was encrypted by a simple shift (i.e., a Caesar cipher), and each has a frequency distribution revealing information about that shift. Note that the plan is to use frequency analysis to figure out these six shifts. Even though there are only 26 possible Caesar ciphers, the difficulty here is that the keyword has length 6, so we're looking at $26^6 = 308,915,776$ keys to try. We don't want to have to try that many keys (and, of course, it would be far worse if a larger keyword were used).

Below are the frequency distributions for each of these six groups.

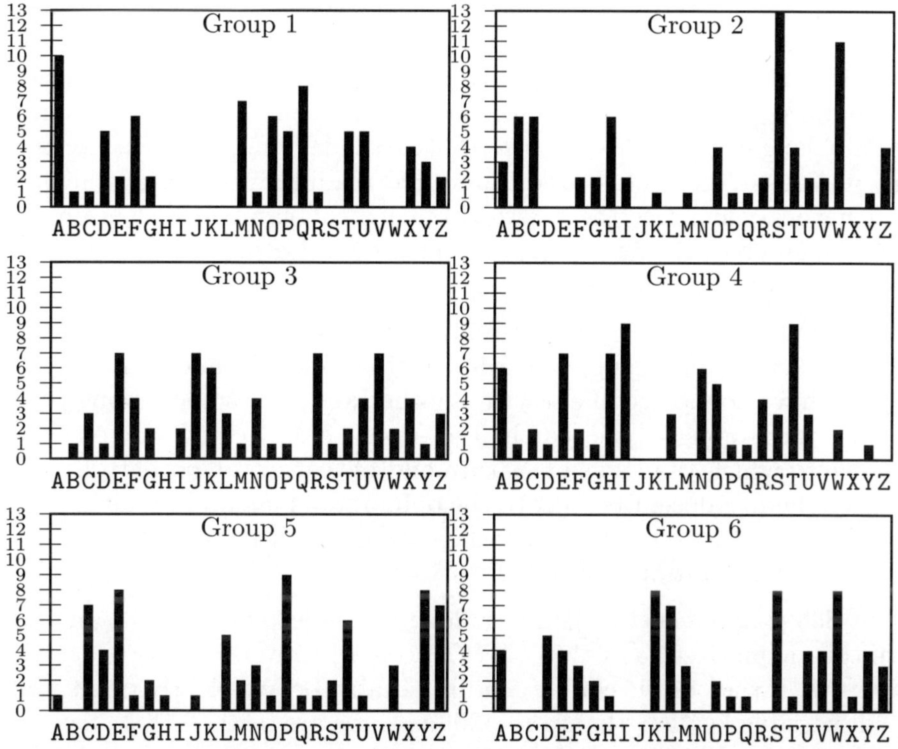

Each of these distributions gives information on the shift used for its respective groups of letters. For example, the fact that `Q` is common in the distribution for Group 1 is due to the fact that `E` is common in the plaintext and corresponds to the letter `Q` under a shift by 12 (which corresponds to `M`, the first letter of the keyword). Similarly, `S` is common in the distribution for Group 2 because it's the shift of `E` by 14 (corresponding to `O`, the second letter of the keyword); `V` is common in Group 3 because it's the shift of `E` by 17 (`R`); `E` and `T` are common in Group 4 because they weren't shifted at all (a shift of 0 corresponds to A in the keyword); the common letter P in Group 5 corresponds to E under a shift of 11 (L); and finally the common letter W in Group 6 corresponds to E under a shift of 18 (S).

This analysis is predicated upon the assumption that the eavesdropper can somehow figure out the length of the keyword. Of course, an eavesdropper with enough time could simply try lots of possible keyword lengths, until one worked. There is, however, a much better way of efficiently determining the keyword length. It's known as the **Kasiski test**, named after Friedrich Kasiski who in 1863 published the first attack on the Vigenère cipher (though the method was known to Charles Babbage 10 or 15 years earlier).

4.7. The Kasiski Test

For hundreds of years, the Vigenère cipher provided enormous protection. Its great strength was the need to guess the keyword. While frequency analysis is an effective tool to break it, it requires knowing the length of the keyword. In actual implementations that length could be enormous. While a long keyword requires both the sender and recipient to know a long phrase, if they agree upon a source readily available to them, they don't have to memorize the keyword but instead can quickly obtain it. For example, maybe both parties agree that they'll use the first sentence of the Declaration of Independence:

> When in the Course of human events it becomes necessary for one people to dissolve the political bands which have connected them with another and to assume among the powers of the earth, the separate and equal station to which the Laws of Nature and of Nature's God entitle them, a decent respect to the opinions of mankind requires that they should declare the causes which impel them to the separation.

At 330 characters, this is a very long key indeed. Further, many people have this memorized, and thus there's no need to worry about forgetting it (of course, if they do, it can be readily obtained). We clearly don't want to try all possible key lengths up to 330, as there are over 10^{400} to try!

In 1863 Kasiski devised a method to find the keyword's length, though Babbage had independently discovered this more than a decade earlier. The **Kasiski test** determines the length of the keyword used by the Vigenère cipher by taking advantage of repetitions in the ciphertext. For example, an observant cryptanalyst might notice that the strings ATKHP and NHTUT both appear twice in the ciphertext:

```
FVVRP AEBFP ZKEWS IWAFM FFEZU BBIYY ATRNJ LTWEG LLMZC IYLTS NOCDP
CIEGW ZCLTZ XUHNH TUTQR NMWDS XACVQ RRSRG ARNIE ZAIKQ FSXWW INSFW
FNPPO SGTLY ACUWT DXWET PDXWX EYUQK ZTUMP UDEYL MBUWS SFSME CLMZV
```

```
NEKAT KHPEU BUOYW YWXHE OMBKT ZFMAV ACWPC LBEDQ GJIYE MBPRP KBSTT
DYACU AYVPS JICSN ZVADS DSJUN ZCIRL TLUSJ OQLQA GECSY SETLK OCLRL
YQFVS ZDGHZ OYHQF JEGWD OECPT GHKHP QOOEA WKAPV CZEQS OTCWY SCYMS
POEDS SDAWU WARHY EHAXZ NHTUT WJTZE MYVUD WATKH PKQUZ FEKAT EAEMD
SRNOO TWTHT FUHJS AWOWR LNGZG KIEMF WFNTK OOCLP VOVRR LUFSI IDFAH
XOZV
```

How can it happen that long words are repeated in the ciphertext like this? If we look at the encryption calculation on page 35, we see that these instances of ATKHP and NHTUT correspond to the words OFTHE and WHICH, respectively, which were repeated in the plaintext. These aren't the only long strings repeated in the plaintext. The string GOODWI occurs both near the end of the second line of the plaintext (as the beginning of "good without qualification") and near the beginning of the third line (as the beginning of "good will"). However, the repetition of the string GOODWI in the plaintext doesn't lead to a repetition in the ciphertext since the first instance encrypts to RGARNI and the second to YACUWT.

Why do some repeated phrases encrypt the same while others do not? The explanation is the distance between the repetitions. Notice that in the two instances of WHICH in the plaintext that were encrypted to NHTUT, both occurrences of WHICH lined up with the keyword in the same way. In each the string WHICH was added to the segment RALSM of the running keystream ...MORALSMORALSMORALS...). Similarly, the two instances of OFTHE which encrypt to ATKHP both line up with the segment MORAL from the keystream. The situation is very different for GOODWI; the first time it's above LSMORA while the second time it's above SMORAL. They almost line up, but almost isn't good enough. The fact that they are off by one means that the letters are shifted by different amounts, and thus the encrypted texts differ.

Kasiski's great observation is that two instances of a repeated string line up with the key stream in the same way exactly when *the distance between the first letters of the instances of the repeated string is a multiple of the keyword length.* In our example, we find the distance between the start of the two instances of ATKHP in the ciphertext is 198, and between the start of the two instances of NHTUT is 282. Kasiski's test tells us that we expect both of these distances to be a multiple of the keyword length, since we expect that both of these repetitions happened because some repetitions in the plaintext lined up in the same way with the keystream.

Another way of phrasing this is that if k is the length of the keyword, then k should divide both 198 and 282. The only common divisors of 198 and 282 are 1, 2, 3, and 6. Which should we try? It's safest to try the largest. Imagine the keyword was just the three letter word SOX. As the keystream is the keyword repeated, we can't tell the difference from using the keyword SOX or using the keyword SOXSOX; both give the same ciphertext. Obviously some six letter keywords are not the same as a three letter keyword repeated (REDSOX is very different than SOX repeated); if we use the largest common divisor we'll never miss a shorter keyword, while if we tried a shorter keyword

TABLE 4.3. Repeated strings of length at least 3 in the intercepted ciphertext.

Repeated string	Distance between occurrences
VOGYVQBND (×2)	50
CCBPDBEL (×2)	120
KTPCZ (×2)	190
HLNL (×2)	70
OGH (×2)	334
ZVS (×2)	120
VTA (×2)	165
TAG (×2)	135
NLC (×2)	89
PPH (×2)	60
AKT (×2)	112
AJE (×2)	85
HBU (×2)	227
VOG (×4)	50, 55, 15
LEO (×2)	85
GHB (×2)	47

we could miss the actual keyword. There is, of course, a drawback to using a larger keyword. The longer the word, the more frequency charts we must make. This lessens the number of data points per chart, which weakens the effectiveness of a frequency attack, but overall it's worth having a little less data and being sure the keyword is correct.

We end this section by showing how to use the Kasiski test and frequency analysis to decrypt a message. Imagine we intercept the following ciphertext, which we assume we know was encrypted with the Vigenère cipher:

```
KTPCZ NOOGH VFBTZ VSBIO VTAGM KRLVA KMXAV USTTP CNLCD VHXEO CPECP
PHXHL NLFCK NYBPS QVXYP VHAKT AOLUH TITPD CSBPA JEAQZ RIMCS YIMJH
RABPP PHBUS KVXTA JAMHL NLCWZ VSAQY VOYDL KNZLH WNWKJ GTAGK QCMQY
UWXTL RUSBS GDUAA JEYCJ VTACA KTPCZ PTJWP VECCB PDBEL KFBVI GCTOL
LANPK KCXVO GYVQB NDMTL CTBVP HIMFP FNMDL EOFGQ CUGFP EETPK YEGVH
YARVO GYVQB NDWKZ EHTTN GHBOI WTMJP UJNUA DEZKU UHHTA QFCCB PDBEL
CLEVO GTBOL EOGHB UEWVO GM
```

The first step is to identify some long repeated strings in the ciphertext so that we can apply the Kasiski test, which we do in Table 4.3. For example, the strings KTPCZ, HLNL, CCBPDBEL, VOGYVQBND are all repeated, at the positions underlined above.

Fortunately, it's not necessary to find all repetitions to apply the Kasiski test. Typically three or four repetitions will be suffice, and in some cases two repeated pairs may even be enough to give a good guess for the keyword's length.

Unfortunately, there's a problem in using the Kasiski test with the data from Table 4.3. The Kasiski test tells us that the keyword length should be a divisor of the distances between repeated strings, but the only common divisor of all of these differences is 1! It's unlikely that our message was encrypted with a keyword of length one, since that means the sender used a simple Caesar cipher. What went wrong?

The problem is that it's possible for some strings to be repeated just by chance, and not because they correspond to a repeated plaintext word for which the keystream has lined up in a consistent way. This is particularly true for short repeated strings which are only repeated once. On the other hand, we should be confident that the repetitions of the strings `CCBPDBEL`, `VOGYVQBND`, `KTPCZ`, `HLNL`, and `VOG` are not just coincidences, since all of these examples are either longer than three characters or are repeated several times (in the case of `VOG`). The greatest common divisor of the distances separating the instances of these sets of repeated strings is 5, so we'll guess that 5 is the keyword's length.

Since the Kasiski test gives a keyword length of 5, the next step is to consider the ciphertext as five groups of letters, according to how each one lines up with the (as of yet unknown) keyword, and do frequency analysis on each group separately. For example, the first group consists of the letters `K`, `N`, `V`, `V`, ... (the first letter from each group of five), the second consists of the letters `T`, `O`, `F`, ... (the second from each group of five), and so on. We just need to count the frequency of letters from each of these groups, which we do in the table below.

	A	B	C	D	E	F	G	H	I	J	K	L	M	N	O	P	Q	R	S	T	U	V	W	X	Y	Z
1	1	0	6	1	4	1	8	1	0	3	8	1	0	6	0	5	3	3	0	1	5	9	2	0	3	0
2	4	0	3	5	7	3	0	7	4	1	0	3	2	4	5	1	0	1	4	9	2	2	1	0	3	0
3	6	12	3	0	2	2	3	1	0	1	0	3	8	2	1	2	0	1	1	5	1	2	3	7	2	2
4	2	1	10	2	3	2	4	3	1	2	4	1	0	0	3	6	5	0	0	7	3	8	2	0	1	0
5	7	5	0	2	0	0	0	5	2	2	4	10	1	1	6	9	1	0	4	1	1	1	0	0	2	6

We see that the ciphertext letter `A` lines up with the first letter of the keyword one time, the second letter of the keyword four times, and so on.

We now start the frequency analysis. Let's look at the first group. The most common letters in this group are `V`, `G`, and `K`, occurring nine, eight, and eight times, respectively. A first guess is that `V` corresponds to the plaintext letter `E`. If this is the case, then this group was encrypted by a shift of 17. To check if this is reasonable, we examine what this means for the frequencies of other plaintext letters. For example, we expect `T` to be common in the plaintext, and under a shift of 17 this corresponds to the letter `K`, which occurs eight times as a ciphertext letter in this group. Thus, this *seems* reasonable, but is it? Let's check an uncommon letter: `Z` encrypts to `Q` under a shift of 17, which occurs three times in the first ciphertext group, which is a lot for this letter. While not impossible, it *suggests* that 17 is not the correct shift and prompts us to try another possibility. If we instead assume that the plaintext `E` corresponds to the ciphertext `G`, this means that this group was encrypted by a shift of two. This seems to check out okay: `T` encrypts to `V`, which is common (occurring nine times), `Z` encrypts to `B`, which doesn't occur at all, `A` encrypts to `C`, which is relatively common, and so on. Thus it seems likely that the first letter of the keyword is `C` (which gives a shift of two).

For the second group, the most common letters are `T`, `E`, and `H`. These are common letters in English text overall, and a quick glance at group 2's row in the plaintext shows that common English letters are common in this group, while uncommon English letters are uncommon. It's likely that this group wasn't shifted at all, meaning that the second letter of the keyword should be `A`. You're asked to finish the decryption in Exercise 4.9.42.

4.8. Summary

In this chapter we saw the power of frequency analysis. While it was not needed for the Caesar or the affine ciphers, it was indispensable in attacking substitution alphabet and the Vigenère ciphers. We do not need to check the over 10^{26} possible substitution alphabet keys, which is a good thing as we don't have the computer power to do so!

We end this chapter with some suggestions to lessen the effectiveness of attacks using frequency analysis. Remember, the way frequency analysis works is that we look for common letters or letter pairs, and then guess other letters by seeing what words fit the partially decrypted message.

- A natural defense against frequency attacks is to dilliberatly mispell words. For example, `Iqf we adyd, rymouve ore chxangq xax phew litturs, txeh myyningg xis ophtyn stil clur.` Of course, doing this means more time is spent in both writing the message and then understanding it when decrypted, and there is a danger of misinterpretation.
- Another defense against frequency analysis is to carefully choose your words. Consider the following excerpt from *Gadsby* by Ernest Vincent Wright:

  > As Gadsby sat thinking thus, his plan was rapidly growing: and, in a month, was actually starting to work. How? You'll know shortly; but first, you should know this John Gadsby; a man of "around fifty"; a family man, and known throughout Branton Hills for his high standard of honor and altruism on any kind of an occasion for public good. A loyal churchman, Gadsby was a man who, though admitting that an occasional fault in our daily acts is bound to occur, had taught his two boys and a pair of girls that, though folks do slip from what Scriptural authors call that "straight and narrow path," it will not pay to risk your own Soul by slipping, just so that

> you can laugh at your ability in staying out of prison; for Gadsby, having grown up in Branton Hills, could point to many such man or woman.

Do you notice anything? There are no *e*'s! The full title is *Gadsby: A Story of Over 50,000 Words Without Using the Letter "E"*. The full story is on-line at `http://onlinebooks.library.upenn.edu/webbin/book/lookupid?key=olbp37063`.

- The following text has been circulating on the internet since at least 2003.

  > Cdnuolt blveiee taht I cluod aulaclty uesdnatnrd waht I was rdanieg. The phaonmneal pweor of the hmuan mnid. Aoccdrnig to rscheearch at Cmabrigde Uinervtisy, it deosn't mttaer in waht oredr the ltteers in a wrod are, the olny iprmoatnt tihng is taht the frist and lsat ltteer be in the rghit pclae. The rset can be a taotl mses and you can sitll raed it wouthit a porbelm. Tihs is bcuseae the huamn mnid deos not raed ervey lteter by istlef, but the wrod as a wlohe. Amzanig huh? Yaeh and I awlyas tghuhot slpeling was ipmorantt!

  This scrambling should make the permutation cipher more secure, though the fact that "the", "and", "a", and other short words are always unchanged will lessen its power.

As a good exercise, create some rules for applying the defensive methods discussed in the summary. Try them on some of the long messages of this chapter and see how well a friend does at decrypting them.

4.9. Problems

- **From §4.1: Breaking the Caesar Cipher**

EXERCISE 4.9.1. *Break these Caesar ciphers:*
(a) `PAXG LAHNEW B KXMNKG`
(b) `QUCN ZIL U JBIHY WUFF`
(c) `GUR ENOOVG PENJYRQ BHG BS VGF UBYR`
(Hint: What three-letter words are likely to appear at the beginning of an English sentence?)

EXERCISE 4.9.2. *Encrypt the message* `BEAM ME UP` *with a Caesar cipher of shift* 16, *and then encrypt the output of that with a Caesar cipher of shift* 10. *Are you surprised at the output? Why or why not?*

EXERCISE 4.9.3. *In this section we met frequency analysis. Though this isn't needed for the Caesar cipher, it'll be very important later. Look up the letter frequencies in different languages. For languages with similar alphabets, are the frequencies similar?*

EXERCISE 4.9.4. *The letter frequencies depend on the type of work. We don't expect Shakespeare's tragedies to have exactly the same behavior as movie scripts, or text messages or twitter posts. Look up the letter frequencies of different types of English writing, and see how stable the percentages are.*

EXERCISE 4.9.5. *Break the following message (which was encrypted with the Caesar cipher) using frequency analysis.*

`MAXLX TKXGM MAXWK HBWLR HNKXE HHDBG ZYHK`

- **From §4.2: Function Preliminaries**

EXERCISE 4.9.6. *Let* $f(x) = 2x + 5$. *Find an inverse function* g *to* f.

EXERCISE 4.9.7. *Let* g *be the inverse function of* $f(x) = 2x + 5$ *from Exercise* 4.9.6. *Find the inverse function to* g.

EXERCISE 4.9.8. *Let* $f(x) = x^2 + 3$ *be defined for all real numbers. Why can't there be an inverse function to* f*?*

EXERCISE 4.9.9. *Let* $f(x) = (x - 5)^3$. *Find an inverse function* g *to* f.

EXERCISE 4.9.10. *Let* $f(x) = 3x \pmod{26}$. *Find an inverse function to* f *modulo* 26 *(this means find a* g *such that for all* x, $g(f(x))$ *is equivalent to* x *modulo* 26*).*

EXERCISE 4.9.11. *Let* $f(x) = 3x+5 \pmod{26}$. *Find an inverse function to* f *modulo* 26.

- **From §4.3: Modular Arithmetic and the Affine Cipher**

EXERCISE 4.9.12. *Show that if* x *and* n *have a common nontrivial divisor (in other words, their greatest common divisor is greater than* 1*), then* x *cannot be invertible modulo* n.

EXERCISE 4.9.13. *Let* p *be a prime number. Show that if* x *is relatively prime to* p, *then* x *is invertible modulo* p.

EXERCISE 4.9.14. *Consider the affine cipher with key* $a = 8$ *and* $b = 3$. *Find two letters that are encrypted to the same letter.*

EXERCISE 4.9.15. *We claimed that if* a *is relatively prime to* 26, *then* $f_{a,b}(n) = an + b \pmod{26}$ *lists the* 26 *letters of the alphabet in a new order. Verify the claim for* $a = 3, 5,$ *and* 11.

EXERCISE 4.9.16. *Remember* g *is the* ***inverse function*** *of* f *if* $g(f(x)) = x$. *Prove* $g(x) = x^2 + 4$ *is the inverse of* $f(x) = \sqrt{x-4}$ *(where of course* $x \geq 4$*), and* $g(x) = \frac{1}{2}x - \frac{3}{2}$ *is the inverse of* $f(x) = 2x + 3$.

EXERCISE 4.9.17. *Show that* $(x+y \bmod 26)+z \bmod 26 = x+y+z \bmod 26$ *(or, more generally, that this holds for any modulus).*

EXERCISE 4.9.18. *The* ***multiplication cipher*** *is an affine cipher with a shift of zero. The ciphertext* `IASSC GW` *was encrypted using the multiplication cipher with* 4 *as the key, while* `KADDI U` *was encrypted by multiplication with* 5 *as the key. It is possible to decrypt one of these messages. Indicate which can be decrypted, briefly explain why, and give the decryption.*

EXERCISE 4.9.19. *Indicate which of the following key-pairs can be used for the affine cipher:* $(5,6)$, $(13,17)$, $(5,5)$, *and* $(6,6)$. *Explain why.*

EXERCISE 4.9.20. *Indicate all the key-pairs in the following list which can be used for the affine cipher:* $(6,5)$, $(18,19)$, $(17,13)$, *and* $(17,15)$.

EXERCISE 4.9.21. *Encrypt the message* `MATHI SFUN` *using the affine cipher with key* $(7,11)$.

EXERCISE 4.9.22. *Encrypt the message* `CRYPT OISFU N` *with the affine cipher with* $(11,15)$ *as a key.*

EXERCISE 4.9.23. *Decrypt the message* `OAAXG XLCSX YD`*, which was encrypted with the affine cipher using* $(5,6)$ *as a key.*

EXERCISE 4.9.24. *Imagine we have a* 17*-letter alphabet. How many affine ciphers are there? What if we have a* 55*-letter alphabet?*

EXERCISE 4.9.25. *Determine how many affine ciphers there are if our alphabet has* n *letters, with* n *prime. What if* n *is a product of two distinct primes,* p *and* q*? Give an example of a valid affine cipher in each case.*

EXERCISE 4.9.26. *The Caesar cipher corresponds to looking at encrypting functions of the form* $f_s(n) = n + s \pmod{26}$, *with corresponding decryption function* $f_{-s}(n) = n - s \pmod{26}$. *The affine cipher used the encryption-decryption pair* $f_{a,b}(n) = an + b \pmod{26}$ *and* $f_{a^{-1},-b \cdot a^{-a}}(n) = a^{-1}n - ba^{-1}$, *where* a^{-1} *is the multiplicative inverse of* a *modulo* 26. *This suggests trying other polynomial functions. Does there exist a quadratic polynomial* $f_{a,b,c}(n) = an^2 + bn + c \pmod{26}$ *which can be used for encryption (i.e., permutes the order of the* 26 *letters), and if so, what does its decryption function look like? What about higher degree polynomials?*

EXERCISE 4.9.27. (Hard). *Continuing the previous exercise, is there an* a *such that* $f_a(n) = a^n \pmod{26}$ *is a valid encryption function (i.e., permutes the order of the* 26 *letters)? This problem is worth revisiting after reading Chapter* 8*, where we develop more of the theory of modular arithmetic.* Hint: Prove there are only 26 values of a that need to be checked.

- **From §4.4: Breaking the Affine Cipher**

EXERCISE 4.9.28. *Decrypt the message* `B FNPKK D CDI`*, encrypted with the affine cipher using the key* $(3,3)$.

EXERCISE 4.9.29. *Solve the following systems of congruences, or state that there is no solution. Be sure to state if there are multiple solutions.*

(a) $\begin{cases} 6 = 13a + b \pmod{26} \\ 13 = 4a + b \pmod{26}. \end{cases}$

(b) $\begin{cases} 14 = 17a + b \pmod{26} \\ 8 = 7a + b \pmod{26}. \end{cases}$

(c) $\begin{cases} 1 = 15a + b \pmod{26} \\ 10 = 9a + b \pmod{26}. \end{cases}$

EXERCISE 4.9.30. *Decrypt the message*

```
ZVUKE OGDGI HQZIL EUQQV GIFLT UZGLE HUCZZ VUOEX
LAEZV KREUA ZGDGH HIUKX LQGIX LNILM UOUXZ QKTGI
ZAVKZ URUHC GOUQT UDGHU EZ
```

encrypted using the affine cipher with an unknown key. A letter count shows U, Z, *and* G *are the most common letters in this ciphertext, occurring* 14, 12, *and* 104 *times, respectively. You can either use the methods of this section, or try all* 311 *possible ciphers!*

• From §4.5: The Substitution Cipher

EXERCISE 4.9.31. *Decrypt the message* YNTFN WONYY NTF, *which was encrypted using the substitution table at the start of §4.5.*

EXERCISE 4.9.32. *Break the following substitution alphabet cipher. This is made substantially easier by the fact that the original word spacing is intact.*

```
LKZB RMLK X JFAKFDEQ AOBXOV TEFIB F MLKABOBA TBXH
XKA TBXOV LSBO JXKV X NRXFKQ XKA ZROFLRP SLIRJB LC
CLODLQQBK ILOB TEFIB  F KLAABA KBXOIV KXMMFKD
PRAABKIV QEBOB ZXJB X QXMMFKD XP LC PLJB LKB DBKQIV
OXMMFKD OXMMFKD XQ JV ZEXJYBO ALLO Q FP PLJB SFPFQBO
F JRQQBOBA QXMMFKD XQ JV ZEXJYBO ALLO LKIV QEFP XKA
KLQEFKD JLOB
```

For your convenience, here are the letter counts: A, 15; B, 28; C, 3; D, 9; E, 8; F, 19; G, 0; H, 1; I, 8; J, 12; K, 24; L, 22; M, 12; N, 1; O, 19; P, 8; Q, 16; R, 7; S, 3; T, 4; U, 0; V, 9; W, 0; X, 23; Y, 2; Z, 5.

EXERCISE 4.9.33. *The message* CQLEQNMQOGDEUMQOKGNTGJVREZQBEREB *was encrypted with a substitution cipher. Decrypt it if you can; however, as it is a short message and as the spacing is lost, it will be very difficult.*

EXERCISE 4.9.34. *The message*

```
MV OQR NOIJN NQZCWX ISSRIJ ZPR PMEQO MP I OQZCNIPX
KRIJN QZG GZCWX LRP ARWMRBR IPX IXZJR IPX SJRNRJBR
VZJ LIPK ERPRJIOMZPN OQR JRLRLAJIPHR ZV OQR HMOK
ZV EZX
```

was encrypted using a substitution cipher. Decrypt it. This is significantly easier than the previous problem, as the spacing was preserved.

EXERCISE 4.9.35. *Assume you know that the word* SHM *is the result of encrypting* THE. *Knowing this excludes many substitution ciphers; how many substitution ciphers must still be tested? What percentage of all substitution ciphers were eliminated?*

EXERCISE 4.9.36. *Imagine you know the encryption of two different three-letter words. What percentage of substitution ciphers* can *you eliminate? What percentage of substitution ciphers* may *you be able to eliminate?*

- **From §4.6: Frequency Analysis and the Vigenère Cipher**

EXERCISE 4.9.37. *Explain in your own words how the Vigenère cipher can be broken if they keyword length is known. For your attack to be successful, what do you think must be true about the ratio of the message's length to the size of the keyword?*

EXERCISE 4.9.38. *If the keyword in the Vigenère cipher has length* 20, *there are almost* $2 \cdot 10^{28}$ *possibilities! What is wrong with the following approach to winnow that list down? Instead of trying to guess the entire keyphrase, someone suggests trying to just guess the first three or four letters. There are only* 456,976 *possibilities. All we need to do is see what starts off as a reasonable English phrase, and then move on to the next batch of four letters.*

EXERCISE 4.9.39. *The message* BCRRBCQORHKEPSLSLCWRWXXDESPEZMPYQ WCEBCBOSFHCIZHSQWVHCBRWRVLNEGDRCKRRQS *was encrypted with a Vigenère cipher of length* 4. *What is the message?*

EXERCISE 4.9.40. *The message*

LVBOSJCSTKSTLTRELXWTYIIBECXADRWYILOEBXAHROSPWYIX
XZTQIEHLVSEJXTVTYIKSIXKKPRDRVJVWCSXDYTIEYLOEWXDV
TLRWVRDSLPOERVEZAQEACWTRWWHOPXJDYBCIKHNNFQDTPFWA
UGORVILEDXGKIOXGHPEVTFVMYVWPRDSCSZLOIHISSRCOLEXH
LWIHVEJAAMLWPRDKLWAHYYKPRDEELBRKPKWSCBWLOADJDTWH
ZWZLIBXGIMSRGGUSEQEPXIFRVLVYYLACTFFWDICLVISDZILV
SVIWEXOJPWLPZIJRLAEGWAONVWPQAPXZLROWLWIRLF

was encrypted with a Vigenère cipher of length 11. *What is the message?*

EXERCISE 4.9.41. *The keyword for the previous problem, when encrypted by a Caesar cipher, is* KZSCWKHWSJW. *What is the keyword, and what is the message from the previous problem?*

- **From §4.7: The Kasiski Test**

EXERCISE 4.9.42. *Finish decrypting the message from page* 104.

EXERCISE 4.9.43. *For such a short passage, the following ciphertext contains many long repeated strings. Use the Kasiski test to determine the length of the Vigenère keyword used to encrypt the following message. You should find enough repetitions to give you a keyword length of at most* 10.

```
KBPYU BACDM LRQNM GOMLG VETQV PXUQZ LRQNM GOMLG VETQV PXYIM HDYQL
BQUBR YILRJ MTEGW YDQWE GUPGC UABRY ILRJM XNQKA MHJXJ KMYGV ETQVP
XCRWV FQNBL EZXBW TBRAQ MUCAM FGAXY UWGMH TBEJB BRYIL RJMLC CAHLQ
NWYTS GCUAB RYILR JMLNT QGEQN AMRMB RYILR JMPGP BXPQN WCUXT GT
```

EXERCISE 4.9.44. *Continuing the above analysis, find the keystream, and decrypt the beginning of the ciphertext (at least* 20 *characters)*

```
KBPYU BACDM LRQNM GOMLG VETQV PXUQZ LRQNM GOMLG VETQV PXYIM HDYQL
BQUBR YILRJ MTEGW YDQWE GUPGC UABRY ILRJM XNQKA MHJXJ KMYGV ETQVP
XCRWV FQNBL EZXBW TBRAQ MUCAM FGAXY UWGMH TBEJB BRYIL RJMLC CAHLQ
NWYTS GCUAB RYILR JMLNT QGEQN AMRMB RYILR JMPGP BXPQN WCUXT GT
```

EXERCISE 4.9.45. *Use frequency analysis to recover the message from Exercise* 4.9.43*, whose keyword length you determined in that problem. For a passage as short (and unusual) as this one, the most common English letters may not be all that common in the some of the plaintext positions, depending on how our luck goes. In cases like this, it's good to pay close attention to how the* uncommon *English letters line up. The plaintext in this case contains no* j*'s,* q*'s,* x*'s, or* z*'s at all.*

EXERCISE 4.9.46. *It turns out that even if the keyword length was actually not the greatest common divisor, but one of the other common divisors, everything would still work out even if we tried to do frequency analysis on letter groups with the greatest common divisor as our guessed keyword length. Discuss why this is true.*

EXERCISE 4.9.47. *We can successfully crack messages encoded with the Vigenère cipher* without *using the Kasiski test, so long as we have many messages encoded with the same keyword. Imagine we intercept* 10,000 *messages encrypted the same way (this is not unreasonable if we are pulling data off a website that has poor security). The first letter of each of these messages is always encrypted the same, and we can apply frequency analysis here (with the caveat, of course, that the distribution of first letters may not be precisely the same as the distribution of letters elsewhere). Describe how one can continue along these lines to figure out the keyword.*

EXERCISE 4.9.48. *The Kasiski test and frequency analysis combine to provide a powerful attack on the Vigenère cipher. The best way to combat this is to have a very long keyword and* not reuse that keyword! *We can arbitrarily inflate the size of our keyword as follows. Imagine our keyword is* YODA*. We choose as our improved keyword* YODAODAYDAYOAYOD*. We now have a keyword of length* 16 *instead of* 4*. Describe how to use this method for an arbitrary keyword. So long as you remember the method, you and your accomplices can take a short word and greatly magnify its length.*

EXERCISE 4.9.49. *Find variants to the Vigenère cipher to make it more resistant to frequency attacks and the Kasiski test.*

Chapter 5

Classical Cryptography: Attacks II

In Chapter 4 we explored attacks on some of the classical ciphers, in particular the substitution alphabet ciphers. We saw that these are extremely vulnerable to attacks using frequency analysis. The Vigenère cipher was created to avoid these issues, but it too succumbs to frequency attacks. The problem with substitution alphabet ciphers is that a letter always encrypts to the same letter. The Vigenère cipher only partially fixes this; if the keystream has length n, then all letters in the same position modulo n are encrypted the same.

These deficiencies led to the creation of alternative schemes where the same letter was encrypted differently. Note, of course, that this is true of the Vigenère cipher, but only in a very weak way. There the encryption depended solely on the *location* of the letter modulo the keystream's length. In the permutation and Hill ciphers of Chapter 2, however, the encryption depends on nearby letters. Thus, the complexity is greatly increased.

The purpose of this chapter is to see that these methods are also vulnerable and to highlight the need for more advanced encryption schemes. A major issue with all the approaches we've seen is that letters are encrypted either singly or in small blocks. When we reach RSA in Chapter 8, we'll see that either the entire message, or at least large parts of it, are all encrypted together. Further, changing just one letter greatly changes the entire encryption, which is markedly different than what happens with the methods we've seen so far.

There is a real need for systems like RSA. In this chapter we finally find an unbreakable system, but it has a significant drawback; it requires the sender and recipient to meet and exchange keys. Thus while we do construct a system safe from attack, it is impractical to use. You should keep these issues in mind as you read the attacks on the different classical schemes, as the need to be resistant to these attacks, plus the requirement that two people who have not met to exchange a secret should be able to securely communicate, played a large role in the development of the field.

5.1. Breaking the Permutation Cipher

Our first method that is truly impervious to frequency analysis is the permutation cipher. Imagine an eavesdropper intercepts such a message but doesn't know the key. Even if it's a very long message, frequency analysis on the letters can't help break the permutation cipher because *the frequency of letters in the ciphertext is the same as the frequency of letters in the plaintext.* Thus frequency analysis of letters will typically just reveal that `E` and `T` are common letters in the ciphertext, and that `Z` is uncommon. This might be useful in confirming that the original message is English text, but it won't give us any information on the permutation used to encode the message.

On the other hand, knowledge of common pairs and triples of letters in English can be very useful in breaking the permutation cipher. Consider the following message, encrypted with a permutation cipher: `RIBNT HGEES MSGEA TTHOE RODPO IPNRL TH`. The ciphertext is 32 characters long; this already gives us important information, since the length of the permutation must divide the length of the message. In this case, the divisors of 32 are 1, 2, 4, 8, 16 and 32. Let's guess that the permutation has length 4, in which case the cipher works by permuting blocks of length 4:

`RIBN THGE ESMS GEAT THOE RODP OIPN RLTH`

We must figure out how to permute the letters to get English text. We could try a brute force approach as there are only $4! = 16$ possibilities; however, the factorial function grows very rapidly, and slugging away won't work for larger sizes. Figure 5.1 shows how rapidly this function grows; when $n = 100$ already 100! exceeds 10^{150}.

We notice that two of the blocks contain the pattern `TH_E`. It seems likely that this pattern arose from occurrences of the word "the" in the plaintext. If this is true, then the decryption permutation maps the first, second, and fourth letters into consecutive positions; there are two permutations with this property, namely

$$\begin{pmatrix} 1 & 2 & 3 & 4 \\ 1 & 2 & 4 & 3 \end{pmatrix} \quad \text{and} \quad \begin{pmatrix} 1 & 2 & 3 & 4 \\ 3 & 1 & 2 & 4 \end{pmatrix}.$$

Under the first of these permutations, the first few blocks of the message decrypt to `RINB THEG ESSM GETA`, which appears to be nonsense. Decrypting with the second permutation, however, gives `BRIN GTHE MESS AGET OTHE DROP POIN TRLH`. Thus the message was "bring the message to the drop point", padded with `RLH` to make the plaintext to a multiple of 4.

For us, the reason for studying the permutation cipher is that it is our first example which is impervious to a frequency analysis. Unfortunately, it is not secure, and thus our search for a good encryption scheme continues.

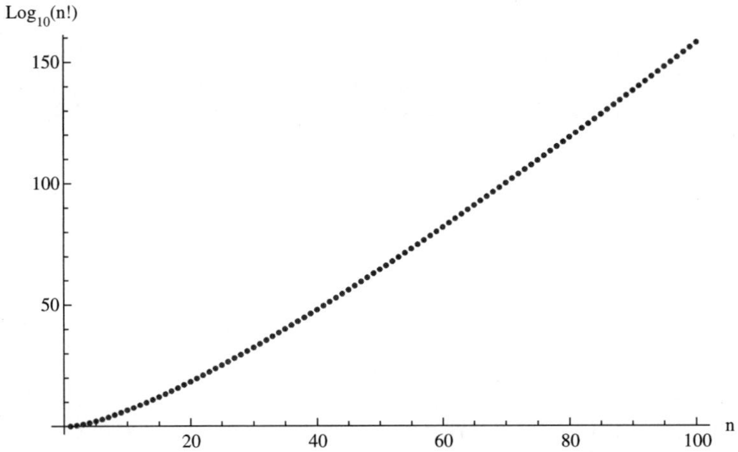

FIGURE 5.1. The extremely rapid growth of the factorial function. Note that the vertical axis is the *logarithm* of the value.

This naturally leads to our next candidate, the Hill cipher. It too is safe from frequency analysis, but we'll find again exploitable weaknesses.

5.2. Breaking the Hill Cipher

Since the Hill cipher encrypts letters in blocks, frequency analysis on individual letters isn't useful. In §2.9, we encrypted the message `ET PH ON EH OM ET` to `WJ JK QB CZ KY WJ`. Like the Vigenère cipher, individual letters aren't always encrypted the same way. The first `E` became a `W` while the second `E` became a `C`. Unlike the Vigenère cipher, where letters were encrypted the same way if they lined up the same way with the keyword, letters can be encrypted differently with the Hill cipher even if they occur at the same place in a block, as is the case with the `E`'s just mentioned. This is one of the great advantages of the Hill cipher, and the reason we're discussing it. For the Hill cipher, the encryption of each letter in each block depends on every other letter in the block, and so there is no direct correspondence between individual letters.

To break the cipher, we have to do a frequency analysis on blocks of letters rather than individual letters. Let's consider the following message, which was encrypted with the Hill cipher using a 2×2 matrix as the key.

```
FRAQR TFRLZ KEFPZ KXYWS XYGZX YZZZF WHPHM SMUMT GZFPK ECUWY BNVZG ZQOVA
TATDJ ROCUQ MTCCI KGZAG ONVZT MDIDJ BNVZF PZKVB BNCXZ KRBMY GZXLV ACLGZ
QOVLL ZNTXL UUVVR XYUCX ZKVMX TGZLZ UUEGN NVMKI QMFPJ CLIGZ EISWV AUGRY
HULMB TGZEI VZCLG ZBNRB XTJIC XTADY VNCSO RPSBV USCCU SVATE UTKSV MKIQM
AGBQU KGZKY UKEOX TGZBT PXRXD DZKCC OKMTJ ICXTA DYVNC SORPS BVUSC CUSVA
TEXOY CXTPS WMUKF JZFWH LFZFX TLSLM XLGZN NUKUT KSNEQ MFPZK GZRBU VZFWH
```

```
PHMSH STPGZ IQNEQ MYSHB ZBIZW KSYFR HDMSU BGZLF MSFEZ EXNJM NEFUD VVOHU
ROALI MLZZU WHZFW HPHMS KOOAK IYEFJ FREYM TMURJ ZKFKB AWHZF WHPHM SBTPS
BVUSC CZBXT HQVLO AGQUV FPYEB TTANW AGSJZ FWHPH MSVZT PSTDZ ZUMTT PYJDZ
ZUMTT PUHHU YGUSW RUSCC BTRCI NBRZK JCRBZ BASZJ BTFWZ JBTZJ BTMUN ZGZZZ
ZFWHP HMSHS VAHYU SBAFP ZKKMK PVCGB HURJZ KNEQM AGNAM TLZFP ZKFKB AWHZF
WHPHM SGZEI XNKTZ KVEWY FPDDZ KNEQM YSTEX OYCRD FPYEB TTANW AGSJZ FWHPH
MSVZT PSTDZ ZUMTT PYJDZ ZUMTT PUHHU YGUSW RUSCC BTRCI NBRZK JCRBZ BASZJ
BTFWZ JBTZJ BTMUN ZGZZZ ZFWHP HMSHS CXZKK MKPKN EIGZL ZNEQM AGYZW GNNPM
HGGZO ACUFR UVZFW HPHMS GZEIX NKTZK VEWYF PDDZK NEQMY SQJNG YCRDF PYEBT
TANWA GVENG TJGZE IXNDW NGIBL QGRNN VCIBZ WGRXO IBFVH UYGUS WRUSC CBTRC
INBRZ KJCRB ZBASZ JBTFW ZJBTZ JBTMU NZGZP HMSWM ZZORZ KKMKP LQGRT JNNKD
DZPCG BHBZK OKBLC ZDDKN CWTOT XGZLZ IZWKN PUSCC HSCXZ KKMKP GKMUR DUUCL
KMKPN NENCU JIZLT JRTFR GZXLW GKTXN WGFLV ZKSSW BTVAG ZLFNG YCNEQ MFPZK
JICXT ADYYI KSGZE IXNHY JIZLR TFRGZ XLWGZ UMTJI RYLSD VBTLV RBAGY ZZKFK
XYCRI EUQMT CRIET JFRZF FJDVU GOWCX TAZBG KHGRI ZKYZZ KKMKP FKNAI BXPUS
MTGZJ RXNAG YZZKZ JKEDI FTCUH QLTTJ TOTXN NDGXL HLDSB VNGTZ XOYZD DJCFS
FPWHY SJMLZ XLNNA LJCBQ UKGZK YMTZB PSXTL SNRTA WMZFU BZKLU VNZTK MKPGZ
HDVXQ RSWGZ ZZZFW HPHMS BTHWH GENIQ MYGZX LLHXL BVOAG QHIUS CCAGX OJZQM
RYSKV ACLUU HBZKG KBRQQ FPFJF RKOYZ ZKKMM ZVCXT RXDDP SLMMU WMZZD SEQUS
KIUXV ZTMNA HGVOD WSPJC WHZFW HPHMS ZKTAD VDIAL JCWHO AQWNA MTZKV YXTJI
ROENV RDYVO NTXLH XHGIX
```

Since the message was encrypted in blocks of size 2, we're interested in finding **bigrams** (pairs of letters) which would be common if this ciphertext were split up into blocks of two (as was done for encryption). The most common blocks of length 2 in this message are GZ and ZK, each occurring 35 times in the ciphertext. Note that not all occurrences of these strings in the ciphertext count towards this total; for example, the instance of ZK starting at the ninth character of the ciphertext runs across two different blocks if the message were split into blocks of length 2, and is just a coincidence. The next most common bigram, ZZ, is significantly more rare, occurring only 22 times. To make use of this information, we need a list of common bigrams, which we give in Table 5.1.

Of course, once we figure out the encryption key we can decode the entire message, and we'll see below that we can typically do this if we can correctly identify two bigrams. While this seems to suggest less security than a substitution alphabet cipher, one must remember that we're just looking at a two-dimensional Hill cipher. The general substitution alphabet cipher has 26 parameters, corresponding to the reordering of the alphabet (actually, we can represent a general substitution alphabet cipher with just 25 parameters, as once we've specified where the first 25 letters go then the last is uniquely determined). To be fair to the Hill cipher, we should look at the security from a larger one, say a 5×5 matrix (which would have 25 free parameters of the same order as a substitution alphabet cipher), but we'll stick to smaller examples. Hill and Louis Weisner made and patented a mechanical device to do a Hill ciphers with a 6×6 key (which had the drawback that the key had to be built into the machine, which is a significant security flaw).

TABLE 5.1. Common English bigrams and trigrams, useful for breaking the two- and three-dimensional Hill ciphers, respectively. (Source: *Cryptographical Mathematics*, by Robert Edward Lewand, Mathematical Association of America, 2000.)

most common bigrams	
1	th
2	he
3	in
4	en
5	nt
6	re
7	er
8	an
9	ti
10	es
11	on
12	at
13	se
14	nd
15	or
16	ar
17	al
18	te
19	co
20	de

common trigrams	
1	the
2	and
3	tha
4	ent
5	ing
6	ion
7	tio
8	for
9	nde
10	has
11	nce
12	edt
13	tis
14	oft
15	sth

Returning to our ciphertext, Table 5.1 suggests that we guess the block GZ corresponds to the plaintext block TH. We can write this in matrix form as

$$\begin{pmatrix} a & b \\ c & d \end{pmatrix}\begin{pmatrix} 6 \\ 25 \end{pmatrix} = \begin{pmatrix} 19 \\ 7 \end{pmatrix} \pmod{26},$$

where we've written $\begin{pmatrix} a & b \\ c & d \end{pmatrix}$ for the unknown decryption key matrix (which is the inverse of the encryption matrix). If we can find this matrix, we can use it to decrypt the rest of the message. By carrying out the matrix multiplication above, we get the following congruence:

$$\begin{pmatrix} 6a + 25b \\ 6c + 25d \end{pmatrix} = \begin{pmatrix} 19 \\ 7 \end{pmatrix} \pmod{26}.$$

Thus our goal is to solve the following pair of congruences, remembering that a, b, c, and d are integers modulo 26:

$$\begin{cases} 6a + 25b = 19 \pmod{26} \\ 6c + 25d = 7 \pmod{26}. \end{cases} \tag{5.1}$$

Unfortunately, we have four unknowns and only two congruences, and thus we don't have enough information to solve the system. We need additional congruences, which we can find by making another guess about a bigram encryption. If we guess that ZK (the other most common bigram in the ciphertext) corresponds to HE in the plaintext (the second most common bigram in English according to Table 5.1), this implies that

$$\begin{pmatrix} a & b \\ c & d \end{pmatrix} \begin{pmatrix} 25 \\ 10 \end{pmatrix} = \begin{pmatrix} 7 \\ 4 \end{pmatrix} \pmod{26},$$

which yields

$$\begin{pmatrix} 25a + 10b \\ 25c + 10d \end{pmatrix} = \begin{pmatrix} 7 \\ 4 \end{pmatrix} \pmod{26}.$$

Combining the two congruences from this with the two congruences from equation (5.1) gives us a system of four equations and four unknowns:

$$\begin{cases} 6a + 25b & = & 19 & \pmod{26} \\ 25a + 10b & = & 7 & \pmod{26} \\ 6c + 25d & = & 7 & \pmod{26} \\ 25c + 10d & = & 4 & \pmod{26}. \end{cases}$$

We can group these four equations into two pairs of two equations such that each pair involves only two of the four unknowns. We begin by solving the pair

$$\begin{cases} 6a + 25b & = & 19 & \pmod{26} \\ 25a + 10b & = & 7 & \pmod{26} \end{cases} \tag{5.2}$$

by subtracting congruences. We could have solved this by substitution to eliminate variables, which we do when we attack c and d so that both approaches to the algebra are shown. To eliminate the b term in each congruence, we multiply each by the coefficient of b in the other and subtract:

$$\begin{array}{rrcll} & 10(6a + 25b) & = & 10(19) & \pmod{26} \\ - & 25(25a + 10b) & = & 25(7) & \pmod{26}, \\ \hline \end{array}$$

which, after simplification and subtraction, gives

$$\begin{array}{rrcrl} & 8a + 16b & = & 7 & \pmod{26} \\ - & a + 16b & = & 18 & \pmod{26} \\ \hline & 7a + \ 0b & = & 15 & \end{array}$$

Since 7 has an inverse modulo 26 (namely, 15), we can solve for a by multiplying both sides by the inverse of $7^{-1} = 15 \pmod{26}$, and find

$$a \ = \ 15 \cdot 15 \ = \ 17 \pmod{26}.$$

We can now find b by plugging our value for a back into one of the congruences from equation (5.2). For example, plugging $a = 17$ into the second one gives

$$25 \cdot 17 + 10b \ = \ 7 \pmod{26}.$$

Simplifying gives

$$10b \ = \ 24 \pmod{26}.$$

Unfortunately, since 10 is not relatively prime to 26, 10 has no multiplicative inverse, and this congruence does not have a unique solution. By looking at Table 4.2 (page 87) we see that $10 \cdot 5 = 24 \pmod{26}$ and $10 \cdot 18 = 24 \pmod{26}$ are both true. Thus we have only determined that $b = 5$ or $18 \pmod{26}$. In some cases this might be the best we could do without further guesswork, in which case we might have to try both possibilities in the final decryption matrix to see which works out (by decrypting the ciphertext and seeing if we get something recognizable). In our case, however, plugging $a = 17 \pmod{26}$ into the first congruence from equation (5.2) instead of the second gives

$$6 \cdot 17 + 25b = 19 \pmod{26},$$

which simplifies to

$$25b = 21 \pmod{26}.$$

This time we get a congruence that has a unique solution, as $25^{-1} = 25 \pmod{26}$ (this is easily seen by recognizing $25 = -1 \pmod{26}$). We thus find that $b = 25^{-1} \cdot 21 = 5 \pmod{26}$.

We'll solve the remaining two congruences for c and d by substitution, so that both methods of solving congruences are shown. We start with the system

$$\begin{cases} 6c + 25d = 7 \pmod{26} \\ 25c + 10d = 4 \pmod{26}. \end{cases} \tag{5.3}$$

We'll look at the second congruence, as there is a problem in the first. The difficulty is that 6 isn't invertible modulo 26, which means the first congruence cannot be uniquely solved for c. Sometimes both congruences have this problem, in which case subtraction of congruences is easier to apply than substitution, although the final result still involves multiple solutions which have to be tried independently. Using the second congruence to solve for c in terms of d gives

$$c = 10d + 22 \pmod{26}.$$

Substituting this value of c into the second congruence gives

$$6(10d + 22) + 25d = 7 \pmod{26},$$

which simplifies to

$$7d = 5 \pmod{26}.$$

Multiplying by 15 (the inverse of 7) gives

$$d = 23 \pmod{26}.$$

Plugging this back into the second of the congruences from equation (5.3) gives

$$25c + 10 \cdot 23 = 4 \pmod{26},$$

which yields

$$c = 18 \pmod{26}.$$

We've found the decryption matrix! It is

$$\begin{pmatrix} 17 & 5 \\ 18 & 23 \end{pmatrix}.$$

We can now use the decryption matrix to decode the ciphertext and see if our bigram guesses were correct. The beginning of the ciphertext FR AQ RT FR LZ KE FP ZK ... corresponds to the matrices

$$\begin{pmatrix} 5 \\ 17 \end{pmatrix}, \begin{pmatrix} 0 \\ 16 \end{pmatrix}, \begin{pmatrix} 17 \\ 19 \end{pmatrix}, \begin{pmatrix} 5 \\ 17 \end{pmatrix}, \begin{pmatrix} 11 \\ 25 \end{pmatrix}, \begin{pmatrix} 10 \\ 4 \end{pmatrix}, \begin{pmatrix} 5 \\ 15 \end{pmatrix}, \begin{pmatrix} 25 \\ 10 \end{pmatrix}, \dots$$

which, upon multiplication by the decryption matrix gives

$$\begin{pmatrix} 14 \\ 13 \end{pmatrix}, \begin{pmatrix} 2 \\ 4 \end{pmatrix}, \begin{pmatrix} 20 \\ 15 \end{pmatrix}, \begin{pmatrix} 14 \\ 13 \end{pmatrix}, \begin{pmatrix} 0 \\ 19 \end{pmatrix}, \begin{pmatrix} 8 \\ 12 \end{pmatrix}, \begin{pmatrix} 4 \\ 19 \end{pmatrix}, \begin{pmatrix} 7 \\ 4 \end{pmatrix}, \dots.$$

This corresponds to the plaintext ON CE UP ON AT IM ET HE ..., and it seems that the encrypted message is a fairy tale. Thus, while frequency analysis cannot crack the Hill cipher, it is vulnerable to other attacks.

5.3. Running Key Ciphers

The purpose of this section is to introduce a nice twist to the Vigenère cipher to greatly increase its security. Suppose we encrypt the message

WEWIL LINFI LTRAT ETHEI RTREE HOUSE ATDAW N

with the Vigenère cipher using the key

THISISTHESUPERSECRETPASSWORDTHATHEYDONTKNOW

We add the letters of the keystream to our message and obtain the ciphertext.

```
WEWIL LINFI LTRAT ETHEI RTREE HOUSE ATDAW N
THISI STHES UPERS ECRET PASSW ORDTH ATHEY D
PLEAT DBUJA FIVRL IVYIB GTJWA VFXLL AMWHA L
```

If an eavesdropper intercepted the encrypted message

PLEAT DBUJA FIVRL IVYIB GTJWA VFXLL AMKEU Q,

it appears that they would have a very hard time breaking it. Imagine their disgust at learning that the keyword is the same length as the message. It would appear that frequency analysis is useless, as each group now consists of exactly one letter. In fact, it might seem that whenever the password used for the Vigenère cipher is at least as long as the message, the cipher is unbreakable.

Running key ciphers: A running key cipher is a Vigenère cipher where the length of the keystream equals the length of the message.

Although it seems such a system should be unbreakable, it can be broken by hand. Though it is indeed substantially more secure than a Vigenère cipher with a short keyword, its weakness is the fact that the keyword is typically *not* random letters, but actually meaningful text. We can exploit this bias to find the original message.

For example, the most common letter in English is "E", which occurs about 12.58% of the time in typical text. Suppose our attacker simply subtracts E from every letter in the ciphertext:

```
PLEAT DBUJA FIVRL IVYIB GTJWA VFXLL AMWHA L
EEEEE EEEEE EEEEE EEEEE EEEEE EEEEE EEEEE E
LHAWP ZXQFW BERNH ERUEX CPFSW RBTHH WISDW H
```

Since the keyword was English text and "E" is so common, this subtraction should yield a high fraction (around 12.58% to be precise) of correct letters from the plaintext. We do well in our example, getting nearly 14% of the letters correct (correct letters are underlined above). While 14% isn't bad, it's not clear how we could possibly figure out the original message using just this technique, since we don't have a reliable way of distinguishing correct letters from incorrect ones.

The solution is to subtract not just a letter from the ciphertext, but blocks of letters. Specifically, common words. For example, "that" is a very common English word, and the attacker could try guessing it appears at some point in the keystream. By subtracting THAT from each possible position in the ciphertext, the attacker can decide which partial decryptions make the most sense. For example, the first few subtractions give

```
PLEA  LEAT  EATD  ATDB  TDBU  DBUJ  BUJA  UJAF
THAT  THAT  THAT  THAT  THAT  THAT  THAT  THAT
WEEH  SXAA  LTTK  HMDI  AWBB  KUUQ  INJH  BCAM
```

Here's the complete list of four-letter blocks resulting from subtracting THAT from different positions of the ciphertext:

WEEH, SXAA, LTTK, HMDI, AWBB, KUUQ, INJH, BCAM, QTFP, HYIC, MBVY, PORS, CKLP, YEIC, SBVF, POYP, CRII, FBBN, PUGA, IZTQ, NMJD, ACWH, QPAC, DTVM, HOFE, CVXS, MQLS, EELH, SEAT, STMD, HFWO, TPHH, DAAS.

Of these, most of these are unlikely to arise in an English test. Depending on how creative you're willing to be, there are either many or few exceptions. Some possibilities are BCAM (which could be *Bob. Camera please*), PORS (perhaps a waiter is asking us *soup or salad*), and of course, SEAT (not only *seat*, but also, as in the case of our plaintext, *house at*).

The diligent attacker then builds on these discoveries with further guesswork. For example, if she decides that SEAT is likely to occur in the plaintext, she has decided on the partial decryption

```
PLEAT DBUJA FIVRL IVYIB GTJWA VFXLL AMWHA L
                                 TH AT
                                 SE AT
```

At this point, she could try subtracting some other common words from other parts of the message. If she tried subtracting THE from different parts of the message, for example, she might find that

```
PLEAT DBUJA FIVRL IVYIB GTJWA VFXLL AMWHA L
                                 TH ATTHE
-------------------------------------------
                                 SE ATDAW
```

was a likely decryption (especially since the word "that" is often followed by "the"). At this point, lucky guessing might lead her to

```
PLEAT DBUJA FIVRL IVYIB GTJWA VFXLL AMWHA L
                                 TH ATTHE Y
-------------------------------------------
                                 SE ATDAW N
```

and then to

```
PLEAT DBUJA FIVRL IVYIB GTJWA VFXLL AMWHA L
                              WO RDTH ATTHE Y
-------------------------------------------
                              EH OUSE ATDAW N
```

and she's well on her way. It should be emphasized, of course, that this kind of attack requires a *lot* of trial and error, and cracking running key ciphers by hand is very labor intensive and dependent on luck. This is another example of why linguists were in such demand for early cryptosystem attacks. Nowadays, computers do very well when programmed to take advantage of more sophisticated information about the language of the message and keystream (such as which words are likely to come after which other words), and there is sophisticated software which can be used to break running key ciphers.

5.4. One-Time Pads

The weakness of running key ciphers is that information about the likely properties of keystreams (for example, they're likely to contain common words, such as "that") can be used to deduce likely decryptions of the ciphertext. The **one-time pad** is a slight modification of the running key cipher which eliminates this weakness.

> **One-time pads.** A one-time pad is a running key (or Vigenère) cipher where the the keystream is a *random* stream of letters.

For example, assume we again want to encrypt the message

WEWIL LINFI LTRAT ETHEI RTREE HOUSE ATDAW N

The one-time pad demands that we generate a random keystream. The letter sequence YYIVFQPUBVKCPKDGYJDSWFRTSGOMDXWXXVHR was generated "randomly" with the help of a computer. We can use it as the key for a one-time pad to encrypt our message:

```
 WEWIL LINFI LTRAT ETHEI RTREE HOUSE ATDAW NR
+YYIVF QPUBV KCPKD GYJDS WFRTS GOMDX WXXVH R
 UCEDQ BXHGD VVGKW KRQHA NYIXW NCGVB WQAVD E
```

None of the techniques for cracking the running key cipher can help, as there are no letters or blocks of letters which are more likely to appear than others in the keystream. Since it was generated randomly, THAT is exactly as likely to appear in the keystream as are ZZZZ and MQPX.

Indeed, the one-time pad ***cannot be broken***, because the randomness of the key used for one-time pad encryption means that ***each plaintext is equally likely to be the source of an observed ciphertext***. For example, even though our message WEWIL LINFI LTRAT ETHEI RTREE HOUSE ATDAW N was encrypted to the ciphertext UCEDQ BXHGD VVGKW KRQHA NYIXW NCGVB WQAVD E, the plaintext

THEPE OPLEI NTHET REEHO USEAR EOURF RIEND S

encrypts to the same ciphertext under the random key

```
 THEPE OPLEI NTHET REEHO USEAR EOURF RIEND S
+BVAOM NIWCV ICZGD TNMAM TGEXF JOMEW FIWIA M
 UCEDQ BXHGD VVGKW KRQHA NYIXW NCGVB WQAVD E
```

In fact, if a message was intercepted with a one-time pad and someone claimed to know the message's contents, we couldn't even verify their claim! By subtracting their claimed message from the ciphertext, we get the keystream used for the encryption. If the running key cipher had been used, we could verify their claim by seeing that the keystream used was English text. For the one-time pad, however, all keystreams are equally likely to occur, and thus no inference about the plaintext can be made from ciphertext, even with very lucky guessing. Because of this, the one-time pad is said to have **perfect security**, which means that the ciphertext gives the cryptanalyst no information at all about the plaintext, since any plaintext gives rise to any ciphertext with equal probability.

Our search is finally over—we've discovered a perfectly secure encryption scheme! Unfortunately, while the one-time pad is perfectly secure, it is not easy to use. This claim requires some comment, as it seems no harder to use than the Vigenère cipher; all we have to do is add letter by letter. The difficulty is not in the act of encryption, but in the act of agreeing on the key. Somehow, we have to have the sender and recipient agree upon a key. This is where the difficulty lies.

We'll talk later in the section about the issues in exchanging a key. Right now, there's another issue that merits discussion. As the name suggests, the one-time pad is meant to be used *once*! For the one-time pad to be secure, we can never reuse a keystream. Suppose the keystream YYIVFQPUBV... used earlier to encrypt the message WEWIL LINFI LTRAT

`ETHEI RTREE HOUSE ATDAW N` was also used to encrypt the message `BEGIN PREPA RATIO NSIMM EDIAT ELY`:

```
 BEGIN PREPA RATIO NSIMM EDIAT ELY
+YYIVF QPUBV KCPKD GYJDS WFRTS GOM
 ---------------------------------
 ZCODS FGYQV BCISR TQRPE AIZTL KZK
```

Let's assume an eavesdropper has intercepted both ciphertexts, but doesn't know the keystream. Because she knows two messages, she can subtract one from the other.

```
 ZCODS FGYQV BCISR TQRPE AIZTL KZK
-UCEDQ BXHGD VVGKW KRQHA NYIXW NCG
 ---------------------------------
 FAKAC EJRKS GHCIV JZBIE NKRWP XXE
```

Let's think about what the resulting text represents. It's

$$(\texttt{BEGIN PR}\ldots + \texttt{YYIVF QP}\ldots) - (\texttt{WEWIL LI}\ldots + \texttt{YYIVF QP}\ldots)$$
$$= (\texttt{BEGIN PR}\ldots - \texttt{WEWIL LI}\ldots).$$

It's the difference (subtraction) of the two plaintexts. This means that the result is essentially a running key cipher, using one of the plaintexts as a keystream to encrypt the other, and using subtraction instead of addition! We've now reduced the problem of breaking a twice-used one-time pad cipher to breaking a slightly modified running key cipher, which we've seen how to do.

This example should serve as a warning. It's very easy to make small errors and have some extra structure remain that a clever attacker can exploit. We saw other examples of this when we studied Enigma and Ultra in Chapter 3.

There's another issue that we need to address for the one-time pad cipher. How can we generate a random keystream? Consider the following random sequence of letters, which was generated by banging on a keyboard haphazardly and then removing nonletters and grouping what remained into blocks of five:

```
LKAJS DFPOI UZXBP UIOYQ WERMN YWERU YZXVT YIWER BNZVX MNZCX LKAHS DIOPY
QWERU YTWEQ IOPAJ KHASD NVZXC MNASJ KLHAS DFHJQ WEPIO UYQTR JKLSF DGZNX
ZXVCB NSDLA FFHGH ASDFY QOWER IOPER JKHAS DKLVB ZXCLH JASFD UYQWE RQTWR
EPUIO HBVAS DFGJL KQETU IOQWR EYPAJ KLSDF BNZGL ASDOF IUYQW ERKLJ HASDF
MBCNZ JOASK LJASG ASOIU YQWEL RKJHA MBCLV KJASH DFOIU YQWER HFVJD KASOI
UYQWE RHAJM ZXMZX CBVKL JHASO UIYQW ERLHA SMNZX BVALK JASDI OPQWE OIUYQ
WEROY LASDF PIOQW ERUIO XTRBV BVRME VBNRE WQGHJ HLQWR EOQPW REUIO PEUUI
OPASD FZVCX MBASD FUYOQ WREHL AGUYA SFDTQ WREYI OAFAS FDBMZ XVCLA SFDOQ
ETQWR EIASD BMZVK GJASF DQWRE OYIAS FGASF DZXVC ASFDF GFFGW REQWE UIOQW
EUIPO ASFDJ KLASF DHASF DZXCX VCHJB FDBMS RFDWR ETUWR EEQWU IOPQW EJKLA
SDHJA SFDGH ZXCZX CVXBN VCZXV CMASD FJKLQ WEUIO PQWRE ASFDY QWREP UIOJK
```

Though we tried to hit the typewriter randomly, the resulting stream is far from random. Let's look at the letter distribution:

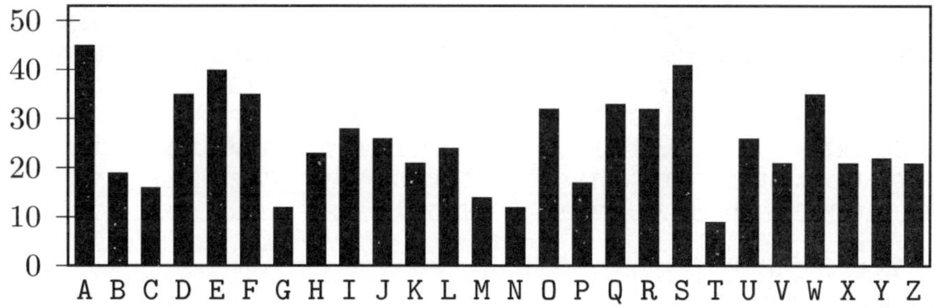

The distribution is clearly not uniform. For example, letters accessible from the "home position" on the keyboard (*a*, *s*, *d*, *f*, *j*, *k*, *l*) are among the more common letters. While not quite as nonuniform as English text (see the graph on page 99), there's still substantial bias which an attacker can exploit. For example, if an attacker who is aware of our bias in choosing random numbers is simply trying to decide whether an intercepted message `YWW` corresponds to the message `YES` or `NO_` (where _ represents any letter, being used here as padding), they might observe that a plaintext of `YES` implies a keystream of `ASE` was used, which consists of letters which are commonly produced by our random letter generation, suggesting that this plaintext is likely, while the plaintext `NO` would imply a keystream beginning with `LI`, consisting of relatively fewer common letters. Try randomly typing letters on your keyboard, and see what you get.

Looking at long blocks, we see an even greater bias in our random stream. The stream has several long repeated blocks: for example,

```
JKHASD, HJASFD, UYQWER, WREPUIO, IUYQWE, KLJHAS,
ASOIUYQ, OIUYQW, YQWERH, IOPQWE, ASFDZX, QWEUIO,
UIOPQW
```

are all repeated in the random stream, and some several times. (It's instructive to sit at a keyboard and type some of these strings. They feel very "natural" and are easy to type, never requiring a finger to change position in the middle of a block. It's perhaps not surprising that blocks like this appear frequently when one tries to generate random streams using a keyboard.) An attacker who knows these biases in the generation of our key material could use it very effectively to break a one-time pad that used this keystream. The same techniques discussed in the running key ciphers section would be very effective, the only difference is that now we use strings like `UYQWER` in place of common words like `THAT`.

To highlight the bias of our key generation by randomly hitting keys, we compare our frequency distribution to that of a more random stream of letters. Below is a long stream of letters randomly chosen by computers, so that each letter is chosen independent of the others and each has the same probability of being any English letter.

```
BRAGZ QGMDB TNZKH LVFWQ OTHFX SYKNL GYYOY HOIAQ ODJOC PEKFW RDDOH QIGXP
SAFFX FXNOR GOWTO KTODH FACOQ DPJBB CTZFC XAHAO SUEXS BEUWR JINLJ IINTI
FCKFN BFTYZ YVHLN FOQFQ KSISB CATRJ ZAILH GGFVW YTCCD OZTNJ HYLNA MLUWF
CDLYS IOONH ZDZOP FISPM LEZSM EJXWR OMJLH EZXPW AOPFE ZECDK AZZIV MJDXV
ACZKX OGBYN WAIWW PKVLZ XYIEG MIRYX WNNBK WPJNI XKOAD UVBLA NANFT SJSHN
FSIML MJSSH SLRDK REYHB VPNZY EDDQY JGWMS ZBUFJ DVBAO XDDFS TVOZW HBXWW
TOVEX INFYB QGUAF QXMRV JGWHR QHGSX DSCIT NIXHF GCPZH ISALB BWMKC NCWOP
EIEOU IHZQE VTIXM DDSHD PRDGC IMKYS ZGOAF FJQDD UOPRL GLXPZ YXYJY TGZSH
WMRHU FAOHQ LKIDR FTVZQ KJQXI KBSZE OIUFO HNUID FBYWW EKYWF ZSKKZ LTUJJ
RPKRU LEUSY EPYCC AMXXP EKQTK ZIVXA XHWHC DFIBZ YRPTD TVWZQ IGRNZ YNQZX
SZCYI IDBIY CYMTJ UWKHF PVVAD TNQRJ XZCHH VHQXK TUPRF DJNWW UGZXE PILYI
```

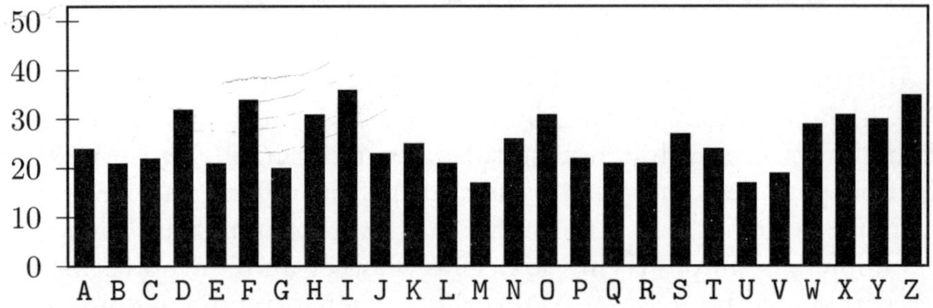

Note that the distribution of letters, shown above, is noticeably more uniform than what we got from hitting the keyboard. If we were to take a longer string of text it would flatten out even more (this is essentially the Central Limit Theorem). Even more striking is the data on repeated strings. The above block of random text contains *no* repeated strings of length 7, 6, 5, or even 4. There are just ten repeated strings of length 3, all of which occur just twice in the text. This is starkly different than the stream generated by banging on the keyboard, which contains 79 repeated blocks of length 3, many of which occur ten or more times.

The issue of random letter (and number) generation is actually a very deep one. How can one program a computer to produce random numbers? Computers and some calculators have functions to generate a random number, for example. These functions can't actually generate a truly random number. They carry out deterministic operations which should nevertheless produce sequences which have the properties of a random sequence. Designing good functions for this purpose can actually be quite difficult, and many software random number generators have weaknesses which would make them unsuitable to use for the generation of one-time pad key material. For an amusing story of how Mohan Srivastava, a geological statistician, cracked the random algorithms for a Canadian scratch lottery, go to

`http://www.wired.com/magazine/2011/01/ff_lottery/all/1`.

Finally, we need to discuss a major drawback that limits the practicality of the one-time pad. Apart from the problem of generating truly random keys, the key used has to be at least as long as the message to be sent, and can never be reused. This means that parties wishing to exchange messages need to arrange for the secure exchange of key material, and need to exchange key material which is at least as long as the total length of messages to be sent. Of course, if it's possible to securely exchange so much key material, it seems like it might make more sense to securely exchange unencrypted messages, and avoid encryption altogether!

The one-time pad has found real practical use by spies, however. In the past, spies obtained key material (books of random letters to be used for encryption) from their intelligence agencies before departing on their assignments. The pages from the books would be used to encrypt messages, and then destroyed so that they couldn't be discovered or used again (key material was sometimes printed on special highly flammable paper to make it easy to burn them without a trace). When the key material was sufficiently random, this provided complete cryptographic security through a method which was easy to carry out by hand with nothing more than pencil and paper. For most modern applications, such as encrypting an email, there is no practical way for parties to securely exchange enough key material in advance for the one-time pad to be practical. In Chapters 6 and 8, we discuss modern encryption schemes which attempt to provide excellent security without requiring huge amounts of key material being exchanged.

5.5. Summary

Of all the encryption schemes that we've studied, only the one-time pad was impervious to our attempts to break it "from scratch"; i.e., without using anything other than the ciphertext we're trying to read. This kind of attack on a cipher is called a **ciphertext-only attack**, because the cryptanalyst only has access to the ciphertext. These kinds of attacks typically rely on statistical information about likely messages, such as letter or bigram frequencies. Obviously, the fact that the classical ciphers are vulnerable to ciphertext-only attacks is a serious flaw, and this was the impetus for developing many of the modern ciphers discussed in Chapters 6 and 8.

There's another kind of attack on a cipher, called a **known-plaintext attack**. In this situation, the attacker has a ciphertext she wants to decrypt, but also has some ciphertext (which was encrypted with the same key) that's already decrypted. It's not surprising that, in this case, the cryptanalyst can break the ciphers we've discussed!

For example, suppose we intercepted the message `NOKBT YRXCO XNSDC YYX`, encrypted with the Caesar cipher. If we know that the first word of the plaintext is `DEAR`, we can mount a known-plaintext attack: this tells us that `D` was encrypted to `N`, and so the shift is 10. We can immediately decrypt the entire message and recover the plaintext, which reads `DEARJ OHNSE NDITS OON`. A similar attack works for many of the other ciphers we've studied.

Other than the one-time pad, every cipher in this chapter is extremely vulnerable to this kind of attack. Even a small amount of known plaintext can allow one to completely break the classical ciphers.

In fact, it's perhaps difficult to imagine how an encryption scheme could possibly be resistant to a known-plaintext attack. Intuition might suggest that if one knows a plaintext and the corresponding ciphertext and the encryption method being used, that the key which would transform the given plaintext into the given ciphertext can be "figured out" in some way.

It turns out, however, that it's possible to develop encryption schemes that are resistant to attack even in this situation. It's essential that these methods exist, because, actually, opportunities for plaintext attacks arise in all sorts of situations in modern cryptography. If someone encrypts a computer file, for example, then the type of file they are encrypting may have standard identifying information that occurs regardless of the actual file content. For example, if someone wants to encrypt a webpage for transmission, the file source will start with tags like `<DOCTYPE... >` and `<HTML>` that come at the beginning of html format pages. (If you use the "view source" option in your web browser, you can see that even completely different web pages share lots of structure.) For this reason, it's of paramount importance for modern cryptographic needs to have encryption systems that are secure even against known-plaintext attacks. Such systems are the subject of Chapter 6.

5.6. Problems

• From §5.1: Breaking the Permutation Cipher

EXERCISE 5.6.1. *The factorial function's rapid growth makes a brute force test of all possible permutations impractical. Plot the first ten values of the factorial function.*

EXERCISE 5.6.2. *Show* $n! > (n/2)^{n/2}$ *and deduce a lower bound for* $100!$*; this can be done with an entirely elementary calculation. While the true size of* $n!$ *is about* $n^n e^{-n}\sqrt{2\pi n}$*, this simple bound is enough to see very rapid growth.*

EXERCISE 5.6.3. *Determine the number of permutations of length* 5 *that leave the middle letter unchanged.*

EXERCISE 5.6.4. *Decrypt the message*

`HTESE RCHTE SAEEB PNMRE TUTDE,`

encrypted with an unknown permutation of length 5.

EXERCISE 5.6.5. *Decrypt the message*

`MANA PALA AACN APLN NMAA QEWA,`

encrypted with an unknown permutation of length 5.

EXERCISE 5.6.6. *Imagine we have a permutation cipher with blocks of length* 5*, and we frequently see the letters* T, H *and* E *in positions* 2, 1*, and* 5*. How many different permutations are there such that these three positions* *are sent to three consecutive positions, with* T *first, then* H*, and finally* E*. What if we had blocks of length n instead of* 5*?*

EXERCISE 5.6.7. *The length of a message encrypted with a permutation cipher provides a powerful clue, as the length of the permutation must divide the length of the message. Imagine the sender knows this and decides to add a couple of garbage letters at the end of the message so that the new message's length is no longer a multiple of the true key. Discuss the value of this method. Would something like this help protect a message encrypted with a Vigenère cipher from the Kasiski test?*

EXERCISE 5.6.8. *A friend of ours wants to send us a message. Worried that a permutation cipher may not provide enough security, she decides to apply* two *permutations. Which provides more security, two permutations of length* 10 *or one permutation of length* 10 *and one of length* 7*?*

EXERCISE 5.6.9. *Our friend from the previous exercise has a new idea; to increase the security of the permutation cipher, she first encrypts the message with a Caesar cipher and then applies a permutation cipher. Has she increased the security?*

- **From §5.2: Breaking the Hill Cipher**

EXERCISE 5.6.10. *Decrypt some more of the message from the section (i.e., the one that begins* FRAQR TFRLZ KEFPZ KXYWS*); enough to identify which fairy tale it is.*

EXERCISE 5.6.11. *The following ciphertext was encrypted with a* 2×2 *Hill cipher:*

```
HOOYH BTXGW KMVDH KKBMC HOHOM TMIJN FBCPY PTRAW SCTCK XEZHO APUCT
MGYKS MTMCT CETQF ZTDJC YNVFW PPBMM GUURG PPGSX OEZHO MTIWY IQDPC
CMUBH OGEXN TCQID QPPPP QBIGK AZEYP HOAPE BEKWX HOCUW XKBAP MIMRI
JMTTL UEYPH OIJIL UDNZE IHNZR HOWXN
```

A quick count shows that the most common bigrams occurring in this ciphertext are HO *(occurring ten times),* PP *(occurring four times), and* MT *(occurring four times). Break the cipher.*

EXERCISE 5.6.12. *The key for the Hill cipher is an $n \times n$ matrix which is invertible modulo* 26*.*

(1) *How many* 2×2 *invertible matrices (modulo* 26*) are there?*
(2) *What if instead we used an alphabet with p letters, with p a prime number? In other words, how many invertible* 2×2 *matrices (modulo p) are there?*
(3) *Let p and q be two primes (they may or may not be distinct). How many* 2×2 *invertible matrices (modulo pq) are there?*

EXERCISE 5.6.13. *Discuss the security of first applying a Vigenère cipher with keystream of length n followed by a Hill cipher with matrix size m. Do you want $n = m$, or $n \neq m$, or does it not make a difference? Discuss your reasons for your choice.*

EXERCISE 5.6.14. *Table* 5.1 *gives common English bigrams and trigrams. Note the percentage of the kth most popular bigram is higher than that of the kth most popular trigram. Are you surprised by this? Why or why not?*

EXERCISE 5.6.15. *Read about Hill's life, and write a short report about him and applications of his method.*

• From §5.3: Running Key Ciphers

EXERCISE 5.6.16. *Charlie gives Alice a message which he claims comes from Bob. The encrypted message is*

HNKUTFNYHTYQGFIUWVIEZHJQXHSMQCPVNJSIQDEANDWVMEVTES.

Charlie claims the message says HI ALICE THIS IS BOB I NEED YOU TO DO WHATEVER CHARLIE SAYS BYE, *and says he knows this because he knows Bob used the running key*

AFKJLDJFALGIOEUTOIEAWJVWETPYUVPCJOOROWEJCVSDMGDSGO.

Can Alice trust Bob?

EXERCISE 5.6.17. *Consider the encrypted message from the previous exercise. Find a running key cipher so that it decrypts to* GREETINGS ALICE FROM BOB DO NOT TRUST CHARLIE AT ALL GOODBYE.

EXERCISE 5.6.18. *Assuming the original message and the keystream have typical English letter frequencies, what percentage of the* e*'s in the ciphertext correspond to an* e *in the plaintext?* Hint: The letter frequencies from Table 4.1 might be useful for this problem.

EXERCISE 5.6.19. *Generalizing the previous problem, we say a position in the plaintext is* good *if the letter there is the same as the ciphertext letter at that position. For each letter of the alphabet, we can see how often it's involved in a good position. Do you think "E" is more likely to be involved in a good position than "Q"? Why or why not?*

• From §5.4: One-Time Pads

EXERCISE 5.6.20. *This exercise assumes familiarity with some basic concepts from probability and statistics. The following Mathematica code randomly generates* num $= 660$ *integers from* $\{0, 1, \dots, 25\}$, *with each integer equally likely to be chosen each time.*

```
randomlist = {};
num = 660;
For[n = 1, n <= num, n++,
  randomlist = AppendTo[randomlist, Random[Integer, {0,25}]]];
Histogram[randomlist, 26]
```

Explore the letter frequencies as num *increases (either in Mathematica or in another environment). We expect each letter to occur approximately* num/26 *times in the random string. Investigate the deviations from the expected values, in particular how many times does the most likely letter occur, and how many times does the least likely letter occur. Is there a rule that governs the behavior of these outliers? What do numerical simulations suggest?*

EXERCISE 5.6.21. *Write a short report about Mohan Srivastava's cracking of the Canadian scratch lottery. Do you think a similar method could work elsewhere? Be careful....*

EXERCISE 5.6.22. *The greatest disadvantage of the one-time pad is that you need to have a large text. As it can be impractical to carry a large text on you, there are several options. One is to use an agreed upon document, which may be accessible through the internet. Here, however, is another idea. Let* a_1 *be any integer in* $\{1, 2, \ldots, 25\}$ *and shift the first letter of the text by* a_1. *Let* $a_2 = a_1^2 \pmod{26}$, *and shift the second letter by* a_2. *We continue, and let* $a_k = a_{k-1}^2 \pmod{26}$ *and shift the* k^{th} *letter by* a_k. *We can create as long of a pad as needed, simply by taking more and more terms. We get different pads by choosing different starting values, and now all we need to do is remember that starting value and the fact that we square modulo 26 to get the next shift. Is this a good system? Why or why not?*

- **From §5.5: Summary**

EXERCISE 5.6.23. *The ciphertext* `HQGDJ GTQEL HGVQL BQGMQ` *was encrypted with the affine cipher. Find the original message, which begins with the word* `DEAR`.

EXERCISE 5.6.24. *The ciphertext* `UENZH ZIMPW EPEVZ PETJR NI` *was encrypted with the Vigenère cipher. Find the original message, which begins with* `JANEDOE`.

EXERCISE 5.6.25. *The ciphertext* `GENMA NCMNJ WQHF` *was encrypted with the* 2×2 *Hill cipher. Find the original message, which begins with the name* `KARLA`.

Chapter 6

Modern Symmetric Encryption

Unfortunately (or fortunately, depending on whether you're the sender or the attacker), the classical encryption schemes are vulnerable to a variety of attacks. The purpose of this chapter is to introduce new schemes. As always, the goal is to be able to encrypt and decrypt quickly and easily if and only if you know the secret key. There are issues with these methods as well, but when combined with more secure public channel cryptosystems such as RSA (discussed in Chapters 7 and 8), we'll see that they play an important role in modern encryption.

In the course of studying these systems, we'll encounter several mathematical concepts that will resurface later in this book. These include representing messages by numbers, binary expansions, and generating seemingly random streams of 0's and 1's. These are key ingredients in designing modern encryption schemes, and thus you should remember this motivation as you read on.

6.1. Binary Numbers and Message Streams

In all the classical ciphers covered in the previous chapters, we encrypted messages written in the English alphabet. Up to some slight modifications for different languages (and possibly the inclusion of punctuation), this was the alphabet of cryptography for thousands of years, well into the twentieth century. In modern times, however, cryptography takes place almost entirely on computers, and the internal "alphabet" of computer systems is the binary number system.

The **decimal number system** we are all used to gets its name from the fact that it's based on powers of 10. For example, given a number like 3,725, each digit indicates increasing powers of 10, from right to left. The right-most digit is the 1's place, the second right-most digit is the 10's place, the third right-most digit is the 100's place, and the fourth right-most digit is the 1000's place. Thus we have

$$3,725 \;=\; 3\cdot 10^3 + 7\cdot 10^2 + 2\cdot 10^1 + 5\cdot 10^0.$$

There are ten possible values for each digit (0 through 9), since larger digits aren't needed (for example, $10\cdot 10^0$ is the same as $1\cdot 10^1$).

The **binary number system**, on the other hand, is based on powers of 2. For example, the number 10011 in binary represents

$$1\cdot 2^4 + 0\cdot 2^3 + 0\cdot 2^2 + 1\cdot 2^1 + 1\cdot 2^0 \;=\; 16+2+1 \;=\; 19.$$

Rather than there being a 1's place, 10's place, 100's place, and so on, there is a 1's place, a 2's place, a 4's place, etc. The powers of 2 take the place of the powers of 10. Moreover, in binary, we use only *two* possible values for each digit, 0 and 1, since it's unnecessary to use any more. For example, the number 2 is 10 in binary, where here 10 means $1\cdot 2^1 + 0\cdot 2^0$.

We quickly describe how to find decimal and binary representations. We'll revisit this topic in greater detail in §7.4, as binary expansions play a central role in fast exponentiation, which is the reason many modern cryptosystems can be implemented in real time. The reader wishing more details is encouraged to read that section now.

If you haven't used base two before, it can be unpleasant to make the switch from decimal expansions, but it is a switch that must be made. We use base 10 because we have ten fingers; there is nothing "natural" about 10. Base two, however, is different. This can be viewed as on/off, yes/no, have/don't have. There are enormous dividends once we gain fluency in binary; it turns out that many of the most important algorithms involve binary expansions, and thus being comfortable with these expansions allow us to compute needed quantities quickly.

Thus, it's important to learn how to convert a number from a decimal expansion to a binary one. Let's do an example. If we want to represent a decimal number, say 35, in binary, we subtract the largest power of 2 less than 35, which is 32, giving 3, and repeat. The largest power of 2 less than 3 is 2, and subtraction gives us 1, which is itself a power of 2. The binary representation has a 1 for each power of 2 subtracted in this process, and a 0 when no power of 2 was removed. For 35 we get 100011, since we have a 1 in the 32's place, the 2's place, and the 1's place. Sometimes people put a subscript after the number to indicate the base to avoid confusion; thus 35 in binary might be written 100011_2 (to be truly pedantic, we could write $35_{10} = 100011_2$). *If the base is understood, however, it's often suppressed; this is usually the case with decimal expansions.*

```
We'll only work in base 2 (i.e., binary) in this
chapter, so henceforth all numbers are in binary.
```

The fact that any number can be represented with just two distinct digit values is what makes binary so attractive for use in computers. Two digit types can be much more easily represented electronically than ten. As mentioned above, we can think of this as power on or off, magnetized or demagnetized, etc.

Of course, for binary numbers to be useful to computers, we must be able to do arithmetic operations on them. It turns out that the ways you learned to do arithmetic with decimal numbers carry over directly to binary numbers. For example, consider the addition problem

$$\begin{array}{r} 10011 \\ +101011 \\ \hline \end{array}$$

We carry out the addition just like we would with decimal numbers, keeping in mind that in binary 1+1=10. Remember, in this chapter all numbers are written in binary, and thus 10 is really $1 \cdot 2 + 0 \cdot 1$ or two (which is 1 plus 1). Thus, for example, addition of the right-most digits results in a carry:

$$\begin{array}{r} {\scriptstyle 1} \\ 10011 \\ +101011 \\ \hline 0 \end{array}$$

Thus in the second right-most digit, the addition is 1+1+1, which is 11 in binary, resulting in 1 and a carry of 1:

$$\begin{array}{r} {\scriptstyle 1\,1} \\ 10011 \\ +101011 \\ \hline 10 \end{array}$$

The rest of the addition is shown below:

$$\begin{array}{r} {\scriptstyle 1\,1} \\ 10011 \\ +101011 \\ \hline 111110 \end{array}$$

We can check that we get the same answer through decimal arithmetic: the binary number 10011 represents the decimal number 19, the binary number 101011 represents the decimal number 43, and the binary number 111110 represents the decimal number 62, which is the sum of 19 and 43. We discuss binary expansions in great detail in §7.4.

Of course, computers need a way to convert between the binary used for internal calculations and the human-readable character set presented to users. This is done with a character encoding scheme, the most famous being the **American Standard Code for Information Interchange**, or **ASCII**. ASCII converts characters to 7-**bit** (or digit) binary numbers (see Figure 6.1). Newer character-encoding schemes such as UTF-8 allow the encoding of more objects, including accented characters, other non-English characters, and many special symbols.

USASCII code chart

				b_7 b_6 b_5 →	0 0 0	0 0 1	0 1 0	0 1 1	1 0 0	1 0 1	1 1 0	1 1 1
b_4	b_3	b_2	b_1	Row ↓ Column →	0	1	2	3	4	5	6	7
0	0	0	0	0	NUL	DLE	SP	0	@	P	`	p
0	0	0	1	1	SOH	DC1	!	1	A	Q	a	q
0	0	1	0	2	STX	DC2	"	2	B	R	b	r
0	0	1	1	3	ETX	DC3	#	3	C	S	c	s
0	1	0	0	4	EOT	DC4	$	4	D	T	d	t
0	1	0	1	5	ENQ	NAK	%	5	E	U	e	u
0	1	1	0	6	ACK	SYN	&	6	F	V	f	v
0	1	1	1	7	BEL	ETB	'	7	G	W	g	w
1	0	0	0	8	BS	CAN	(	8	H	X	h	x
1	0	0	1	9	HT	EM	)	9	I	Y	i	y
1	0	1	0	10	LF	SUB	*	:	J	Z	j	z
1	0	1	1	11	VT	ESC	+	;	K	[	k	{
1	1	0	0	12	FF	FS	,	<	L	\	l	|
1	1	0	1	13	CR	GS	-	=	M	]	m	}
1	1	1	0	14	SO	RS	.	>	N	^	n	~
1	1	1	1	15	SI	US	/	?	O	_	o	DEL

FIGURE 6.1. An old ASCII code chart. The column determines the first three binary digits, while the row determines the final four. Thus, for example, < corresponds to 0111100. (Uploaded from Wikipedia, scanner copied from the material delivered with TermiNet 300 Impact Type Printer with Keyboard, February 1972, General Electric Data Communication Product Dept., Waynesboro VA.)

For the purpose of doing cryptography with binary message streams, we don't need to be able to convert all the punctuation marks, and we don't need to distinguish between upper- and lower-case letters. Thus we use the conversion given in Table 6.1, which maps characters to 5-bit binary strings. Note that we have given a character assignment for the 32 possible 5-bit binary numbers. Using the ASCII table, the string `HELLO` becomes 0011100100010110101101110, while the string 011011010101110 breaks up into 01101 10101 01110, giving `NVO`.

We can perform one-time pad encryption and decryption on binary message streams (see §5.4 for a review of the one-time pad). To send the message `HELP`, we first convert it to the binary string 00111001000101101111. If we have generated the random keystream 01010111111011010000 (and given a copy to the intended recipient of the message), we can encrypt the message by adding it to the keystream. When we were working with the English alphabet, the addition was carried out modulo 26 for each character. For binary streams, we add each digit *modulo* 2. There are only four possibilities:

- $0 + 0 \equiv 0 \pmod{2}$,
- $1 + 0 \equiv 1 \pmod{2}$ and $0 + 1 \equiv 1 \pmod{2}$, and
- $1 + 1 \equiv 0 \pmod{2}$.

Thus the encryption is given by

$$\begin{array}{r} 001110010001011011111 \\ \oplus 010101111110110100000 \\ \hline 011011101111101111111 \end{array}$$

We indicate this bitwise addition modulo 2 by the symbol $\oplus$. The operation is also sometimes referred to as **XOR**. XOR stands for eXclusive OR, since it returns 1 if the first digit is 1 *or* if the second digit is 1, but not if both are 1. Note that *no carrying* is done when carrying out bitwise addition; this lack of carrying will be important later.

TABLE 6.1. Binary character conversion table

Decimal Number	Character	5-bit Binary Representation
0	A	00000
1	B	00001
2	C	00010
3	D	00011
4	E	00100
5	F	00101
6	G	00110
7	H	00111
8	I	01000
9	J	01001
10	K	01010
11	L	01011
12	M	01100
13	N	01101
14	O	01110
15	P	01111
16	Q	10000
17	R	10001
18	S	10010
19	T	10011
20	U	10100
21	V	10101
22	W	10110
23	X	10111
24	Y	11000
25	Z	11001
26	.	11010
27	!	11011
28	?	11100
29	☺	11101
30	☹	11110
31	_	11111

As discussed in the section on one-time pads (Section §5.4), producing "random-seeming" streams in a deterministic way is a difficult problem. If we have a way of turning a small key into a large random-seeming binary stream, then we can use such a method as a basis for an encryption method which XORs the stream and the binary message string. Ciphers of this type are called **stream ciphers**. In the next section, we discuss Linear Feedback Shift Registers as a way of producing apparently random streams from short keys.

6.2. Linear Feedback Shift Registers

In this section we describe Linear Feedback Shift Registers. These were created around 1950 and provide a fast way to generate pseudo-random sequences. We start with a problem seemingly unrelated to these issues. There is a connection, though, as will become clear soon.

Imagine an office of imitators, who obsess about whether or not to wear hats to work to satisfy social pressures.

- Jan thinks Alice is the coolest one in the office, and wears a hat to work whenever Alice wore one the previous day.
- Albert thinks Jan is the coolest person in the office, and so wears a hat whenever Jan wore one on the previous day.
- Jeff thinks that Albert is totally awesome, and so wears a hat whenever Albert wore on the previous day.
- Finally, Alice wears a hat whenever either Jeff or Albert wore a hat on the previous day, but not if they both did. (We will not explore the psychology behind such behavior.)

Let's assume that on the first day of the new year, Albert wears a hat and nobody else does. The following table shows how the days will progress in terms of hat-wearing, with 1 indicating that someone is wearing a hat on the given day and 0 that they are not.

day	Alice	Jan	Albert	Jeff
1	0	0	1	0
2	1	0	0	1
3	1	1	0	0
4	0	1	1	0
5	1	0	1	1
6	0	1	0	1
7	1	0	1	0
8	1	1	0	1
9	1	1	1	0
10	1	1	1	1
11	0	1	1	1
12	0	0	1	1

Consider now the perspective of Jeff's lunch buddy Bobby, who never sees Alice, Jan, and Albert, and does not know about the dedicated attempts being made at imitation. Bobby notices that Jeff sometimes wears a hat and sometimes does not, but he knows nothing about how the decision is made for a particular day. In fact, to Bobby, the sequence $0, 1, 0, 0, 1, 1, 0, 1, 0, 1, 1, 1, \ldots$ indicating when Jeff decides to wear a hat (the right-most column in the above table) might appear to be random. In spite of this appearance, we know that there was actually no randomness involved in the production of this sequence (except perhaps Albert's decision on the first day of the year). Instead, the sequence was produced by the deterministic imitations of Jeff and his coworkers.

Linear Feedback Shift Registers can be used to build random-seeming sequences in the same way that Jeff's hat-wearing sequence was produced. Consider, for example, the following "transition rules":

$$b_3 \leftarrow b'_4, \qquad b_2 \leftarrow b'_3,$$
$$b_1 \leftarrow b'_2, \qquad b_4 \leftarrow b'_1 + b'_2.$$

These rules tell us how to transform one 4-bit string into another. By convention, the indices are written right-to-left, so we send the string (b'_4, b'_3, b'_2, b'_1) to the string (b_4, b_3, b_2, b_1). For example, the string 1011 corresponds to the assignment $b'_4 = 1$, $b'_3 = 0$, $b'_2 = 1$, and $b'_1 = 1$. Transforming this string according to the rules given above produces the sequence 0101 (corresponding to $b_4 = 0$, $b_3 = 1$, $b_2 = 0$, and $b_1 = 1$).

In fact, these rules represent the rules of imitation practiced by Jeff and his coworkers. In this case, b_4, b_3, b_2, b_1 are the variables indicating whether Alice, Jan, Albert, and Jeff, respectively, wear hats on a given day. The transition rule $b_2 \leftarrow b'_3$, for example, implies that Albert wears a hat whenever Jan did the day before. Notice the last transition rule $b_4 \leftarrow b'_1 + b'_2$: adding the bits b'_1 and b'_2 modulo 2 gives 1 whenever exactly one of them is 1, but not both. Thus $b_4 \leftarrow b'_1 + b'_2$ is exactly Alice's hat-wearing rule: she wears a hat whenever Jeff *or* Albert—but not both—wore one the day before.

Linear Feedback Shift Register (LFSR): In general, a Linear Feedback Shift Register (LFSR) is specified by a single formula, such as

$$b_4 \leftarrow b'_1 + b'_2.$$

The subscript of the first variable tells us that the register has four bits. It is understood that the three unspecified rules are the rules $b_3 \leftarrow b'_4$, $b_2 \leftarrow b'_3$, and $b_1 \leftarrow b'_2$. Similarly, the LFSR specified by the rule

$$b_5 \leftarrow b'_1 + b'_2 + b'_4$$

is a 5-bit LFSR whose complete transition rules are

$$\begin{aligned} b_4 &\leftarrow b'_5 \\ b_3 &\leftarrow b'_4 \\ b_2 &\leftarrow b'_3 \\ b_1 &\leftarrow b'_2 \\ b_5 &\leftarrow b'_1 + b'_2 + b'_4. \end{aligned}$$

The table of output values can be filled out beginning with a **seed**, which is just an initial condition for the register. For example, if we seed this 5-bit register with the string 10110, the output begins with

	b_5	b_4	b_3	b_2	b_1
1	1	0	1	1	0
2	1	1	0	1	1
3	1	1	1	0	1
4	0	1	1	1	0
5	0	0	1	1	1
6	0	0	0	1	1
7	0	0	0	0	1
8	1	0	0	0	0
9	1	1	0	0	0
10	1	1	1	0	0
11	1	1	1	1	0
⋮	⋮	⋮	⋮	⋮	⋮

The output stream of any LFSR is taken to be the sequence of states of the bit b_1. For the 5-bit sequence whose table of states is shown above, this sequence would be 01101110000.... LFSRs are commonly used as part of schemes to generate "pseudo-random" streams of bits because they are quite easy to implement on hardware. In spite of the fact that they produce streams that seem very random, the only operations involved in producing the streams are simple bitwise addition, which can performed very efficiently even on extremely simple computer chips.

The output of an LFSR can be used for the keystream of a stream cipher. For example, let's encrypt the message HELP with the output from the LFSR specified by $b_4 \leftarrow b_1' + b_2' + b_4'$ and seeded with the string 0110. The first step is to use Table 6.1 to translate the message HELP into the binary message stream 00111001000101101111. Next, we compute the output stream of the LFSR with the given stream. Here is the table of output states:

	b_4	b_3	b_2	b_1
1	0	1	1	0
2	1	0	1	1
3	1	1	0	1
4	0	1	1	0
5	1	0	1	1
6	1	1	0	1
7	0	1	1	0
8	1	0	1	1
9	1	1	0	1
10	0	1	1	0
11	1	0	1	1
⋮	⋮	⋮	⋮	⋮

The output sequence (the b_1 column) appears to be repeating: 011011011.... We can see that the seventh state of the register is the same as the first state; thus, the eighth state will be the same as the second state, the ninth the same as the third state, and so on. Thus we do not need to continue filling out the table to produce a long keystream; we can infer that the output will be 011011011011011....

To encrypt the binary message stream, we add it digit-by-digit to the keystream modulo 2:

$$\begin{array}{r} 00111001000101101111 \\ \oplus 01101101101101101101 \\ \hline 01010100101000000010 \end{array}$$

We can convert this back to characters using Table 6.1 to get the ciphertext `KSQC`. The recipient, knowing the LFSR register and seed used, would convert this ciphertext back into binary, find the keystream using the LFSR, and subtract the keystream from the binary ciphertext modulo 2 to get the binary message stream (which is then converted back to characters using Table 6.1). Notice that *subtracting modulo 2 is the same as adding!* For example, $1 - 1 \equiv 0 \pmod 2$, but also $1 + 1 \equiv 0 \pmod 2$. This means that the decryption operation is actually just adding the keystream to the binary ciphertext, and so is the same as the encryption operation. The decryption for our message is

$$\begin{array}{r|c|c|c} \texttt{K} & \texttt{S} & \texttt{Q} & \texttt{C} \\ 01010 & 10010 & 10000 & 00010 \\ \oplus 01101 & 10110 & 11011 & 01101 \\ \hline 00111 & 00100 & 01011 & 01111 \\ \texttt{H} & \texttt{E} & \texttt{L} & \texttt{P} \end{array}$$

Recall that in the example above where `HELP` was encrypted to `KSQC`, the keystream produced by the LFSR being used eventually repeated. In fact this will *always* happen eventually, no matter what LFSR is used and no matter the seed. This is simply because there are only a finite number of possible states for the LFSR. For example, consider the 3-bit LFSR register given by $b_3 \leftarrow b_2' + b_1'$, seeded with the string 101. The table of states begins with

	b_3	b_2	b_1
1	1	0	1
⋮	⋮	⋮	⋮

There are only finitely many possible states (in fact, only 8: 111, 110, 101, 100, 011, 010, 001, 000). Therefore, eventually some state is repeated as we fill out the table. Once that happens, the table repeats that section over and over again, since the next row of the table is completely determined by the previous row. The same thing will happen for any LFSR, although with more bits there are more possibilities.

It should be noted that, while LFSR streams *seem* random, they are not really random at all. For one thing, as already noticed, they eventually repeat over and over again. For another, it turns out that given enough of an LFSR stream, it is possible to figure out the rest of it. This is crucial for the application of LFSRs to cryptography, as we will see shortly.

6.3. Known-Plaintext Attack on LFSR Stream Ciphers

Suppose we have intercepted the message `WIMSUJ`, encrypted with an LFSR stream cipher. As is often the case with LFSRs in cryptographic settings, we assume the formula for the LFSR is public knowledge. This is often the case because the LFSR formula is hardcoded (built-into) the hardware used for encryption, and cannot be changed as part of the key. Let's assume the formula is $b_4 \leftarrow b_4' + b_1'$, and that only the seed is kept secret as the key. To mount a known-plaintext attack on the stream cipher, we need to have some known plaintext. In this case, let's assume we know that the intercepted message is a name and begins with the letters `MR`.

Amazingly, this is all we need to uncover the keystream! Here's how. The message `WIMSUJ` corresponds to the binary stream 1011001000..., while the letters `MR` correspond to 0110010001 in binary. By subtracting the second string from the first modulo 2, we can get the beginning of the keystream used for encryption. Since subtraction and addition are the same modulo 2, we are just adding the two strings modulo 2:

$$\begin{array}{r} 1011001000 \\ \oplus 0110010001 \\ \hline 1101011001 \end{array}$$

The result must be the beginning of the keystream used for encryption, thus we know that this string forms the beginning of the right-hand column of the table of LFSR states:

	b_4	b_3	b_2	b_1
1				1
2				1
3				0
4				1
5				0
6				1
7				1
8				0
9				0
10				1
11				
12				

To decrypt more of the ciphertext, we need to be able to fill in more of the b_1 column. To do this, we need to know the other missing bits in the table. We can figure out many of these bits from what we know about the right-most column because we know the transition rule. For example, the fact that $b_1 = 1$ in the second state means that $b_2 = 1$ in the first state (because of the transition rule $b_1 \leftarrow b_2$), as we have indicated in table below on the left:

	b_4	b_3	b_2	b_1
1			1	1
2				1
3				0
4				1
5				0
6				1
7				1
8				0
9				0
10				1

	b_4	b_3	b_2	b_1
1		0	1	1
2			0	1
3				0
4				1
5				0
6				1
7				1
8				0
9				0
10				1

Similarly, the fact that $b_1 = 0$ in the third state means that $b_2 = 0$ in the second state, and, moreover, that $b_3 = 0$ in the first state, as we have indicated in the table above on the right. Continuing in this way, we can fill in much more of the table:

	b_4	b_3	b_2	b_1
1	1	0	1	1
2	0	1	0	1
3	1	0	1	0
4	1	1	0	1
5	0	1	1	0
6	0	0	1	1
7	1	0	0	1
8		1	0	0
9			1	0
10				1

At this point, we can start working down the table from the transition rules (in this case, $b_1 \leftarrow b_2'$, $b_2 \leftarrow b_3'$, $b_3 \leftarrow b_4'$, and $b_4 \leftarrow b_4 \oplus b_1$). For example, the eighth state of the bit b_4 is given by $1 \oplus 1$, which is 0. This, in turn, is then the ninth state of the bit b_3, the tenth state of the bit b_2, and the eleventh state of the bit b_1:

	b_4	b_3	b_2	b_1
1	1	0	1	1
2	0	1	0	1
3	1	0	1	0
4	1	1	0	1
5	0	1	1	0
6	0	0	1	1
7	1	0	0	1
8	0	1	0	0
9		0	1	0
10			0	1
11				0

We can now fill in the ninth state of the bit b_4, and so on. Continuing in this manner we produce

	b_4	b_3	b_2	b_1
1	1	0	1	1
2	0	1	0	1
3	1	0	1	0
4	1	1	0	1
5	0	1	1	0
6	0	0	1	1
7	1	0	0	1
8	0	1	0	0
9	0	0	1	0
10	0	0	0	1
11	1	0	0	0
12	1	1	0	0
13	1	1	1	0
14	1	1	1	1
15	0	1	1	1
16	1	0	1	1
17		1	0	1
18			1	0
19				1

Since the entire message we are trying to decrypt consists of six characters, and so corresponds to a binary string of length 30, we need the first 30 bits of output to decrypt the stream. We can stop filling in the table at this point since we have encountered a repeated state: The sixteenth state in the above table is the same as the first states. This means that the first 15 digits of the output stream from this LFSR repeat over and over again, giving an output stream of

$$110101100100011\,110101100100011\,110101100100011\ldots.$$

Subtracting (adding) this to the binary ciphertext string gives:

	`W`	`I`	`M`	`S`	`U`	`J`
	10110	01000	01100	10010	10100	01001
$\oplus$	11010	11001	00011	11010	11001	00011
	01100	10001	01111	01000	01101	01010
	`M`	`R`	`P`	`I`	`N`	`K`

We've found the plaintext: `MRPINK`! Even with just a tiny amount of known plaintext, a stream cipher based on a single LFSR can be completely broken by a known-plaintext attack. The problem, roughly speaking, was that it is too easy to work backwards from a given part of an output stream and determine previously unknown states of the registers. Nevertheless, the simplicity of LFSRs make them very attractive for many forms of cryptography, especially in cases where the encryption or decryption is to be carried out by a dedicated hardware device. There are sophisticated ways of designing stream ciphers which make use of LFSRs, but which try to overcome their weakness towards known-plaintext attacks. We cover one such type of cipher in the next section.

6.4. LFSRsum

We've already seen that encryption schemes based on a single LFSR are vulnerable by a known-plaintext attack. They can be easily broken once we know a little of the plaintext. By combining multiple LFSRs, however, we can attempt to create a system which is more resilient to plaintext attacks. As a first example of this, we consider a scheme, which we call LFSRsum, which forms an encryption keystream as the modulo 2 sum of two different LFSR output streams.

LFSRsum: The LFSRsum system uses a 3-bit LFSR and a 5-bit LFSR in tandem to create a single pseudo-random stream. For example, we may choose the 3-bit output stream defined by the formula $b_3 = b_2' + b_1'$, and a 5-bit LFSR defined by $c_5 = c_4' + c_2' + c_1'$. We refer to these registers as LFSR-3 and LFSR-5, respectively.

- The LFSRsum system works by adding these two streams together modulo 2 to form a keystream, which is then added (modulo 2) to the binary message stream to encrypt the message.
- Decryption is done by adding (or subtracting) the keystream from the ciphertext.
- The key for the LFSRsum cipher is six binary digits. The first two digits give the first two digits of the seed for the LFSR-3 register, and the last four digits of the key give the first four digits of the LFSR-5 register seed. The last digit of the seed for each register is always 1: this means that no matter what key we choose, the registers will never get set to all 0's.

For example, if we seed LFSR-3 and LFSR-5 with the strings 011 and 10101, respectively, their output is

b_3	b_2	b_1
0	1	1
0	0	1
1	0	0
0	1	0
1	0	1
1	1	0
1	1	1

c_5	c_4	c_3	c_2	c_1
1	0	1	0	1
1	1	0	1	0
0	1	1	0	1
0	0	1	1	0
1	0	0	1	1
0	1	0	0	1
0	0	1	0	0

To encrypt the message Y using these seeds, the encryption stream is

$$\begin{array}{r} 11001 \\ \oplus 10101 \\ \hline 01100 \end{array}$$

This is then added to the plaintext to encrypt:

$$\begin{array}{r} \texttt{Y} \\ 11000 \\ \oplus 01100 \\ \hline 10100 \\ \texttt{U} \end{array}$$

Notice that for the example above, the key was 011010. Just as with single LFSR streams, the decryption operation is *exactly the same* as the encryption operation, since adding is the same as subtracting modulo 2.

At first glance, this new scheme might seem secure against a known-plaintext attack. Assume, for example, that using a known plaintext we have discovered that the beginning of an encryption stream produced by the LFSRsum system using some unknown key is 011110011. Can we figure out the key used, and calculate more of the decryption stream? It seems like we can't since, for example, the fact that the first digit of this keystream is a 1 is either because the first digit of the LFSR-3 output stream is a 1 and the first digit of LFSR-5 is a 0, or because the first digit of the LFSR-5 output stream is a 1 and the first digit of the LFSR-3 output stream is a 3. While it doesn't look like there's an obvious way to figure out which of these possibilities is correct, we can completely determine how each register contributes to the encryption stream by solving a system of equations as we did in Section 5.1. We'll see this idea of solving systems of linear equations to crack messages again when we study the Perfect Code cryptosystem in §7.1.

Since the 6-bit key used to produce the encryption stream 011110011 is unknown to us, let's represent it with the six variables $k_1k_2k_3k_4k_5k_6$, each either a 0 or a 1. The LFSR registers used for the LFSRsum cipher are initialized with these bits:

b_3	b_2	b_1
k_1	k_2	1
$\vdots$	$\vdots$	$\vdots$

c_5	c_4	c_3	c_2	c_1
k_3	k_4	k_5	k_6	1
$\vdots$	$\vdots$	$\vdots$	$\vdots$	$\vdots$

In spite of the fact that we don't know the values of these seed bits, we can nevertheless use the transition rules of the LFSR-3 and LFSR-5 registers to fill in subsequent rows of the tables. Note that modulo 2, we have $1+1 \equiv 0 \pmod 2$, but also, for example, $k_2 + k_2 \equiv 2 \cdot k_2 \equiv 0 \pmod 2$. Thus, for example, in the LFSR-3 table below, the fifth state of the bit b_3 is simplified to $1 + k_1$ from $(1 + k_2) + (k_2 + k_1)$:

b_3	b_2	b_1
k_1	k_2	1
$1+k_2$	k_1	k_2
k_2+k_1	$1+k_2$	k_1
k_1+1+k_2	k_2+k_1	$1+k_2$
$1+k_1$	k_1+1+k_2	k_1+k_2
1	$1+k_1$	k_1+1+k_2
k_2	1	$1+k_1$
k_1	k_2	1
$\vdots$	k_1	k_2
$\vdots$	$\vdots$	k_1

c_5	c_4	c_3	c_2	c_1
k_3	k_4	k_5	k_6	1
$1+k_6+k_4$	k_3	k_4	k_5	k_6
$k_6+k_5+k_3$	$1+k_6+k_4$	k_3	k_4	k_5
$1+k_5+k_6$	$k_6+k_5+k_3$	$1+k_6+k_4$	k_3	k_4
$k_4+k_5+k_6$	$1+k_5+k_6$	$k_6+k_5+k_3$	$1+k_6+k_4$	k_3
$k_3+k_4+k_5$	$k_4+k_5+k_6$	$1+k_5+k_6$	$k_6+k_5+k_3$	$1+k_6+k_4$
$\vdots$	$k_3+k_4+k_5$	$k_4+k_5+k_6$	$1+k_5+k_6$	$k_6+k_5+k_3$
$\vdots$	$\vdots$	$k_3+k_4+k_5$	$k_4+k_5+k_6$	$1+k_5+k_6$
$\vdots$	$\vdots$	$\vdots$	$k_3+k_4+k_5$	$k_4+k_5+k_6$
$\vdots$	$\vdots$	$\vdots$	$\vdots$	$k_3+k_4+k_5$

Thus we know that, in terms of the key bits $k_1, k_2, k_3, k_4, k_5, k_6$, the keystream generated by the LFSRsum cipher should be the (modulo 2) sum of the two output streams found above; namely, $(1+1)$, (k_2+k_6), (k_1+k_5), $(1+k_2+k_4)$, $(k_1+k_2+k_3)$, $(k_1+1+k_2+1+k_6+k_4)$, $(1+k_1+k_6+k_5+k_4), \ldots$. On the other hand, we *know* (from a known-plaintext attack) that the actual encryption stream is 0111100. This allows us to set up a system of congruences.

For example, we know from the tables that the first digit of the encryption stream should be congruent to $1 + 1 \pmod 2$, and from the known-plaintext attack that it should be 0. This isn't helpful as 2 is always equivalent to 0 modulo 2, no matter what values we assign to the variables.

What about the second digit? For the second digit of the encryption stream, the tables tell us that the value should be $k_2 + k_6$ (modulo 2), while the known-plaintext attack tells us the digit should be 1. This gives us the congruence

$$k_2 + k_6 \equiv 1 \pmod 2.$$

Unlike our investigation of the first digit, this is no longer trivial and provides real information. Continuing, we can create such a congruence for each digit of the encryption stream that we know: for the third digit, we get $k_1+k_5 \equiv 1 \pmod 2$; for the fourth, we get $1+k_2+k_4 \equiv 1 \pmod 2$. Continuing in this fashion, we obtain the system

$$\begin{cases} k_2+k_6 & \equiv 1 \pmod 2 \\ k_1+k_5 & \equiv 1 \pmod 2 \\ 1+k_2+k_4 & \equiv 1 \pmod 2 \\ k_1+k_2+k_3 & \equiv 1 \pmod 2 \\ k_1+k_2+k_6+k_4 & \equiv 0 \pmod 2 \\ 1+k_1+k_6+k_5+k_3 & \equiv 0 \pmod 2. \end{cases}$$

At this point, we have six unknowns and six congruences. As we have as many equations as unknowns, there's reason for optimism and hope that we can solve and find the solution. While this may look daunting at first, it's really not that bad. For now, we'll proceed by looking at the equations and seeing what operations help reduce the complexity. There are lots of ways to do the algebra; with enough patience we'll reach the solution.

Let's begin by reordering the variables in each congruence. It's convenient to have only variables on the left-hand side and only numbers on the right. To do this, we add 1 to both sides of the third and sixth congruences. This eliminates the constants on the left, as $1+1 \equiv 0 \pmod 2$.

$$\begin{cases} \quad k_2 \qquad\qquad\qquad\qquad + k_6 \equiv 1 \pmod 2 \\ k_1 \qquad\qquad\qquad + k_5 \qquad \equiv 1 \pmod 2 \\ \quad + k_2 \qquad + k_4 \qquad\qquad\quad \equiv 0 \pmod 2 \\ k_1 + k_2 + k_3 \qquad\qquad\qquad\quad \equiv 1 \pmod 2 \\ k_1 + k_2 \qquad + k_4 \qquad\quad k_6 \equiv 0 \pmod 2 \\ k_1 \qquad + k_3 \qquad + k_5 + k_6 \equiv 1 \pmod 2. \end{cases} \tag{6.1}$$

The system can be solved very efficiently by adding congruences. There are lots of ways to do the algebra. A guiding principle is that, unlike normal arithmetic, here anything multiplied by 2 is zero as 2 is equivalent to zero modulo 2. This suggests adding all the congruences together, as we'll have an even number of k_1's, an even number of k_2's, and so on. All those sums will vanish, and we'll be left with $3k_6$, which is k_6 modulo 2. We prove the last claim, as the others follow similarly. This is because $3k_6 = 2k_6 + k_6$, and $2k_6$ is 0 modulo 2. One way to see this is to try the two possible values for k_6; if k_6 is 0, then $2k_6$ is 0, while if it is 1, then $2k_6$ is 2, which is also equivalent to 0 modulo 2.

So, adding *all* of the congruences together (modulo 2) gives

$$\begin{array}{r} k_2 \qquad\qquad\qquad\qquad + k_6 \equiv 1 \pmod 2 \\ k_1 \qquad\qquad\qquad + k_5 \qquad \equiv 1 \pmod 2 \\ + k_2 \qquad + k_4 \qquad\qquad\quad \equiv 0 \pmod 2 \\ k_1 + k_2 + k_3 \qquad\qquad\qquad\quad \equiv 1 \pmod 2 \\ k_1 + k_2 \qquad + k_4 \qquad\quad k_6 \equiv 0 \pmod 2 \\ + k_1 \qquad + k_3 \qquad + k_5 + k_6 \equiv 1 \pmod 2 \\ \hline k_6 \equiv 0 \pmod 2. \end{array} \tag{6.2}$$

Again, this is because only the k_6 column has an odd number of the variable, and an even number must be zero modulo 2. Thus $4k_1$, $4k_2$, $2k_3$, $2k_4$, and $2k_5$ are all zero modulo 2.

We now know k_6! Now that we know k_6, we can substitute its value in the congruences in (6.1). While we can look at these in any order, we see this information is most valuable in the first congruence. Since $k_2 + k_6 \equiv 1 \pmod{2}$ and $k_6 \equiv 0 \pmod{2}$, we have $k_2 \equiv 1 \pmod{2}$. In other words, we've just discovered that k_2 is 1.

We have now determined two of the six variables (k_2 is 1 and k_6 is 0). Of the four remaining equations, the best one to turn to is the third as that's just $k_2 + k_4 \equiv 0 \pmod{2}$. Substituting k_2 is 1 gives $1 + k_4 \equiv 0 \pmod{2}$, so k_4 must be 1. We now have three of the six variables.

We're left with three unknowns. Notice the fifth equation involves k_1, k_2, k_4, and k_6. We know the last three, so substituting these values gives

$$k_1 + 1 + 1 + 0 \equiv 0 \pmod{2},$$

so k_1 must be 0 (the right-hand side is an even integer, so the left must be even too).

Only two variables are still unknown: k_3 and k_5. Note the second congruence involves k_5 and variables we know, the fourth congruence has k_3 and variables we know, and the sixth has k_3 and k_5 and variables we know. Thus it's better to use either the second or the fourth congruence next. To show it doesn't matter, we'll use the slightly harder fourth congruence now. Substituting for k_1 and k_2 gives

$$0 + 1 + k_3 \equiv 1 \pmod{2}.$$

As the right-hand side is odd, the left-hand side must be odd as well, so k_3 is 0.

All that remains is k_5. As we know every other variable, we can substitute those values into any congruence involving k_5. The simplest choice is the second. As $k_1 = 0$, we find

$$0 + k_5 \equiv 1 \pmod{2},$$

so k_5 is 1.

We found the key! It's 010110, and armed with this information we can decipher the rest of the LFSR encryption stream and decrypt the message.

Congruences (and equations) in which variables are never multiplied by each other (only added or subtracted) are called *linear*. Thus $y - 3x = 0$ is a linear equation, since x is not multiplied by y. Note that the congruences we needed to solve to break LFSRsum system were all linear. It turns out that this can always be done efficiently. Even though it seemed like there was a bit of trial and error involved in solving the congruences in the above example (for example, in deciding which congruences to add together), the process can be carried out by the method of **Gaussian elimination**.

Gaussian elimination is discussed in other mathematics classes, and is a systematic way of efficiently carrying out the appropriate additions and subtractions of congruences needed to reach a solution. Linear systems can be solved quickly even when there are lots and lots of congruences; a computer can solve a linear system containing millions of congruences (and variables) in a matter of seconds. We discuss this a bit in Exercise 6.11.36, where we show that a brute force attack (of trying all possible keystreams) takes far too long, but a linear equation approach is feasible. This means that even if LFSRsum was modified to use gigantic LFSRs (and thus take a gigantic key), it would still not be secure against known-plaintext attacks. We'll see another success of linear algebra in attacking Perfect Codes in §7.1.

We're now in a very common situation in cryptography. We had a very good idea, and we created a very difficult problem where there were so many possibilities that a brute force attack is completely infeasible, requiring more time than the life of the universe! Unfortunately, it turned out that there were powerful methods available to the attacker, and our system was insecure. This will happen again when we investigate KidRSA in §7.2. What do we do? We look at our system and see if there's any way we can make it more complicated.

Our system was based on adding two LFSR streams. This is a linear process, and led to a system of linear equations. This observation suggests the solution: we need a more complicated way to combine the two streams. It turns out that there is no good, general method to solve nonlinear systems of congruences. This suggests that to make a secure cipher based on LFSRs, we need to introduce some "nonlinearity" to prevent known-plaintext attacks such as the one shown above. We discuss one such approach in the next section.

6.5. BabyCSS

BabyCSS is in many ways similar to LFSRsum. It was developed around 1970, and uses the same LFSR-3 and LFSR-5 registers (given by the formulas $b_3 \leftarrow b_2' + b_1'$ and $c_5 \leftarrow c_4' + c_2' + c_1'$, respectively), and again uses a 6-bit key, whose first two bits give the first two bits of the LFSR-3 seed, and whose final four bits give the first four bits of the LFSR-5 seed. As we did for the LFSRsum system, we take the last seed-bit of each register to be 1 to prevent getting all 0's. There are two differences: we do addition in blocks of length five with carrying.

Notice there are *two* new items here. The first is that we break the sequences into blocks of length five. We chose the length to be five as our characters in Table 6.1 are 5-bit binary numbers. The second new element is that we are doing addition with carrying (both within and between blocks). This is responsible for the nonlinearity, which will frustrate a Gaussian elimination attack on the system of congruences.

BabyCSS: BabyCSS has the same set-up as LFSRsum. A 3-bit LFSR and a 5-bit LFSR are used in tandem to create a single pseudo-random stream. For example, we may choose the 3-bit output stream defined by the formula $b_3 = b_2' + b_1'$, and a 5-bit LFSR defined by $c_5 = c_4' + c_2' + c_1'$. We refer to these registers as LFSR-3 and LFSR-5, respectively.

- The BabyCSS system works by adding these two streams together in blocks of length 5 through addition (modulo 2) with carrying to form a keystream, which is then added (modulo 2) to the binary message stream to encrypt the message. The carrying is done both within and between blocks (if there is a carry after summing two blocks of length 5, that 1 is carried to the next block sums).
- Decryption is done by adding (or subtracting) the keystream from the ciphertext.
- The key for the BabyCSS cipher is six binary digits. The first two digits give the first two digits of the seed for the LFSR-3 register, and the last four digits of the key give the first four digits of the LFSR-5 register seed. The last digit of the seed for each register is always 1: this means that no matter what key we choose, the registers will never get set to all 0's.

To see how BabyCSS works, let's use it to encrypt the message `HI` with the cipher key 011010 (the same used in the first example in the LFSRsum section). Since the binary message stream for `HI` has ten digits, we need to compute ten digits of the BabyCSS encryption stream. The output tables of the LFSR-3 and LFSR-5 registers are:

b_3	b_2	b_1
0	1	1
0	0	1
1	0	0
0	1	0
1	0	1
1	1	0
1	1	1
0	1	1
0	0	1
1	0	0

c_5	c_4	c_3	c_2	c_1
1	0	1	0	1
1	1	0	1	0
0	1	1	0	1
0	0	1	1	0
1	0	0	1	1
0	1	0	0	1
0	0	1	0	0
0	0	0	1	0
1	0	0	0	1
1	0	0	0	0

We now add the streams in blocks of five **with carrying**, as opposed to the bitwise addition modulo 2 we did for LFSRsum. For example, adding the first five bits of each output stream gives

$$\begin{array}{cccccc} \textcircled{1} & & & & 1 & \\ & 1 & 1 & 0 & 0 & 1 \\ + & 1 & 0 & 1 & 0 & 1 \\ \hline & 0 & 1 & 1 & 1 & 0 \end{array}$$

Notice that the first five values of b_1 are 1, 1, 0, 0, 1 and the first five values of c_1 are 1, 0, 1, 0, 1; however, when we add these two blocks, we start the addition *on the right*. Thus, the first addition is really the fifth value of b_1 plus the fifth value of c_1, and *not* the first value of b_1 plus the first value of c_1. If you find this a minor annoyance, take heart in hoping that maybe the attacker will be annoyed as well!

Returning to the addition with carrying, the resulting bits 01110 form the first five output bits of the encryption stream. Notice that there was an "extra" carry digit, which we circled above. Instead of becoming a sixth digit in the sum, we carry this digit over to the next block addition problem. Thus, adding the next five digits gives

$$\begin{array}{cccccc} \textcircled{1} & 1 & 1 & 1 & & 1 \\ & 0 & 1 & 1 & 1 & 0 \\ + & 1 & 0 & 0 & 1 & 0 \\ \hline & 0 & 0 & 0 & 0 & 1 \end{array}$$

There is again an extra carry digit. Since we do not need to compute any more digits of the encryption stream, the extra carry digit from this addition is simply discarded.

`HI` corresponds to the binary message stream 0011101000. The encryption stream found through the additions with carrying between blocks of five above is 0111000001. Encryption works by adding these streams modulo 2:

$$\begin{array}{r|l} \texttt{H} & \texttt{I} \\ 00111 & 01000 \\ \oplus 01110 & 00001 \\ \hline 01001 & 01001 \\ \texttt{J} & \texttt{J} \end{array}$$

Note that in spite of the fact that BabyCSS uses addition with carrying in blocks of five to produce its encryption stream, the resulting stream is still added to the message modulo 2 for encryption. Like other stream ciphers, encryption and decryption are still exactly the same operation. The only difference is in generating the keystream. Adding the encryption stream to the encrypted message recovers the plaintext, since addition and subtraction are the same modulo 2:

$$\begin{array}{r|l} \texttt{J} & \texttt{J} \\ 01001 & 01001 \\ \oplus 01110 & 00001 \\ \hline 00111 & 01000 \\ \texttt{H} & \texttt{I} \end{array}$$

6.6. Breaking BabyCSS

Recall that the LFSRsum system can be completely and quickly broken by a known-plaintext attack by solving a system of linear equations. In this section, we see that BabyCSS does not suffer the same weakness. Our starting point is the same as before. Just like the LFSRsum cipher, the output tables of the LFSR-3 and LFSR-5 registers under a key $k_1k_2k_3k_4k_5k_6$ are

b_3	b_2	b_1
k_1	k_2	1
$1+k_2$	k_1	k_2
k_2+k_1	$1+k_2$	k_1
k_1+1+k_2	k_2+k_1	$1+k_2$
$1+k_1$	k_1+1+k_2	k_1+k_2
1	$1+k_1$	k_1+1+k_2
$\vdots$	$\vdots$	$\vdots$

c_5	c_4	c_3	c_2	c_1
k_3	k_4	k_5	k_6	1
$1+k_6+k_4$	k_3	k_4	k_5	k_6
$k_6+k_5+k_3$	$1+k_6+k_4$	k_3	k_4	k_5
$1+k_5+k_6$	$k_6+k_5+k_3$	$1+k_6+k_4$	k_3	k_4
$k_4+k_5+k_6$	$1+k_5+k_6$	$k_6+k_5+k_3$	$1+k_6+k_4$	k_3
$k_3+k_4+k_5$	$k_4+k_5+k_6$	$1+k_5+k_6$	$k_6+k_5+k_3$	$1+k_6+k_4$
$\vdots$	$\vdots$	$\vdots$	$\vdots$	$\vdots$

Suppose we know from a known-plaintext attack that the first five bits of a BabyCSS encryption stream is 01000 (for example, we have an intercepted ciphertext, and we know what the first letter of the message is). What congruences can we form from this information? In the arguments below, remember that we write numbers from the two streams in blocks of five but add in reverse order. Thus, the first addition is from the fifth bit of each of the two streams. We don't have to worry about any carries here; however, after we add these two numbers there might be a carry that must be brought forward.

We know that the fifth bit of the stream is the sum of the fifth output bits from each of the registers: this gives the congruence

$$k_1 + k_2 + k_3 \equiv 0 \pmod{2}.$$

So far so good. But now it gets tricky. Remember we're adding the two streams in blocks of five. The fourth bit of the output stream is the sum of the fourth output bit from each of the registers, *plus* any carry bit from the sum of the fifth digits. When is there a carry? The only way to have a carry in binary addition is if both digits are 1. Note that the *product* of two binary digits is 1 if and only if both digits are 1. In other words, the carry from adding two digits is the same as the product of the two digits!

This simple observation allows us to write down exact equations for all the congruences. We can represent the value of the carry digit (1 or 0) from the sum of the fifth bits as the product of those bits: $(k_1 + k_2) \cdot k_3$. The fourth bit of the encryption stream is thus given as the sum of this carry bit with the sum of the fourth output bits from each of the registers, giving the congruence

$$(k_1 + k_2) \cdot k_3 + 1 + k_1 + k_4 \equiv 0 \pmod{2}.$$

This congruence is *not linear*, since it involves multiplication of the variables. Subsequent congruences become more and more complicated. The fact that the congruences can't be represented with addition alone means that the congruences can't be efficiently solved through Gaussian elimination; BabyCSS does not suffer the same weakness as LFSRsum.

While we are heartened that BabyCSS is resistant to an attack using Gaussian elimination, this does not mean that it is secure. Perhaps there is another deficiency that can be exploited? The LFSRsum cipher could be completely broken by a known-plaintext attack, in the sense that an attacker who knows some corresponding plaintext and ciphertext (and thus can figure out some of the encryption stream) can quickly deduce the key without any guesswork, allowing her to decrypt the rest of the message. We've just seen that the BabyCSS cipher is not *as* susceptible to a known-plaintext attack. The attacker must solve nonlinear congruences to find the key used to produce a given encryption stream. Nonlinear problems are typically much harder and more time-intensive than linear ones, which suggests the possibility that there is not an efficient way to solve the resulting nonlinear equations.

Nevertheless, a known-plaintext attack is still quite effective against the BabyCSS cipher. The BabyCSS cipher has $2^6 = 64$ possible keys, thus a brute force attack on the cipher would require at most 64 guesses to break the cipher. (While 64 is a small number, in practice we would use a long key of length n, which would result in an exponentially large number of possible keys, namely 2^n.) It turns out that a known-plaintext attack can break the cipher using at most four guesses! After showing how to do this, we'll put 4 and 64 in perspective. As you read on, try and figure out how these are related so that if we replace 64 with n, you would know what to replace 4 with.

To see how this attack works, suppose we have intercepted the message

`FPSM☹ U`

which was encrypted with the BabyBlock cipher using an unknown key. We know the message contains the name of a double agent, and believe it begins with the letters `MR`. We find the 5-bit values for F, P, M, and R in Table 6.1. Since BabyCSS encrypts binary message streams by adding them to the BabyCSS encryption stream modulo 2, we can find the beginning of the BabyCSS encryption stream by computing

$$\begin{array}{rl} 0010101111 & (\texttt{FP}) \\ \oplus 0110010001 & (\texttt{MR}) \\ \hline 0100111110 & \end{array}$$

It's worth noting that, even though we're using BabyCSS, there is no carrying. Carrying is only used to create the pseudo-random keystream; it is not used for encrypting or decrypting. If you're a little confused that we're adding `MR` to `FP` instead of subtracting `MR` from `FP` to find the keystream, remember that addition and subtraction are the same in binary without carrying. For example, $1-1$ and $1+1$ are both 0 modulo 2. As it will be very important later, we remind the reader that *there is carrying between blocks, so all borrows must be paid for!*

We have already seen that knowing part of the encryption stream is not enough to work backwards to find the seeds of the registers by solving a system of linear congruences. This is because carrying makes the congruences nonlinear. Imagine, however, that someone tells us that the LFSR-3 register used for this BabyCSS encryption was seeded with 001. In other words, they tell us that the first two bits of the BabyCSS key are 00. Never mind for the moment how they might know this; we'll return to this later.

In that case, we now can compute the output table of the LFSR-3 register:

b_3	b_2	b_1
0	0	1
1	0	0
0	1	0
1	0	1
1	1	0
1	1	1
0	1	1
0	0	1
1	0	0
0	1	0

Let's recap what we know. The BabyCSS encryption stream begins with 0100111110, while the LFSR-3 output begins with 1001011100. From these two pieces of information, can we figure out the LFSR-5 output stream? Yes! Since the BabyCSS encryption stream is formed by adding the LFSR-3 and LFSR-5 output streams (in blocks of 5 with carrying), the LFSR-5 output stream can be found by subtracting the LFSR-3 output stream from the BabyCSS stream (in blocks of 5 with borrowing). We arrange the two streams into blocks of 5:

$$\begin{array}{ccccc c ccccc} 0 & 1 & 0 & 0 & 1 & \quad & 1 & 1 & 1 & 1 & 0 \\ 1 & 0 & 0 & 1 & 0 & \quad & 1 & 1 & 1 & 0 & 0 \\ \hline \end{array}$$

The top row is the deduced encryption stream, while the second row has the values from the b_1 register. We now subtract the blocks in order to find the values of the LFSR-5 output stream. We start with the left-most block of five. Remember that when we add or subtract two blocks, we start with the digits furthest to the right and move leftwards.

When subtracting, we'll see that we frequently need to borrow. This is very similar to what happens in decimal subtraction. For example, if we had 32 minus 9, we borrow 1 from the 3 and replace the 2 with 12; here

12 is really $1 \cdot 10 + 2$. The process is similar in binary. Note that while $10 - 1$ is 9 base 10, $10 - 1$ is 1 base 2 (if 10 represents a number in binary, then it's $1 \cdot 2 + 0$, or 2 in decimal). The arithmetic becomes a bit confusing. We'll write the answer first, and then slowly go through and explain what everything means.

$$\begin{array}{r} 0\;1\;\;\;\;\; \\ {}^{1}0\;\not{1}\;\not{0}{}^{1}0\;1 \\ -\,1\;0\;0\;1\;0 \\ \hline 1\;0\;1\;1\;1 \end{array}$$

Note that there was an "extra borrow" from the left-most digit of the problem; *this is borrowed from the right-most digit of the next block of* 5.

Let's discuss what we just did above. We're subtracting the binary number 10010 (which is $1 \cdot 2^4 + 1 \cdot 2$ or 18 base 10) from the binary number 01001 (which is $1 \cdot 2^3 + 1$ or 9 base 10). We can't do this subtraction as 9 is less than 18. We therefore need to "borrow", and instead of 01001 we have 101001 (or $1 \cdot 2^5 + 1 \cdot 2^3 + 1$, which equals 41 base 10). The decimal subtraction is $41 - 18 = 23$, and 23 in binary is our claimed answer of 10111 (that's just $1 \cdot 2^4 + 1 \cdot 2^2 + 1 \cdot 2 + 1$ or 23 in decimal).

What's going on is we really have the subtraction problem

$$\begin{array}{cccccc} 1 & 0 & 1 & 0 & 0 & 1 \\ - & 1 & 0 & 0 & 1 & 0 \\ \hline \end{array}$$

where we must remember we borrowed. We're left with what is almost a standard subtraction problem; the only difference is that we're working base 2 rather than base 10. One way to solve is to convert our two numbers to decimal, subtract, and then convert back to binary. Another approach is to modify the standard subtraction method from decimal to binary. We show how to do that now, but if you find this confusing don't worry. Just do $41 - 18 = 23$ in decimal, and then write 23 in binary to get the answer. Of course, if you do convert to decimal and subtract, you're doing essentially the same thing; the only difference is you're doing subtraction base 10 and not base 2. So, as you read on, remember that you already know how to do this. We're just recasting the subtraction lessons from your childhood into base 2.

We start at the right and move out. We can do 1 minus 0, that's just 1.

$$\begin{array}{cccccc} 1 & 0 & 1 & 0 & 0 & 1 \\ - & 1 & 0 & 0 & 1 & 0 \\ \hline & & & & & 1 \end{array}$$

Now we look at the second digits, and we find $0 - 1$. We need to borrow. Unfortunately, the third digit in our top number is 0, so we cannot borrow from that. We must move to the fourth digit, which is 1. We borrow from here. The fourth digit now becomes 0 and we have 10 in the third digit. Remember that 10 in binary is $1 \cdot 2$ or two. We now have 10 or 1 plus 1 in the third digit. We move one of these 1's to the second digit, giving us 10 (i.e., 1 plus 1) there. We can now do the subtraction in the second digits. We have 10 minus 1 in binary, which is just 1. We can also immediately do the subtraction for the third and fourth digits.

$$\begin{array}{cccccc} 1 & 0 & 0 & 1 & 10 & 1 \\ - & 1 & 0 & 0 & 1 & 0 \\ \hline & & 0 & 1 & 1 & 1 \end{array}$$

We're almost done. We can't do the subtraction in the fifth digit, as 0 is less than 1. We therefore borrow from the sixth digit. We carry that 1 over, which becomes 10. We get

$$\begin{array}{cccccc} 0 & 10 & 0 & 1 & 10 & 1 \\ - & 1 & 0 & 0 & 1 & 0 \\ \hline & 1 & 0 & 1 & 1 & 1 \end{array}$$

as 10 minus 1 is 1 in binary (since 10 is 1 plus 1). Thus, our answer is 10111 in binary, exactly as we had before.

Remember that, as remarked above, *there is carrying between blocks.* Thus the second subtraction problem depends on the first, as we had to borrow 1. If we didn't have to worry about that borrow, we would have had the subtraction problem

$$\begin{array}{cccccc} & 1 & 1 & 1 & 1 & 0 \\ - & 1 & 1 & 1 & 0 & 0 \\ \hline \end{array}$$

Unfortunately, however, we do need to pay for the borrow. We borrowed 1 from the first digit. Since the first digit is 0, we need to keep moving to the left until we find a nonzero digit. That happens in the second digit. We replace the 1 there as 0 and carry 1 over to the first digit. As we're carrying 1, it becomes 10 (or $1 + 1$ which is two) when we move it over, and the our subtraction is

$$\begin{array}{cccccc} & 1 & 1 & 1 & 0 & 10 \\ - & 1 & 1 & 1 & 0 & 0 \\ \hline \end{array}$$

We're in luck, as we can do all the subtraction here without carrying. If we weren't so fortunate, we would just attack it as we did in the previous problem (either doing a binary version of the subtraction methods we know, or converting the numbers to decimal, subtracting, and then converting back). We find

$$\begin{array}{cccccc} & 1 & 1 & 1 & 0 & 10 \\ - & 1 & 1 & 1 & 0 & 0 \\ \hline & 0 & 0 & 0 & 0 & 1 \end{array}$$

Thus we have found that the LFSR-5 output stream begins with 1011100001. We can thus fill in the right-hand column of the LFSR-5 output table,

b_5	b_4	b_3	b_2	b_1
1	1	1	0	1
0	1	1	1	0
0	0	1	1	1
0	0	0	1	1
0	0	0	0	1
1	0	0	0	0
	1	0	0	0
		1	0	0
			1	0
				1

which tells us that the seed for the LFSR-5 register was 11101! Thus the BabyCSS key used was 001110.

As seen above, we can break the BabyCSS cipher with a known-plaintext attack *if* we know the seed of the LFSR-3 cipher. What if we don't know the seed? Make a guess! There are only four possible seeds for the LFSR-3 cipher (001, 011, 101, 111). By trying each one, we can see which one gives a sensible decryption. Further, a lot of incorrect guesses can be quickly discarded based on whether or not the resulting LFSR-5 output table is consistent with the transition rules.

6.7. BabyBlock

A **block cipher** is one in which the message is broken up into blocks before encryption, and identical blocks are encrypted to the same expression. For example, the Hill cipher from §2.9 is a block cipher where the blocks have length 2. Modern block ciphers are used in situations where the highest levels of security are required. The only publicly known cipher which is approved for encryption of "Top Secret" information at the government level is the block cipher **Advanced Encryption Standard (AES)**. Block ciphers are also used for secure internet transactions. We will study a simplified example of a block cipher to see how this type of cipher achieves security. This is meant to be just a brief discussion of the possibilities and issues. The algebra below is a bit more involved, and the details can be safely skimmed or skipped.

The **BabyBlock cipher** operates on blocks of size 4, and has a key consisting of four bits. Consider the situation where we are encrypting the message `HELP` with the key 1110. As with stream ciphers, the actual encryption/decryption is done in 0's and 1's, so we begin by converting the message to a binary string using our character table: `H` corresponds to 00111, `E` to 00100, `L` to 01011, and `P` is 01111. We are therefore encrypting the message

0011100100010110111 1

Since the BabyBlock cipher operates on blocks of size 4, we begin by breaking the message into blocks of this size, adding extra bits to the end of the message if needed:

0011 1001 0001 0110 1111

As the BabyBlock cipher encrypts the message one block at a time, we begin with the first block. Remember our key is 1110.

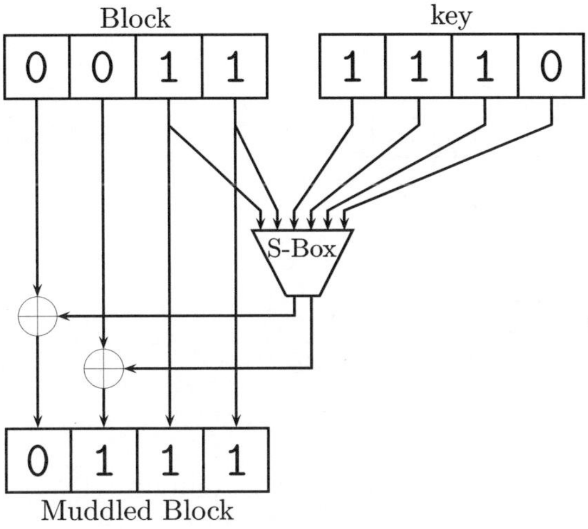

S-box

key	last bits of block: 00	01	10	11
0000	10	10	00	11
0001	00	11	10	00
0010	00	10	01	00
0011	01	00	11	11
0100	11	00	10	01
0101	10	11	01	10
0110	11	01	11	10
0111	10	01	00	01
1000	00	01	11	10
1001	01	00	01	11
1010	11	11	00	00
1011	11	10	01	01
1100	01	11	10	10
1101	00	01	11	00
1110	10	00	00	01
1111	01	10	10	11

FIGURE 6.2. The muddling of block `0011` when the key is `1110`. The $\oplus$'s mean addition modulo 2. The S-box is a table-based substitution.

We use the key to **muddle** the blocks using the following procedure, which is depicted in Figure 6.2. In summarizing the procedure below, we describe what happens to the block `0011` with a key of `1110` and an S-box table as in Figure 6.2.

Muddling procedure: To muddle a four-digit binary block with a given four digit binary key, given an S-box table, perform the following steps.

(1) Take the last two digits of the block and look up the corresponding entry in the row of the S-box table corresponding to the given key. In our example, this generates the two-digit string `01`.

(2) Add (without carrying) the generated two-digit string to the first two digits of our initial block. The resulting sum is the first two digits of the muddled string, and the last two digits of the muddled string are the last two digits of the initial block. In our example, adding `01` to `00` gives `00`, so the muddled string is `0111`.

Note the last two bits of the block are not changed by the muddling operation, but are used with the key to produce a bit pair using an **S-box**. The S-box is just a substitution table.

Notice that the last two bits are never changed by the muddling operation. We'll return to this in a moment. For now, the obvious question is, How do we unmuddle a block? By muddling it! Muddling a muddled block returns the original block (see Figure 6.3). Why is that? The S-Box gives the same output as before because the last two bits haven't changed and the key is the same. Adding modulo 2 is the same as subtracting modulo 2, so when we add the S-Box output to the muddled bits, it changes them back to the original bits.

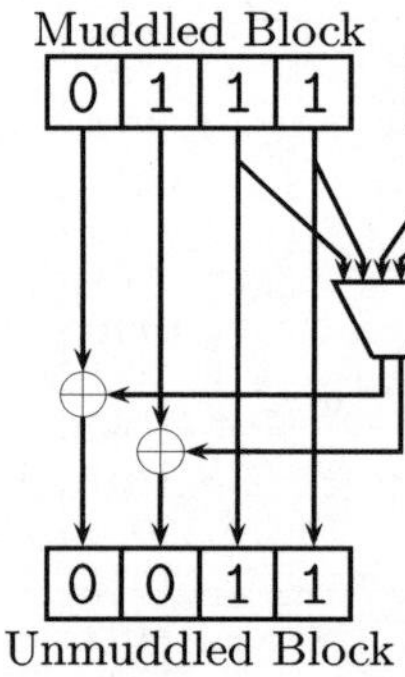

FIGURE 6.3. Unmuddling is the same as muddling!

If we're going to use the muddling operation we've defined to encrypt messages, we have to deal with the fact that the last two bits never change. BabyBlock addresses this problem by muddling a block repeatedly, but swapping the first and last pairs of digits between each muddle. For example, we've seen that the plaintext 0011 is muddled to 1011 with a key of 1110. Now we switch the first pair (10) and last pair (11) of digits in the string 1011 to get 1110. We now muddle this string, but to increase security we use a different key. After each muddling operation, we shift the digits of the key to the left. By having an explicit procedure, once we specify the initial key all subsequent keys are readily determined. BabyBlock does this three times; we depict this in Figure 6.4. The ciphertext for the first block of our plaintext turns out to be 1110.

We've already defined the encryption side of BabyBlock: muddle, swap first and last pairs, muddle, swap first and last pairs, and muddle. Notice that each of these operations reverses itself. Muddling reverses muddling, and swapping pairs reverses swapping pairs. Because of this, encryption and decryption for BabyBlock are exactly the same, except you have to use the keys in reverse order: first use the key shifted to the left by 2, then use the key shifted to the left by 1, then use the original key.

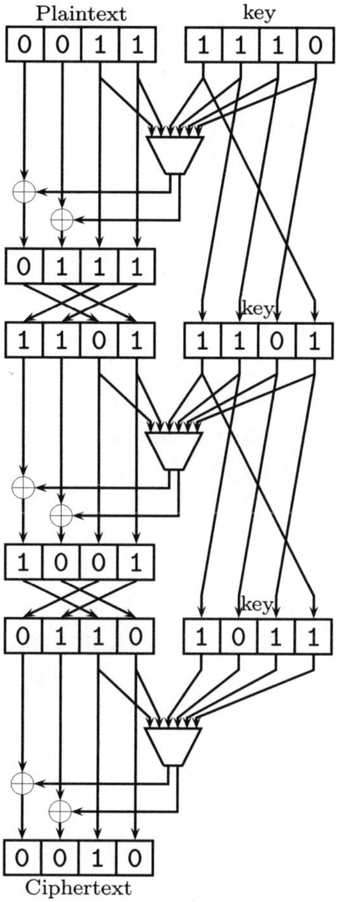

FIGURE 6.4. BabyBlock

6.8. Security of BabyBlock

While we know how BabyBlock's encryption works, we haven't said anything about why it was designed this way. As modern ciphers are expected to be resistant to known-plaintext attacks, let's see how BabyBlock fares in this arena. Imagine we are given a string 1110 and are told that, when muddled with a certain key, it turns into 0110. Can we figure out what the key is?

We know that the input to the S-Box was the (unknown) key and the bits 10, since those are the last two bits of the original string. Since the output of the S-box gives 01 (the first two bits of the muddled string) when added to 11 (the first two bits of the original string), we know that the output of the S-Box is 10. Is this enough information to figure out the key? Looking at the S-Box table, we see that there are four possible keys which could give rise to the output 10 when the last two bits of the block being muddled are 10. The possibilities are 0001, 0100, 1100 and 1111. Even though we knew a plaintext and its corresponding ciphertext, we haven't been able to figure out the key. This is because the S-box is not *invertible.* This means that, even if we know the output and one of the inputs, we still may not be able to figure out the other input. In spite of the fact that the S-Box is not invertible, it is still possible to easily decrypt messages. This is because messages are not encrypted directly with the S-Box; instead, the S-Box is only used to choose a pair of bits which will be added to plaintext bits to encrypt them. Since addition can be reversed, the encryption can be reversed, and it is possible to decrypt messages so long as we know the key.

We've seen that knowing a plaintext isn't enough to figure out the key used for the muddling operation, but it *did* narrow the choices down to just four keys. This is a good reason not to encrypt messages just by muddling them! (Of course, an even better reason is that half the bits remain unencrypted!) When the muddling operation is repeated, however, there's no longer a quick way of narrowing down to four keys. The goal for a well-designed block cipher is that there isn't a way to break the cipher much faster than just trying all possible keys. We emphasize, however, that BabyBlock is *not* a well-designed block cipher. It's just a simplified example showing the general structure of such schemes.

There are a number of advanced techniques developed to attack block ciphers. Some poorly designed block ciphers have turned out to be very insecure, but those in most widespread use (such as AES) appear to be essentially as secure as the size of their keyspace (guessing the key is basically your best option). Most of the advanced techniques used to attack block ciphers are beyond the scope of this course, but in the next section we briefly examine one type of attack which has had important ramifications for the use of block ciphers.

6.9. Meet-in-the-Middle Attacks

The BabyBlock cipher has a 4-bit key, meaning that there are a total of just 16 possible keys. If we want more security, we might try to encrypt a message *twice* with the BabyBlock cipher, using a different key for each step (see Figure 6.5). We'll call this procedure **DoubleBabyBlock**.

Since two 4-bit keys are used for DoubleBabyBlock, we can view DoubleBabyBlock as a cipher with a single 8-bit key which is just the concatenation of the two (if key1 is 1100 and key2 is 1010, the single 8-bit key is 11001010).

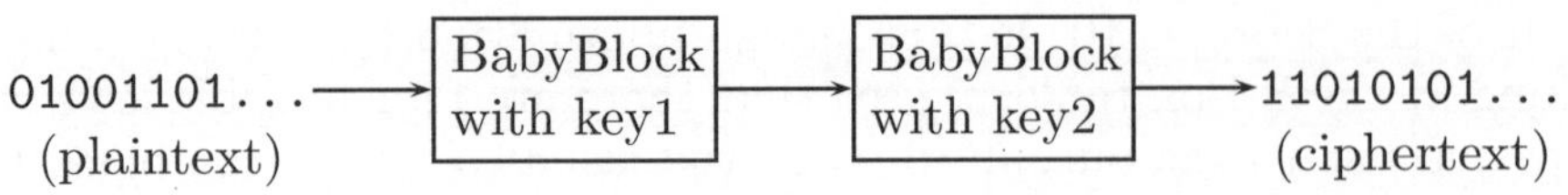

FIGURE 6.5. **DoubleBabyBlock encryption with key** key1key2.

There are a total of 256 possible 8-bit keys, which means that a brute force attack on DoubleBabyBlock would take up to 256 guesses for the key (and 128 guesses on average).

It turns out, however, that DoubleBabyBlock does *not* provide much more security than BabyBlock, as there is a clever way to avoid having to make 256 guesses. The **meet-in-the-middle** attack is a known-plaintext attack. Assume we know that the plaintext HELP became the ciphertext REHN under encryption by the DoubleBabyBlock cipher with unknown key (after conversion to and from binary). The idea of the meet-in-the-middle attack is simple. If a plaintext is encrypted first with a key *key1* and then with a key *key2* to produce a ciphertext, then *decrypting* that ciphertext with *key2* must produce the same result as *encrypting* the original plaintext with *key1*. These operations "meet in the middle", at the halfway point between the original plaintext and the doubly encrypted ciphertext. It turns out that, by checking which pairs of keys meet up in the middle like this, we can break the cipher with far fewer guesses than would be required for a brute force attack.

To illustrate the meet-in-the-middle attack, assume we know that the first letter of the a four-letter plaintext message is H, and the four-letter ciphertext is CYU!. Converting to binary gives that the first block of four of the plaintext is 0011, the first block of the ciphertext is 0001. The brute force attack would now try all 256 possible key-pairs to see which one encrypts 0011 to 0001. Instead, we will try all 16 possible 4-bit keys to *encrypt* the plaintext 0011 using the regular BabyBlock algorithm, and all 16 possible 4-bit keys to *decrypt* 0001 using the regular algorithm, and see which pair of guesses gives the same thing.

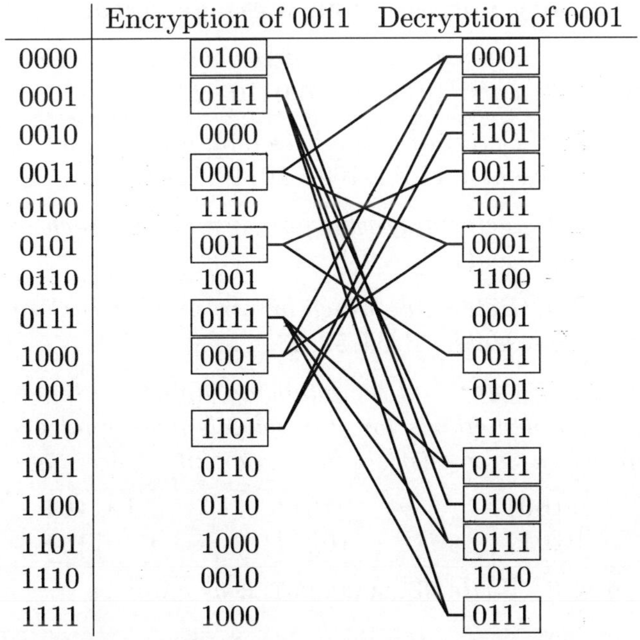

For example, encrypting 0011 with 0000 gives the same result as decrypting 0001 with the key 1100. This means that one possible key combination that would encrypt 0011 to 0001 is 0000 followed by 1100, corresponding to the 8-bit DoubleBabyBlock key 00001100. Altogether, there are 17 8-bit keys formed by matched pairs in the above table; the 8-bit key which will decrypt the rest of the message `CYU!` must be among them. Note that we calculated 32 BabyBlock encryptions/decryptions to make the table, and it will take at most 17 more DoubleBabyBlock decryptions to find the true key and decrypt the message. Since a BabyBlock encryption/decryption takes half as long as a DoubleBabyBlock decryption, we can break the cipher with the equivalent of at most 33 DoubleBabyBlock decryptions. Although this is quite tedious to do by hand, it is much better than having to try all 256 possible decryptions!

6.10. Summary

We discussed several new encryption schemes. Unlike many of the classical methods, here we convert our message to binary numbers. For many of these schemes, the encryption is essentially a Vigenère cipher: we add our keystream letter by letter. The new twist is in the generation of the keystream, which often comes from creating pseudo-random strings of 0's and 1's. The challenge is in creating a pseudo-random stream that is resistant to attack.

6.11. Problems

• From §6.1: Binary numbers and message streams

EXERCISE 6.11.1. *Determine which decimal number each of the following binary numbers represents:* 1011, 101, 100110, 111.

EXERCISE 6.11.2. *Determine which decimal number each of the following binary numbers represents:* 11011, 1001, 10110, 1111.

EXERCISE 6.11.3. *Determine which binary number each of the following decimal numbers represents:* 14, 29, 82, 4.

EXERCISE 6.11.4. *Determine which binary number each of the following decimal numbers represents:* 33, 54, 59, 23.

EXERCISE 6.11.5. *Carry out the following binary arithmetic problems by adapting the standard methods for the equivalent decimal arithmetic problems. (You can check your answers by converting to decimal numbers).*

$$\begin{array}{r} 11001 \\ +01011 \\ \hline \end{array} \qquad \begin{array}{r} 101011 \\ -\ \ 10011 \\ \hline \end{array} \qquad \begin{array}{r} 101101 \\ \times\ \ \ 1101 \\ \hline \end{array}$$

EXERCISE 6.11.6. *Write* `HELLOCOMPUTER` *as a binary stream using Table* 6.1.

EXERCISE 6.11.7. *Convert the following to a string of letters using Table* 6.1:

001110010001011010110111010010100111010000011001000110110011

EXERCISE 6.11.8. *Let x be a positive integer. Show that there cannot be two distinct binary representations of x.*

EXERCISE 6.11.9. *Instead of the ordinary binary representation of numbers, consider* signed *binary representations. The only difference is that we allow positive and negative summands; we denote a negative summand with a negative subscript. Thus, in addition to terms such as* 1011 *(which is* $1 \cdot 2^3 + 0 \cdot 2^2 + 1 \cdot 2^1 + 1 \cdot 2^0 = 11$*), we could also have* 101_-1_- *(which is* $1 \cdot 2^3 + 0 \cdot 2^2 - 1 \cdot 2^1 - 1 \cdot 2^0 = 5$. *Does every number have a unique signed binary representation? Why or why not?*

EXERCISE 6.11.10. *Building on the previous problem, consider signed ternary (base* 3*) representations. Now the possible digits are* $0, \pm 1, \pm 2$. *Does every number have a unique signed ternary representation? Why or why not?*

EXERCISE 6.11.11. *Consider the signed ternary representation from Exercise* 6.11.10, *but now additionally assume each summand can only be* 0 *or* ± 1*; in other words, we no longer allow* ± 2. *Does every number have a unique restricted signed ternary representation? Why or why not?*

EXERCISE 6.11.12. *Multiply the following binary numbers:* 1011 *and* 10101. *You should check your answer by converting to decimal numbers. Notice how easy it is to multiply in binary.*

EXERCISE 6.11.13. *Research the history of ASCII and UTF-8, and write a short note on these.*

EXERCISE 6.11.14. *The following is a fun example of binary representations. It's possible to guess someone's birthday in ten questions, provided that if you guess incorrectly the person says if you are earlier or later in the year. Your first guess should be July* 1st*; if it is earlier your next guess should be April* 1st*, while if it is later go with October* 1st. *Figure out what the next guesses should be, and prove why it works. This is known as the Method of Binary Search, and is used in a variety of problems.*

EXERCISE 6.11.15. *Look up the Method of Binary Search, and write a short report about it.*

- **From §6.2: Linear Feedback Shift Registers**

EXERCISE 6.11.16. *Consider the LFSR given by* $b_4 \leftarrow b_3' + b_1'$ *and seeded with* 1011. *Find the first fifteen terms in the sequence.*

EXERCISE 6.11.17. *Encrypt the message* `HI` *using a stream cipher using the LFSR given by* $b_4 \leftarrow b_3' + b_1'$ *and seeded with* 1011.

EXERCISE 6.11.18. *Encrypt the message* BLUE *using a stream cipher using the LFSR given by* $b_5 \leftarrow b'_4 + b'_1$ *and seeded with* 11011.

EXERCISE 6.11.19. *Decrypt the message* ?SY, *which was encrypted using a stream cipher with a keystream produced by the LFSR with formula* $b4 \leftarrow b'_3 + b'_1$ *and seed* 0110.

EXERCISE 6.11.20. *Consider a 3-bit LFSR. Show the all zero state* 000 *can never arise, unless the register eventually outputs all* 0*s and so has a period of repetition of* 1. *This means that, for a 3-bit register, the longest possible period of repetition is* 7.

EXERCISE 6.11.21. *Consider an* n*-bit LFSR. Assume two states, say the* i^{th} *and* j^{th} *states, are identical. Why must the pattern repeat? How is the length of the period of the pattern related to* i *and* j*?*

EXERCISE 6.11.22. *What is the longest possible period of repetition of a* 4*-bit LFSR register? Of a* 5*-bit LFSR register? Of an* n*-bit LFSR register?*

EXERCISE 6.11.23. *Show that adding and subtracting in binary are the same when there is no carrying.*

EXERCISE 6.11.24. *Can you have an LFSR with just one variable (i.e., only* b_1*)? If yes describe this system.*

EXERCISE 6.11.25. *The LFSRs work by adding the keystream to the message digit by digit, modulo* 2. *What would happen if instead of adding we multiplied? Would this still be a valid encryption scheme?*

• From §6.3: Known-Plaintext Attack on LFSR Stream Ciphers

EXERCISE 6.11.26. *The message* .ZYHM.! *was encrypted with a LFSR stream cipher which uses the formula* $b_4 \leftarrow b'_4 + b'_1$. *Knowing that the message begins with the letters* Mr, *determine the key and decrypt the message.*

EXERCISE 6.11.27. *The message* ☹YWTCQ! *was encrypted with a LFSR stream cipher which uses the formula* $b_4 \leftarrow b'_4 + b'_2 + b'_1$. *Knowing that the message begins with the letters* Mr, *determine the key and decrypt the message.*

EXERCISE 6.11.28. *The two previous exercises involve LFSR with formulas involving four variables. In such a setting, how many of the first few letters of the message must we know in order to be able to crack it?*

EXERCISE 6.11.29. *Consider a LFSR with a formula involving* r *variables. Prove that after some point the output* must *repeat, and bound how long we must wait to see this repetition as a function of* r.

• From §6.4: LFSRsum

EXERCISE 6.11.30. *Encrypt the message* HELLO *using the LFSRsum cipher, using* 001100 *as a key.*

EXERCISE 6.11.31. *Encrypt the message* ROSEBUD *using the LFSRsum cipher, using* 010010 *as a key.*

EXERCISE 6.11.32. *Decrypt the message* DTQRZ_Y, *which was encrypted with the LFSRsum cipher using the key* 001110.

EXERCISE 6.11.33. *Decrypt the message* Y_ZRQTD, *which was encrypted with the LFSRsum cipher using the key* 001110. *Do you expect it to be the message from the previous exercise but written in reverse order? Why or why not.*

EXERCISE 6.11.34. *The message* FCJUWRMX *was encrypted with the LFSRsum cipher.*

(a) The first two letters of the plaintext are Mr. *Use this information to determine the first ten digits of the binary stream used for encryption.*
(b) Using the binary digits found in part (a), mount a known-plaintext attack on the LFSRsum cipher to determine the key used for encryption, and decrypt the rest of the message. Hint: *The equations you need to solve to break the cipher are exactly the same as those in line* (6.1), *except for the 0's and 1's on the right-hand side of each. This means that the same combinations of congruences can be summed to solve for the separate variables!*

EXERCISE 6.11.35. *In the section we considered combining an LFSR-*3 *with an LFSR-*5; *would we have greater or less security if instead we combined two LFSR-*5*s?*

EXERCISE 6.11.36. *If we have a system of n equations in n unknowns, Gaussian elimination takes on the order of n^3 operations. What this means is that there's some constant C (which turns out to be less than* 1) *such that we need at most Cn^3 additions, subtractions, multiplications, and divisions of binary numbers to find the n variables. This should be contrasted to the number of possible n-tuples, which is 2^n. Compare the number of operations if $n = 100$ and if $n = 1000$. The point of this exercise is to show that while it is impractical to try to break LFSRsum by trying all possible keystreams, we can easily beat it by Gaussian elimination.*

• **From §6.5: BabyCSS**

EXERCISE 6.11.37. *Encrypt the message* BLUE *with the BabyCSS cipher using the key* 110011.

EXERCISE 6.11.38. *After reading about how vulnerable LFSRsum is to Gaussian elimination attacks, a classmate suggests the following nonlinearity. Instead of adding the two streams, let's multiply the two streams. Is this a good idea? Why or why not.*

EXERCISE 6.11.39. *Imagine you have two n digit binary numbers and you are going to add them with carrying. About how many carries do you* *expect? What if the numbers were in decimal notation? If you can, write a computer program to gather some data.*

EXERCISE 6.11.40. *BabyCSS has an LFSR-*3 *and an LFSR-*5; *would we have greater, equal or less security if instead we combined two LFSR-*5*s?*

• From §6.6: Breaking BabyCSS

EXERCISE 6.11.41. *You are trying to break a message which was encrypted with the BabyCSS cipher. The ciphertext is* DXMP, *and you know that the plaintext begins with* MR.

(a) What are the first seven digits of the encryption stream?
*(b) If you guess that the LFSR-*3 *register is seeded with* 111, *what does this mean the LFSR-*5 *output stream was?*
*(c) What was the LFSR-*5 *output stream seeded with?*
(d) What was the key?
(e) Decrypt the rest of the message.

EXERCISE 6.11.42. *In the known-plaintext example carried out in this section, we had ten bits of known plaintext (the ten bits corresponding to the known characters* MR*). Would the same attack work if we just knew eight of the plaintext bits? What about four of the plaintext bits? What is the minimum number of known bits necessary for the known-plaintext attack described in this section to work?*

EXERCISE 6.11.43. *Imagine we use BabyCSS with registers LFSR-m and LFSR-n, with $m < n$. The key has $(m-1)+(n-1)$ binary digits (remember the last digit of each must be a* 1*). There are 2^{m+n-2} possible keys. Show, using the methods of this section, that we can break the cipher by exploring at most 2^{m-1} guesses. This tells us how to compare the numbers* 64 *and* 4 *from the example in the chapter. We have $64 = 2^{(3-1)+(6-1)}$ and $4 = 2^{3-1}$. A consequence of this is that we can break BabyCSS by trying at most the square-root of the number of possible keys. While this is a tremendous savings, for large m and n it is not practical.*

• From §6.7: BabyBlock

EXERCISE 6.11.44. *The second* 4*-bit block from the plaintext* HELP *is* 1001. *Muddle this block using the key* 1110.

EXERCISE 6.11.45. *The third* 4*-bit block from the plaintext* HELP *is* 0001. *Muddle this block using the key* 1110.

EXERCISE 6.11.46. *The final two* 4*-bit blocks from the plaintext* HELP *are* 0110 *and* 1111. *Muddle the first with* 1100 *and the second with* 0011.

EXERCISE 6.11.47. *Decrypt the ciphertext* 0010 *with the key* 1110 *and check that you can recover the plaintext* 0011. *(Remember you need to start your decryption with the key rotated two positions to the left!)*

EXERCISE 6.11.48. *Building on the previous problem, are there any other keys which decrypt the ciphertext* 0010 *to* 0011?

EXERCISE 6.11.49. *Encrypt the next block from the plaintext message* `HELP` (1001) *with the key* 1110.

EXERCISE 6.11.50. *Read about AES, and write a short note describing it.*

- **From §6.8: Security of BabyBlock**

EXERCISE 6.11.51. *The S-Box is not invertible; give another example of two different inputs that are mapped to the same output.*

EXERCISE 6.11.52. *In the text we mentioned that if there is repeated muddling, then there's no longer a quick way of narrowing down to four keys. Generalize the example from the text by introducing another muddling, and investigate what happens.*

- **From §6.9: Meet-in-the-Middle Attacks**

EXERCISE 6.11.53. *Assume* 0110 *encrypts to* 0011. *Use a meet-in-the-middle attack to try and deduce the key.*

Chapter 7

Introduction to Public-Channel Cryptography

We've covered many encryption schemes, from ancient systems based on simple arithmetic to toy versions of modern block ciphers which are believed to be secure against all practical attacks. All of these systems have been **symmetric ciphers**, meaning that knowing the key used to encrypt a message allows one to decrypt the message. For example, in the Vigenère cipher to encrypt we repeatedly write the keystream under the message, adding letters modulo 26; to decrypt, we write the keystream under the received message and subtract. Indeed, at this point, this seems like an unavoidable phenomenon: how could we know how to encrypt a message to someone, but not now how to decrypt it? Interestingly, there are several different systems that allow people to securely encrypt messages that they themselves cannot decipher! The point of this chapter is to introduce some of the ideas and the mathematics behind these systems, leaving a detailed analysis of one of the most popular systems (RSA) to the next chapter.

At first, using symmetric ciphers appears fine. Modern block ciphers are indeed very secure: data encrypted with AES with a 256-bit key should be considered safe for the foreseeable future. We feel confident sending credit card information over insecure internet connections, without fear that the information could be decrypted and used by third parties. But there remains a fundamental problem. How can we agree with the merchant on a secret key to use for the transmission? Spies and their agencies have always been able to agree on keys for encryption when they meet "face to face", but we have never met Amazon.com or Target.com in private. If all of our communications have to take place over insecure internet transmission, we never have an opportunity to agree on a secret key to use for transmission without people listening in. Without such an agreement, all of the systems we developed are useless. Thus, we need to find a way to send a secret through insecure channels.

To emphasize the point, suppose Alice and Bob are standing on opposite sides of a room, and Charlie is standing in the middle. Alice and Bob have never met before, but Alice has a message she wants to get to Bob, and only Bob. Alice and Bob each have pen and paper to carry out any calculations they wish, but can only shout to each other across the room, and Charlie hears everything. This situation seems impossible to deal with from a cryptography standpoint. Since Charlie hears whatever Bob hears, can't Charlie just carry out whatever process Bob does to decrypt the message? If we're assuming that Alice and Bob have no shared information that Charlie doesn't also have, what will enable Alice to ensure that her message can only be understood by Bob?

Had you asked cryptographers in the 1960s if it was possible to communicate encrypted messages in such a setting, the consensus would likely have been "no". However, in the 1970s, in one of the most remarkable mathematical developments of the twentieth century, several different cryptographic schemes to effectively deal with this kind of "public-channel" communication were independently developed. We end this introduction by highlighting a few of the topics of this chapter, the issues they raise, and hints of how we solve them in the next chapter.

We begin our discussion of public-channel cryptography with a toy example of a public channel system: the perfect code. The perfect code cryptography system illustrates how it could—in principle—be possible to exchange messages through a public channel with no shared secret information. Unfortunately, just because something appears hard does not mean it actually is hard, and we'll see that this method has a fundamental security defect. We then move on to another system. While it too appears secure, it turns out that it also has an exploitable weakness. This weakness, though, appears to be fixable, and a related system (RSA) appears secure. So, even though the examples here fail in the end, they set the stage for the successes in the next chapter.

To develop a public-channel cryptography system, we essentially need two keys, one key to encrypt the confidential message and a second key to decrypt the message. The messages themselves may be in the form of letters or numbers, or transactions for your credit card, or a variety of other pieces of information that must be kept secret. In general the key used to encrypt the message is out in public (Alice or Bob can say it out loud in the room), and the key for decryption is not. The keys are generally numeric keys, and often the message is converted to numbers. In the perfect code cryptography system, the subject of the next section, we use only one key or secret; when we advance to RSA in Chapter 8, we'll have two, one for each player (i.e., one for the sender and one for the receiver).

7.1. The Perfect Code Cryptography System

The perfect code cryptography system is based on graphs.

> **Graphs:** A **graph** is a collection of points (called the **vertices** or **nodes**) and segments connecting pairs of vertices (called the **edges** or **links**). In this book, we do not allow edges to connect a vertex to itself (such edges are called **loops**), nor do we have multiple edges between the same pair of vertices. If two vertices are connected by an edge, we say the vertices are **adjacent**. The **degree** of a vertex is the number of vertices adjacent to it (or, equivalently, the number of edges emanating from it as we are allowing neither multiple edges nor loops).

For example, the graph in Figure 7.1 has nine vertices and 16 edges. We have labeled the vertices so that we can easily refer to them. For example, notice that the vertex H is adjacent to vertex D, but not to vertex C. Vertex A has degree 1 while vertex C has degree 4.

We give another presentation of the graph from Figure 7.1 in Figure 7.2, drawn a slightly different way. It is the same graph because vertices which were adjacent in the first graph are still adjacent, while vertices not adjacent in the first graph are not adjacent here either.

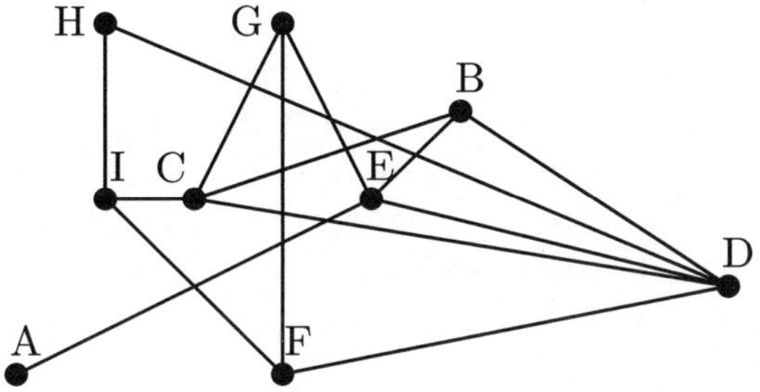

FIGURE 7.1. A graph with nine vertices and 16 edges.

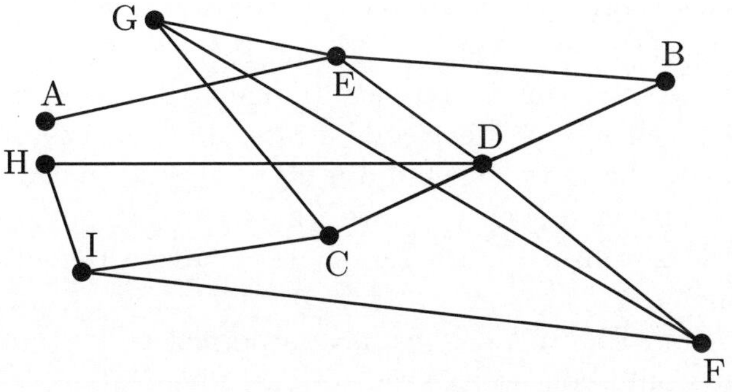

FIGURE 7.2. The graph from Figure 7.1, drawn differently.

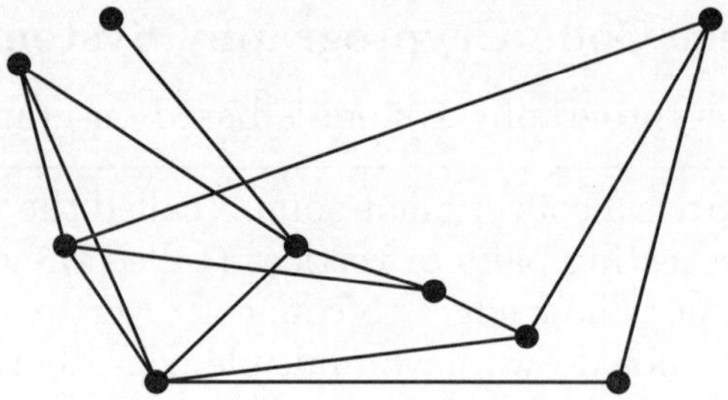

FIGURE 7.3. The same graph as the ones depicted in Figures 7.1 and 7.2, but drawn differently.

It can be surprisingly difficult to tell whether two graphs are actually the same. For example, the graph in Figure 7.3 is the same graph as the previous two already drawn (you are asked to show this in Exercise 7.8.2).

We now state the main ingredient for the cryptographic method of this section.

Perfect code: A **perfect code** of a graph is a set V_{PC} of vertices with the following two properties:

(1) None of the vertices in the perfect code V_{PC} are adjacent to each other (in other words, there are no edges connecting a pair of vertices in V_{PC}).
(2) Every vertex from the graph which is not in the perfect code V_{PC} is adjacent to *exactly one* vertex from V_{PC}.

In the graph from Figure 7.1, vertices E and I form a perfect code. They are not adjacent to each other, and every other vertex is adjacent to exactly one of them: vertices A, B, D, and G are adjacent to E, while C, F, and H are adjacent to I.

Graphs can have more than one perfect code, although it happens that there are no more for this particular graph. For example, the vertices E, F, and H, are all not adjacent to each other, and all other vertices are adjacent to at least one of these. However, some other vertices are adjacent to more than one of these (note that I is adjacent to both F and H), violating the requirement of exactly one. Thus, E, F, and H do not form a perfect code.

At first glance, it seems to be very difficult to find perfect codes in a graph, or to show that no perfect code exists. This is actually a promising development, as perhaps we can build a cryptography method around this graph concept. To get a sense of how difficult it can be to find a perfect code, look at the graph in Figure 7.4. Does it have a perfect code? If yes, can you find one?

The graph from Figure 7.4 does have a perfect code, but it is not easy to see it. What would you do if asked to locate it?

You might start with two nonadjacent vertices, then three, then four, etc. Try it!

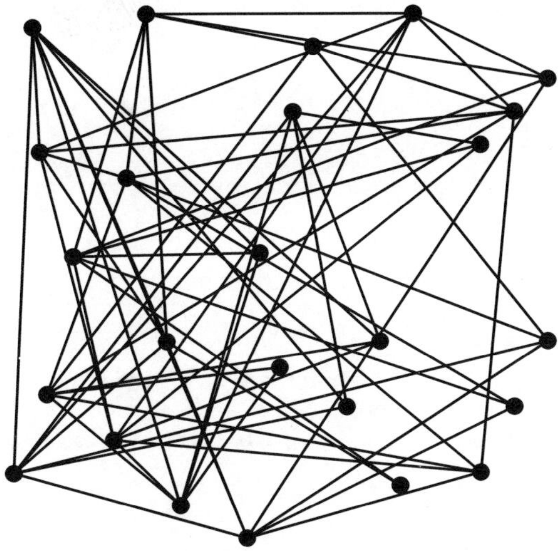

FIGURE 7.4. A graph with a hard-to-find perfect code.

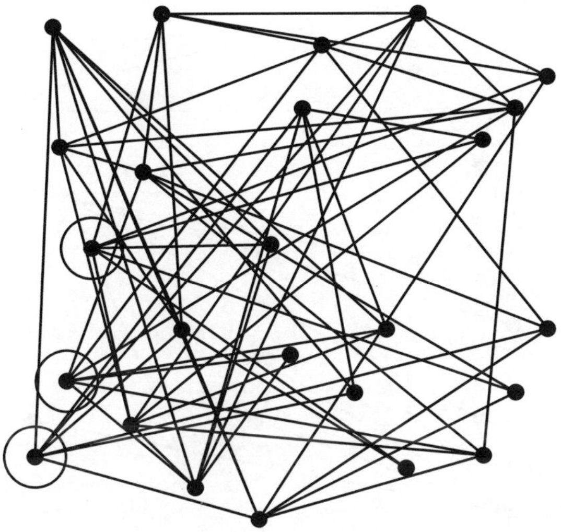

FIGURE 7.5. The graph from Figure 7.5 with the perfect code shown.

It seems you would essentially have to try all possible sets of vertices, checking each one against the perfect code conditions. This would take a long time! We quantify this approach in Exercise 7.8.4. We display the graph again in Figure 7.5 with the perfect code circled. You can check that these vertices are not adjacent to each other, and that every other vertex in the graph is adjacent to exactly one of them.

Even though it can be quite difficult to *find* a perfect code in a graph, it is quite easy to *make* a graph in which we know a perfect code exists. For example, suppose we begin with the set of vertices from Figure 7.6.

FIGURE 7.6. A set of vertices.

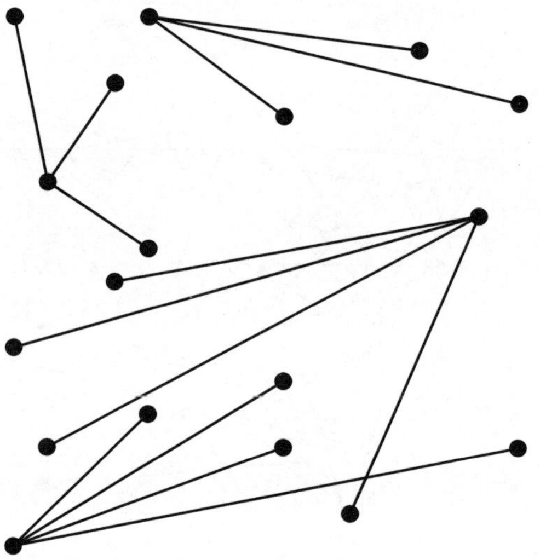

FIGURE 7.7. Adding edges to the vertices from Figure 7.6 to create a perfect code.

We now pick a few vertices arbitrarily, and in Figure 7.7 we connect every other vertex in the graph to *exactly one* of the chosen vertices.

The vertices (four of them in this case) are the perfect code. While we have constructed a graph with a perfect code, it would be foolish to stop here, as the perfect code is highly visible. We now arbitrarily add lots of edges between pairs of the vertices *not in the perfect code* (Figure 7.8).

Now we know a perfect code for the graph, but these vertices wouldn't be obvious to someone who hadn't seen us making the graph. From this, we can create a cryptosystem.

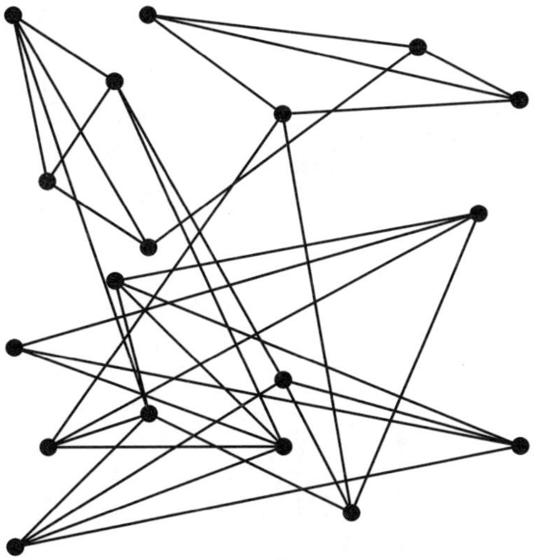

FIGURE 7.8. The final graph, where additional edges have been added to the graph from Figure 7.7. Note the perfect code cannot be easily found.

The perfect code cryptosystem: The perfect code cryptosystem allows Alice to encrypt and send a number to Bob.

(1) Bob creates a graph where he knows a perfect code exists. He keeps this perfect code secret (this is his **private key**), and publishes the graph for all to see.
(2) Alice chooses a number to send to Bob. She writes any integers she wishes at the vertices of the graph, *provided* that they sum to the number she wants to send.
(3) Alice encrypts her message by replacing the number at each vertex with the sum of that number and the numbers of all vertices adjacent to that vertex; this is called **clumping**. She transmits this graph (with the summed values at the vertices) to Bob.
(4) Bob adds the numbers at the vertices in the perfect code. That sum is Alice's message.

Let's go through the steps above, and see why it works. For Alice to send Bob a message using the perfect code cryptography system, Bob first creates a graph in which he knows the location of a perfect code. He can reveal the graph to Alice (or anyone else, for that matter), but he must keep secret the location of the perfect code (the vertices that correspond to the perfect code). To keep things simple, let's assume Bob creates the graph discussed at the beginning of this section, shown again in Figure 7.9.

This graph is Bob's public key: he publishes it for everyone to see. Alice wants to send Bob a message consisting of a single number. She can either convert a text message into a number, or use the number she sends as the key

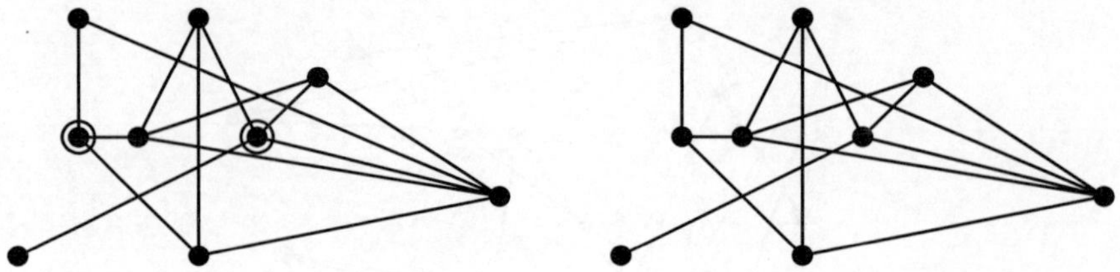

FIGURE 7.9. (Left) Bob creates a graph in which he knows the location of a perfect code (which we have circled). (Right) He sends the graph to Alice, *without revealing the perfect code.*

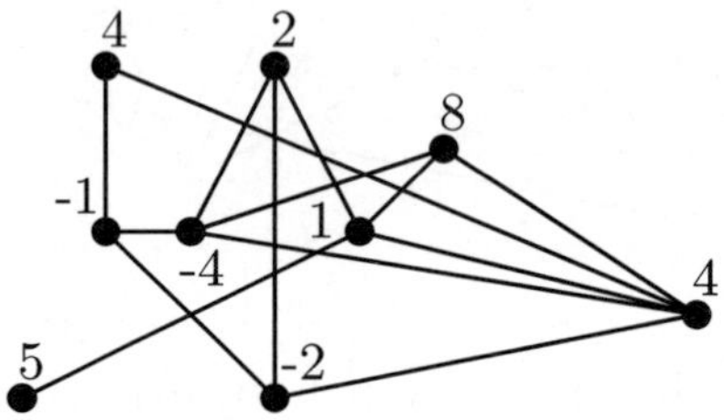

FIGURE 7.10. First Alice "spreads out" the message 17 on the graph.

for a symmetric cipher. Suppose that Alice wants to send Bob the message "17". She begins by "spreading out" the number 17 across the vertices of the graph in any way she wishes. This means that she writes a number next to each vertex of the graph, so that the sum of all of the numbers is 17 (see Figure 7.10).

At this point, the message hasn't been encrypted. Anybody intercepting the graph with these labels can easily recover the original message simply by adding all the numbers together.

To encrypt the message, Alice makes a new copy of the graph. Each vertex gets a new label, which is the sum of its old label, together with the labels of all the vertices adjacent to it. For example, the upper-left vertex is relabeled from 4 to $4 + (-1) + 4 = 7$, while the lower left vertex is relabeled from 5 to $5 + 1 = 6$. We call this operation **clumping**. The complete relabeling is shown in Figure 7.11.

After completing this summing operation, the original message can no longer be recovered by simply summing the labels at all the vertices. In our example, we get 46, which is not the message! Alice now sends this message (the graph with the labels arising from her clumping) to Bob over open channels, without worrying about an eavesdropper being able to deduce the message. So far both the graph that Bob originally sent and the new graph that Alice sends are public.

Bob has special knowledge about this graph: he knows a perfect code (this is his private key). By adding the labels of the vertices in the perfect code, he can recover the original message (see Figure 7.12).

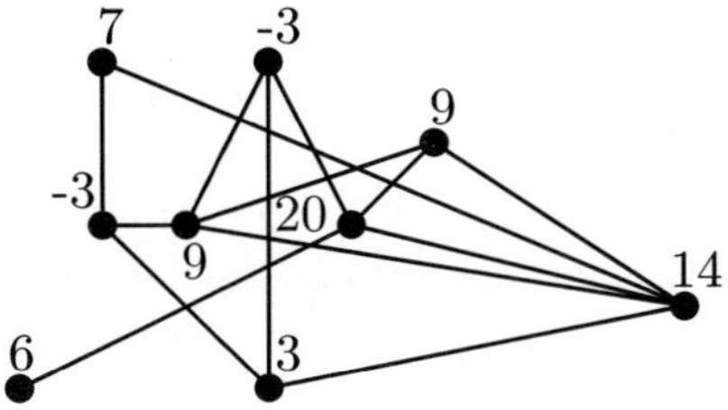

FIGURE 7.11. Alice now creates a new "clumped" labeling of the graph, in which each vertex is labeled with the sum of its original label and the original labels of its adjacent vertices.

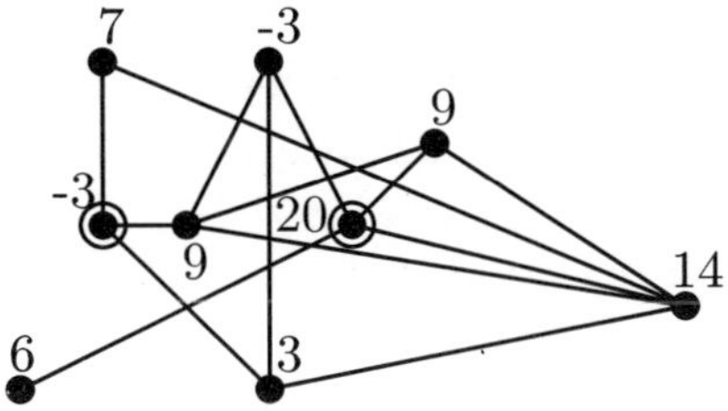

FIGURE 7.12. Bob decrypts the message by adding the labels at the perfect code vertices. In this case, he recovers $-3+20$, which is the original message of 17.

Before reading further, think about why Bob is able to recover the original message.

Remember, the original message was spread out onto the vertices, so that the sum of the labels equals the original number. Alice then clumped the message, replacing each vertex label with the sum of its label and the labels of all adjacent vertices, and transmits that to Bob. Consider what happens when Bob sums the labels at the vertices of the perfect code. The fundamental properties of a perfect code are that no two vertices in the perfect code are adjacent to each other, and every other vertex is adjacent to exactly one vertex in the perfect code. This means that when Bob adds the labels at the perfect code, he gets the sum of the labels Alice deposited on the graph *before* clumping; in other words, he gets the message.

Let's look at our example more closely. There are nine vertices. Using the vertex names from Figure 7.1, the perfect code is the pair of vertices I and F. Vertices H, C, and F are adjacent to I, while A, B, D, and G are adjacent to E. Let $x_A, x_B, \ldots, x_I$ be the numbers that Alice places on the vertices (so $x_A + \cdots + x_I = 17$). If we let y_V denote the clumped value on vertex V, then we see $y_A = x_A + x_E$, $y_B = x_B + x_C + x_D + x_E$ and so on. For us, though, the only values that matter are at the perfect code:

$$y_E \ = \ x_A + x_B + x_D + x_E + x_G, \qquad y_I \ = \ x_C + x_F + x_H + x_I.$$

Notice that

$$y_E + y_I \ = \ x_A + x_B + x_C + x_D + x_E + x_F + x_G + x_H + x_I \ = \ 17,$$

which is our message! The key is that summing the labels at the perfect code gives us the sum of Alice's original labeling.

It is worthwhile to emphasize just how amazing this situation appears to be. It implies that two people can communicate secret messages without sharing *any* secret information ahead of time! Even if one of the parties wishing to exchange messages knew nothing of the perfect code system, it could be completely explained in the presence of eavesdroppers, without the need for any secret keys to be exchanged.

Unfortunately, the perfect code system is a better example of what could be done than what should be done. It illustrates a *possibility* for secure communication, but we have not proved it is secure. The reason we haven't proved it is secure is that it is not! We describe an attack in Exercise 7.8.6. Interestingly, the attacker is able to obtain the message *without* finding the perfect code!

Even though the perfect code system is a failure from the point of view of implementation, it has several good features that ultimately will lead us to a workable system, RSA. The method began by taking a process that was hard one way but easy in the other direction. For example, it is hard to find a perfect code in a general graph, but it is easy to create a graph with a perfect code. RSA is based on the same principle. There, the math problem is factoring. It is currently very hard to factor numbers, but it is easy to multiply two primes together to generate a composite number.

7.2. KidRSA

While the perfect code cryptosystem is promising, it sadly doesn't meet our needs as it is easily broken. Thus, the search continues. In this section we describe another attempt to devise a system to allow two people to communicate in secret without having to previously meet and exchange a key. This is called KidRSA, and is due to Neal Koblitz. Like RSA discussed in the next chapter, KidRSA is a public-key cryptography system where the public and private keys are pairs of integers. As it shares many of the features and requires many of the same mathematical prerequisites as RSA, it's a natural next step.

Unfortunately, like the perfect code system, KidRSA is not actually secure. Unlike the perfect code system, however, the problem with KidRSA is that the "hard problem" it is based on is not actually hard, though it appears so at first. KidRSA assumes that it is difficult to find multiplicative inverses in modular arithmetic. While it is easy to solve $17593x = 1$ in the rational numbers, it seems to be significantly harder to solve

$$17593 \cdot x \ = \ 1 \pmod{104659}.$$

Appearances, however, can be deceiving. Just because a problem looks hard does not mean that it is hard. While we currently don't know how to solve this problem save for trial and error, we'll see in §7.3 that these modular inverses can be found very rapidly through the Euclidean algorithm. Even though this means our search must continue, it is still worthwhile exploring KidRSA. First, it serves as a warning that what we believe is hard may not actually be hard. Second, even though modular inverses are easy, a related operation appears hard. There are currently no known fast ways to solve this related problem, which serves as the basis of RSA.

We now describe KidRSA. As usual, Bob, as the recipient, has to do all the legwork at the beginning by creating a public/private key-pair.

The KidRSA cryptography system: The KidRSA Cryptography System first has the recipient generate a public key (e, n) (e for encryption) and a private key (d, n) (d for decryption) as follows.

(1) Choose integers a, b, a', b', and set $M = ab - 1$.
(2) Let $e = a'M + a$, $d = b'M + b$, and set $n = (ed - 1)/M$. Then (e, n) is the public key, and is displayed for all to see, while (d, n) is the private key, which is kept secret.

A message is any positive integer m that is less than n and relatively prime to n. The encrypted message is $e \cdot m \pmod{n}$. Decryption is done by multiplying the encrypted message by d and reducing modulo n.

Let's see why the above works. First, we need to make sure n is an integer. We have

$$\begin{aligned} ed - 1 &= (a'M + a)(b'M + b) - 1 \\ &= a'b'M^2 + (ab' + a'b)M + ab - 1 \;=\; (a'b'M + ab' + a'b + 1)M \end{aligned}$$

as $M = ab - 1$. Thus M divides $ed - 1$, and n is thus an integer. Let m be a positive integer less than n and relatively prime to it. Bob receives $m \cdot e \pmod{n}$, which is $me - kn$ for some k. Bob now multiplies this by d and reduces modulo n. Since

$$d(me - kn) \;=\; m \cdot ed - kdn \;=\; m \cdot (nM + 1) - kdn \;=\; m + (M - kd)n,$$

we see its reduction modulo n is m, as desired.

Let's consider an example. Suppose Bob begins his key generation with the choices $a = 5, b = 3, a' = 7, b' = 5$. He quickly gets $M = 14$, $e = 7 \cdot 14 + 5 = 103$, $d = 5 \cdot 14 + 3 = 73$, and $n = (103 \cdot 73 - 1)/14 = 537$. Thus $(103, 537)$ is the public key and $(73, 537)$ is the private key. Bob publishes $(103, 537)$. If Alice wants to encrypt the message "30", she computes $103 \cdot 30 \pmod{537}$, which gives 405. Upon receiving the message 405, Bob computes $405 \cdot 73 \pmod{537}$, which recovers 30.

The "point" of the key generation steps for KidRSA is just that they allow us to pick, in a systematic way, numbers e and d for which we know some nontrivial divisor of $ed - 1$, namely M. We then choose n to be the complementary divisor of $ed - 1$, $(ed - 1)/M$. This ensures that $ed = 1 \pmod{n}$ holds. The utility of this is that if Alice encrypts the message by multiplying by e, Bob can undo the encryption by multiplying by e. The *hope* is that it is hard for an eavesdropper to figure out d from the publicly available information: e and n. Again, we're trying to build a cryptosystem out of a problem where we know more information than the attacker (here, it's the numbers a, b, a', and b'). Thus the central question is, How difficult is it for an attacker to determine d *without* knowing these four special numbers? In the next section, we see the answer is not hard at all!

7.3. The Euclidean Algorithm

The KidRSA system is based on our ability to create pairs of numbers which are multiplicative inverses of each other to a given modulus. It would be trivial to do this if we could always invert numbers to a given modulus. If we could do this, all we would have to do is choose any d modulo n, and then take its inverse. We succeed in KidRSA because we are searching for a pair. Thus, we don't have to start with a specific d, but rather will accept any d so long as we can find its inverse.

In the last section, Bob produced the numbers 103, 73, and 537 satisfying $103 \cdot 73 = 1 \pmod{537}$. It is not obvious how Alice, knowing only the numbers 103 and 537, could *quickly* solve the congruence $103 \cdot x = 1 \pmod{537}$ to recover the secret number 73. The keyword is "quickly". There's of course a trivial approach to finding the inverse: try all candidate numbers! This brute force search, of course, is entirely impractical if n is large.

There is, however, a very fast and efficient way to find modular inverses, called the Euclidean algorithm. We first describe the Euclidean algorithm and show how it can compute greatest common divisors. After we get a good feel for it, we then show it provides more information than was apparent at first, giving an extremely fast method to *construct* modular inverses.

Greatest common divisor (gcd): The gcd, d, of two nonnegative integers x and y is the largest integer dividing both (so $d|x$, $d|y$, and any other number dividing both is less than d). We often denote this by writing $\gcd(x, y)$ for d.

As any positive integer a divides 0, we have $\gcd(a, 0) = a$.

At first glance, computing gcd's seems like a straightforward task. The gcd of 15 and 25 is 5, for example, since clearly this is the greatest divisor common to both the numbers. Things look less rosy when numbers get bigger, however. Suppose, for example, that we want to compute $\gcd(707, 980)$. Running through all possible divisors "by hand" now seems like an unpleasant task, and the future looks bleak when we get to numbers with hundreds of digits. However, a simple but elegant observation saves the day.

The key to finding greatest common divisors: Let $x \leq y$ be nonnegative integers. The greatest common divisor of x and y equals the greatest common divisor of x and $y - x$; mathematically, we write $\gcd(x, y) = \gcd(x, y - x)$.

Proof. Why is this true? Let's write d for $\gcd(x, y)$ and d' for $\gcd(x, y - x)$. We must show $d = d'$. We use a nice trick to show equality: we show $d \leq d'$ and $d' \leq d$, which forces them to be equal.

Since $d|x$ and $d|y$, we have $d|y - x$, and thus $d \leq d'$. For the other direction, since $d'|x$ and $d'|y - x$, d' divides $y - x$ plus x. Thus $d'|y$ (you are asked to prove this in Exercise 7.8.23), and hence $d' \leq d$, completing the proof. □

Returning to our example, the important observation above implies that the gcd of 707 and 980 is the same as the gcd of 707 and $(980-707)$; since the gcd divides both 707 and 980, it divides $(980 - 707)$ as well. Thus we have reduced the problem of finding $\gcd(707, 980)$ to one of finding $\gcd(707, 273)$. Why stop now, though? We can use the subtraction trick again. In fact, 273 can be subtracted from 707 multiple times, reducing the problem to finding $\gcd(707 - 273 - 273, 273) = \gcd(161, 273)$. By subtracting 161 from 273, we have reduced the problem to finding $\gcd(161, 112)$. The next reduction gives $\gcd(49, 112)$. Now 49 can be subtracted twice from 112, giving $\gcd(49, 14)$. While it might be clear what the gcd is, it's worthwhile to see how the process ends. We have 14 comes out 3 times from 49, leaving $\gcd(7, 14)$. Finally, we can subtract two 7's from 14, leaving $\gcd(7, 0)$, which is 7.

The procedure of subtracting to find the gcd is called the **Euclidean algorithm**. We need the following notation to formally state it: $\lfloor r \rfloor$ is the largest integer at most r. Thus $\lfloor\sqrt{2}\rfloor = 1$, $\lfloor\pi\rfloor = 3$, $\lfloor 6 \rfloor = 6$, and $\lfloor -4.2 \rfloor = -5$. We call this the **floor** function.

The Euclidean algorithm: Let $x < y$ be two nonnegative integers. The following procedure generates the **greatest common divisor (gcd)**.

(1) If $x = 0$, then the gcd is y, and STOP; otherwise, proceed to Step 2.

(2) If $x < y$, note that $\lfloor y/x \rfloor$ is the largest multiple of x that is at most y. We have $\gcd(x, y) = \gcd(x, y - \lfloor y/x \rfloor x)$.

(3) Let $y_{\text{new}} = x$ and $x_{\text{new}} = y - \lfloor y/x \rfloor$. Note $x_{\text{new}} < x$ and $y_{\text{new}} < y$. If $x_{\text{new}} = 0$, then the gcd is y_{new}. If $x_{\text{new}} = 1$, then the gcd is 1. If $x_{\text{new}} \geq 2$, then relabel, setting $x = x_{\text{new}}$ and $y = y_{\text{new}}$, and return to Step 2 with these values.

Note that in Step 3, y_{new} is x_{old} (i.e., it's the value of x we started with). This suggests a convenient way to record the output of these steps. In a column, we first write y. Underneath that we write x. Each time we perform Step 3, we write x_{new} underneath the last entry in the column.

For example, for gcd(707, 980) we begin by putting the bigger number above the smaller one.

980
707

At each step, we subtract the last number from the one immediately preceding it as many times as possible, writing the result at the bottom of the column. For example, 980 − 707 was 273, so the column becomes

980
707
273

We now continue until we reach zero.

980
707
273
161
112
49
14
7
0

The number immediately before 0 is the gcd; in this case it is 7. It works because the subtractions don't change the gcd's. This column represents a chain of gcd's which are all equal: $\gcd(980, 707) = \gcd(707, 273) = \gcd(273, 161) = \cdots = \gcd(14, 7) = \gcd(7, 0)$.

While finding gcd's is fun (and often important!), breaking KidRSA involves being able to find modular inverses. A slight extension makes the Euclidean algorithm up to this task. All we need to do is keep track of the multiples $\lfloor y/x \rfloor$ of x that we subtracted in Step 2. Not surprisingly, the resulting procedure is called the **extended Euclidean algorithm**.

Of course, we cannot set our sights too high. We can't always find inverses, as some inverses do not exist. We can't hope to find the inverse of 707 modulo 980, since it doesn't have one (we just showed both are divisible by 7). Let's try to find the inverse of 707 modulo 979. We carry out the Euclidean algorithm as before, but do some more "bookkeeping" on the side. We introduce two more columns to our table, and initialize them with 1's and 0's as shown below.

	979	707
979	1	0
707	0	1

The two extra columns keep track of how to write the numbers produced by the Euclidean algorithm as the sum (or difference) of multiples of 979 and 707. For example, the first row of this table should now be interpreted as saying that $979 = 1 \cdot 979 + 0 \cdot 707$, while the second means that $707 = 0 \cdot 979 + 1 \cdot 707$.

We now carry out the Euclidean algorithm exactly as before, except that *whatever we do to the numbers in the left column, we do to the right two columns as well.* For example, the next step of the Euclidean algorithm at this point is to subtract 707 from 979 and write the difference beneath. We now subtract each element of the second row from each element of the preceding row and write the result beneath. We get $979 - 707$ gives 272, $1 - 0$ gives 1, and $0 - 1$ gives -1, so the table becomes

	979	707
979	1	0
707	0	1
272	1	−1

This new row can be interpreted as saying that $272 = 1 \cdot 979 - 1 \cdot 707$ (which is true!) The next step was to take the remainder after dividing 707 by 272. Since 272 goes into 707 two times, this is the same as subtracting $2 \cdot 272$ from 707. We therefore subtract 2 times each element of the last row from each element in the previous row. We get $707 - 2 \cdot 272 = 163$, $0 - 2 \cdot 1 = -2$, and $1 - 2 \cdot (-1) = 3$.

	979	707
979	1	0
707	0	1
272	1	−1
163	−2	3

The new row tells us that $163 = -2 \cdot 979 + 3 \cdot 707$. As 163 goes once into 272, to continue we subtract this row once from the previous row.

	979	707
979	1	0
707	0	1
272	1	−1
163	−2	3
109	3	−4

The new row means that $109 = 3 \cdot 979 - 4 \cdot 707$. Next 109 goes once into 163, giving

	979	707
979	1	0
707	0	1
272	1	−1
163	−2	3
109	3	−4
54	−5	7

Now 54 goes twice into 109, so we subtract two times the final row from the previous row.

	979	707
979	1	0
707	0	1
272	1	−1
163	−2	3
109	3	−4
54	−5	7
1	13	−18

Since we reached 1, we find 979 and 707 are relatively prime, and we don't need to continue on to 0. We can see this from the expansion in the last row: $1 = 13 \cdot 979 - 18 \cdot 707$. Rearranging things, this means that $(-18) \cdot 707 = 1 + (-13) \cdot 979$. In other words, $(-18) \cdot 707$ is 1 plus a multiple of 979, and thus -18 is a multiplicative inverse of 707 modulo 979! Of course, we normally prefer to write our inverses as positive numbers at most 979. Since $-18 = 961 \pmod{979}$, so we have that the inverse of 707 modulo 979 is 961.

The extended Euclidean algorithm gives us multiplicative inverses. After completing the extended table with inputs $x \le y$, the final row (if the x and y are relatively prime) gives us an equation of the form $1 = ax + by$, where 1, a, and b are the numbers from the last row. We see that a is the inverse of x modulo y, and b is the inverse of y modulo x. Note that, although the extended Euclidean algorithm often gives negative numbers as inverses (as in -18 in the above case), these are always equivalent to some positive number in the modular system.

Why does the extended Euclidean algorithm work? We now briefly discuss its proof.

Proof that the extended Euclidean algorithm works. When constructing our table, it suffices to show that whenever we add a new row, say r, a_r, b_r, that $r = a_r x + b_r y$, where $x \le y$ are our initial positive integers. Let's look at the previous two rows:

	y	x
$\vdots$	$\vdots$	$\vdots$
p	b_p	a_p
q	b_q	a_q
r	b_r	a_r

Technically, we're proceeding by what is called *mathematical induction* (see §3.7 for a detailed description of this method). We already showed the first two rows satisfy the desired property; we just need to show that this property propagates downward. Since we're assuming the previous two rows satisfy the claim, we have

$$p = a_p x + b_p y \quad \text{and} \quad q = a_q x + b_q y.$$

We subtract $k := \lfloor p/q \rfloor$ copies of q from p, which gives r. How are a_r and b_r constructed? We subtract k copies of a_q from a_p, and k copies of b_q from

b_p. Thus

$$a_r = a_p - ka_q \quad \text{and} \quad b_r = b_p - kb_q.$$

The rest of the proof is just collecting algebra. We have

$$\begin{aligned} r &= p - kq \\ &= (a_p x + b_p y) - k(a_q x + b_q y) \\ &= (a_p - ka_q)x + (b_p - kb_q)y = a_r x + b_r y, \end{aligned}$$

precisely as claimed! □

We codify what we have done. We only need the case when the gcd is 1, but a similar result holds more generally (see Exercise 7.8.26).

The extended Euclidean algorithm: When applying the Euclidean algorithm to nonnegative integers $x < y$, add two more columns. Label the first y and the second x, and have the table begin

	y	x
y	1	0
x	0	1

Each time you add a number in the first column by subtracting k times the number above from the number 2 above, do the same to the other two columns. Letting $k = \lfloor p/q \rfloor$, we have

	y	x
$\vdots$	$\vdots$	$\vdots$
p	b_p	a_p
q	b_q	a_q
$p - kq$	$b_p - kb_q$	$a_p - ka_q$

If we ever get the row $1, b, a$, then the greatest common divisor is 1 and $1 = ax + by$, which implies that a (or $a \pmod{y}$) is a multiplicative inverse of x modulo y.

It would be criminal if we left the extended Euclidean algorithm without discussing its run-time (how long it takes to execute). Remember, a huge part of cryptography is being able to do something quickly. We've just seen how to use the extended Euclidean algorithm to compute modular inverses. An immediate consequence is that we can crack KidRSA, as given e and n we can now find d such that $ed = 1 \pmod{n}$. Of course, in order for this to be an effective attack, we must be able to compute the modular inverse *quickly*. Thus, it is essential that we know how long the algorithm takes.

At first glance, it appears the extended Euclidean algorithm might be quite slow. As always, we start with two integers $x \le y$. You should think of these as large numbers, say 200 digits. When we run the algorithm, each new row is less than the previous. Unfortunately, if it is less by just 1, then it could take x steps to terminate! If $x \approx 10^{200}$, this is far too long.

Fortunately, the extended Euclidean algorithm is quite fast. The number of rows is at most $2\log_2 x$. This means that even if x is around 10^{200}, the number of steps is at worst about 1329, which is very manageable. We sketch a proof of this critical fact in Exercise 7.8.27.

To recap: The extended Euclidean algorithm allows us to compute modular inverses extremely rapidly (the run-time is proportional to the logarithm of our numbers). In particular, KidRSA is not secure, and we need a better method. The rest of this chapter is devoted to the remaining mathematical prerequisites needed to improve KidRSA.

7.4. Binary Expansion and Fast Modular Exponentiation

In KidRSA we encrypted our message m by multiplying it by the public key e, and then reduced it modulo n; decryption was handled similarly, with the receiver multiplying the encrypted message by d and then reducing modulo n. We saw that this method was insecure, as an attacker could use the extended Euclidean algorithm to quickly compute the secret d from the publicly available e and n. Perhaps, though, all is not lost. What we need to do is make the encryption more complicated and hide more information. If we do this well, hopefully we can frustrate the attacker.

Building on KidRSA, instead of doing *one* modular multiplication, what if we instead do many? In RSA, instead of multiplying the message by the public key e, we exponentiate by e. Thus we transmit not $e \cdot m \pmod{n}$ but $m^e \pmod{n}$, and then somehow use d to recover m. Of course, this can't be the only change; since e is public and an attacker could deduce d and decode the message as fast as the desired recipient. We therefore must change how we choose d, e and n as well.

There is an additional concern. Remember we want a method where encryption and decryption is fast. Typically e and d have hundreds of digits, as does the message m. For definiteness, say $m = \lfloor\sqrt{2} \cdot 10^{185}\rfloor$ and $e = \lfloor\sqrt{3} \cdot 10^{180}\rfloor$ (remember $\lfloor x\rfloor$ is the **floor** function, returning the largest integer at most x). The way we've been taught to exponentiate is through repeated multiplications. In this case, we'd compute $m, m^2 = m \cdot m$, $m^3 = m^2 \cdot m$, $m^4 = m^3 \cdot m$, and so on, all the way to $m^{e-1} \cdot m$. This requires $e - 1$, or over 10^{180} multiplications! To put this in perspective, imagine we have a supercomputer capable of doing one *googol* of these multiplications a second (that's 10^{100}, a number that is so large it even dwarfs the net worth, in dollars, of the entire planet, a number greater than all the subatomic particles in the universe!). It would take 10^{80} seconds to finish the multiplication, or over 10^{72} years. This clearly is impractical—the universe is only around 14 billion years old (i.e., about $14 \cdot 10^9$ years).

If the only way to exponentiate is through repeated multiplication as above, exponentiation is too expensive to use. Fortunately, there's a faster way. It involves the **method of repeated squaring**, and allows us to compute $m^e \pmod{n}$ in at most $2\log_2 e$ steps. To appreciate the power of this, in our example above that means at most 1198 multiplications; this is well within the power of modern computers (and slightly smaller numbers are doable with paper and pen!).

In order to explain the method of repeated squaring, we need to first explain base expansions. These are generalization of the familiar decimal (base 10) notation.

Base B expansions: Let $B \geq 2$ be a positive integer. Given any nonnegative integer x, there is a nonnegative integer n and integers $a_0, a_1, \dots, a_n \in \{0, 1, \dots, B-1\}$ such that $a_n \neq 0$ and

$$x = a_n B^n + a_{n-1} B^{n-1} + \cdots + a_1 B + a_0.$$

We call this the **base B expansion** of x, and the a_i the **digits**. Special choices are $B = 10$ (**decimal**) and $B = 2$ (**binary**). We often write a subscript B after the number to indicate the base, but suppress the subscript if the base is 10. To find the base B expansion of x:

- Find the largest integer n such that $B^n \leq x < B^{n+1}$. The base B expansion has $n+1$ digits.
- The leftmost digit a_n is the largest $k \in \{1, 2, \dots, B-1\}$ such that $kB^n \leq x < (k+1)B^n$.
- To find the next digit, a_{n-1}, find the largest $\ell \in \{0, 1, \dots, B-1\}$ such that $\ell B^{n-1} \leq x - kB^n < (\ell+1)B^{n-1+1}$.
- To find the next digit, a_{n-2}, find the largest $m \in \{0, 1, \dots, B-1\}$ such that $mB^{n-2} \leq x - kB^n - \ell B^{n-1} < (m+1)B^{n-2+1}$.
- Continue as above until you get the rightmost digit, a_0. The base B expansion is $a_n B^n + a_{n-1} B^{n-1} + \cdots + a_0$.

Thus

$$144 = 1 \cdot 5^3 + 0 \cdot 5^2 + 3 \cdot 5 + 4 = 1034_5, \quad 11 = 1 \cdot 2^3 + 0 \cdot 2^2 + 1 \cdot 2 + 1 = 1011_2.$$

If x is our number in base 10, then the number of digits base B is $\lceil \log_B x \rceil$, where $\lceil y \rceil$ is the **ceiling** function, giving the smallest integer at most y.

We justify the claims above for $B = 2$, as this is the base we need (and the other bases follow similarly). Imagine we are given the number $x = 101$, and we want to write it in base 2. In base 10, we use powers of 10: 1, 10, $100 = 10^2$, $1000 = 10^3$, and so on. In binary we use powers of 2: 1, 2, $4 = 2^2$, $8 = 2^3$, $16 = 2^4$, $32 = 2^5$, $64 = 2^6$, $128 = 2^7$, and so on. We look at the largest power of 2 less than our number and subtract that. For $x = 101$, that would be $64 = 2^6$; we can't have $128 = 2^7$ appearing in the binary expansion, as this is larger than our number! We now have

$$101 - 2^6 = 101 - 64 = 37.$$

Thus, the binary expansion of 101 starts as $1 \cdot 2^6$. We're left with 37. The largest power of 2 less than 37 is $32 = 2^5$. We subtract that off, and find

$$101 - 2^6 - 2^5 = 101 - 64 - 32 = 5,$$

giving us a bit more information about the binary expansion of 100: it starts $1 \cdot 2^6 + 1 \cdot 2^5$. We continue. We can't subtract 16 from 5, nor can we subtract 8; both of these digits are therefore 0. The highest power of 2 we can subtract is $4 = 2^2$, giving

$$101 - 2^6 - 2^5 - 2^2 \ = \ 101 - 64 - 32 - 4 \ = \ 1,$$

so the binary expansion of 101 starts $1 \cdot 2^6 + 1 \cdot 2^5 + 1 \cdot 2^2$. We can't subtract 2 from 1, and thus the final subtraction is subtracting $1 = 2^0$, and we find

$$101 - 2^6 - 2^5 - 2^2 - 2^0 \ = \ 101 - 64 - 32 - 4 - 1 \ = \ 0.$$

In other words,

$$101 \ = \ 1 \cdot 2^6 + 1 \cdot 2^5 + 1 \cdot 2^2 + 1 \cdot 2^0 \ = \ 1100101_2.$$

All that's left is the claim on the number of digits. We have $B^n \leq x < B^{n+1}$, so

$$n \ = \ \log_B(B^n) \ \leq \ \log_B x \ < \ \log_B(B^{n+1}) \ = \ n + 1.$$

Thus $\lceil \log_B x \rceil = n + 1$, the number of digits.

We'll see now why we were so obsessed with the number of digits in a base B expansion. It turns out that the number of multiplications needed to find m^x is not $x - 1$, but only $2 \log_2 x$. This is an absolutely remarkable savings. It makes exponentiation feasible. We can't do 10^{200} multiplications, but we can do $2 \log_2(10^{200}) \approx 1329$.

In RSA, we only need to exponentiate modulo n. We'll constantly use the following important fact (which you are asked to show in Exercise 7.8.38):

Let x and y be integers, and let n a positive integer. Then

$$x \cdot y \pmod n \ = \ (x \pmod n) \cdot (y \pmod n) \pmod n.$$

In other words, we can multiply and then reduce modulo n, or we can reduce modulo n, multiply, and then reduce modulo n; both give the same answer.

We illustrate **fast modular exponentiation** by an example, and then isolate the algorithm. Imagine we want to find $107^{101} \pmod{113}$. The normal, brute force approach requires 100 multiplications. We now show how to find it with just *nine* multiplications. We start by finding the base 2 expansion of 101, the exponent. Not coincidentally, we just computed this; it's

$$1100101_2 \ = \ 1 \cdot 2^6 + 1 \cdot 2^5 + 1 \cdot 2^2 + 1 \cdot 2^0 \ = \ 64 + 32 + 4 + 1.$$

Note the highest power of 2 in the decomposition is 6. We now compute $107^{2^k} \pmod{113}$ for $k = 1, 2, \dots, 6$ recursively. This means we use the results of the previous step to get the next value.

For example, $107^2 \pmod{113} = 36$. To find $107^4 \pmod{113}$ we don't do $107 \cdot 107 \cdot 107 \cdot 107 \pmod{113}$, but rather $(107^2 \pmod{113}) \cdot (107^2 \pmod{113}) \pmod{113}$. The advantage of this grouping is that we've just calculated

TABLE 7.1. Computing $m^e \pmod{n}$ via fast modular exponentiation. Here $m = 107$, $e = 101$, $n = 113$, and the a_k's are the digits in the binary expansion of e.

k	2^k	$m^{2^{k-1}} \cdot m^{2^{k-1}} \pmod{n}$	$m^{2^k} \pmod{n}$	a_k	Running Product
0	1		107	1	107
1	2	$107 \cdot 107 \pmod{113}$	36	0	107
2	4	$36 \cdot 36 \pmod{113}$	53	1	21
3	8	$53 \cdot 53 \pmod{113}$	97	0	21
4	16	$97 \cdot 97 \pmod{113}$	30	0	21
5	32	$30 \cdot 30 \pmod{113}$	109	1	29
6	64	$109 \cdot 109 \pmod{113}$	16	1	12

$107^2 \pmod{113}$; that's 36. Thus $107^4 \pmod{113}$ equals $36 \cdot 36 \pmod{113}$, or 53. We continue in this way until we've found $107^{64} \pmod{113}$.

Now that we know $107^{2^k} \pmod{113}$ for $k \leq 6$, we can quickly find $107^{101} \pmod{113}$. We use the laws of exponents to write

$$107^{101} = 107^{64+32+4+1} = 107^{64} \cdot 107^{32} \cdot 107^4 \cdot 107^1,$$

or better yet

$$\begin{aligned} 107^{101} \pmod{113} \quad &= \quad (107^{64} \pmod{113}) \cdot (107^{32} \pmod{113}) \\ &\quad \cdot (107^4 \pmod{113}) \cdot (107^1 \pmod{113}). \end{aligned}$$

We recorded the process in Table 7.1. The only new item is the final column, which we've labeled the running product. The a_k are the binary digits of e. If a_k is 1, then $m^{2^k} \pmod{113}$ occurs in the decomposition, while if a_k is 0 it does not. This is a fast and clean way to keep track of the exponentiations. The running product in the kth row is the product of the running product from the $(k-1)^{\text{st}}$ row with $m^{2^k} \pmod{n}$ if $a_k = 1$, and 0 otherwise.

We now formally isolate this procedure.

Fast modular exponentiation: To quickly compute $m^e \pmod{n}$, use the method from Table 7.1:

- Find the base 2 expansion of e, and let its digits be $a_0, a_1, \ldots, a_\ell$.
- Using repeated squaring, compute modulo n the following: $m \cdot m$, $m^2 \cdot m^2$, $m^4 \cdot m^4$, $\ldots$, $m^{2^{\ell-1}} \cdot m^{2^{\ell-1}} = m^{2^\ell}$.
- Form the running product, multiplying together all the values from the repeated squarings corresponding to nonzero digits.

The number of multiplications is at most $2 \log_2 e$.

Apart from the fact that it works, another absolutely important property of fast exponentiation is that it works quickly. It's similar to the extended Euclidean algorithm in that it too runs in logarithmic time. This means that if we want to use e as an exponent, the number of operations is not of the order of e, but of the order of $\log_2 e$. As the logarithm is so much smaller, this makes exponentiation feasible. We can compute by hand numbers that would overflow typical graphing calculators and popular programs such as Microsoft Excel.

7.5. Prime Numbers

As we prepare to finally tackle the RSA system in the next chapter, our journey may have convinced you of the difficulty of finding a workable public-key cryptography system. Before we began this chapter, you might have been skeptical that such a system was possible at all, mirroring the view of just about everybody before the crucial breakthroughs in the 1970s.

One of the elegant aspects of public-key cryptography systems is that each one is constructed around a fundamental mathematical idea. This stands in contrast to symmetric ciphers like AES, where there generally is no essential mathematical relationship around which the whole system is developed; the mantra of a symmetric cipher is more or less "do something complicated" in a very careful, efficient way.

In the case of RSA the essential mathematical relationship comes from prime numbers.

Prime numbers, composites, unit: Prime numbers are positive integers with exactly two positive divisors. The first ten primes are 2, 3, 5, 7, 11, 13, 17, 19, 23 and 29. We call 1 the **unit**; all other positive integers are called **composite**, and have at least three positive divisors. If n is a positive integer and $n = xy$ with $x, y \geq 2$, then n is composite, while if $n > 1$ and there is no such factorization, then n is prime.

For example, 13 is prime because its only positive divisors are 1 and 13, but 6 is composite as it has four factors (1, 2, 3 and 6), or because we can write 6 as $2 \cdot 3$.

For RSA, essentially all we need to know is that there are infinitely many primes, and that they are fairly plentiful. This is only part of the story. We collect a few amusing properties about primes as an appendix to this section.

The practical role that primes play in systems like RSA is particularly surprising; before these modern developments in cryptography, the study of prime numbers had essentially no practical applications. Mathematicians often pursue problems whose practical uses are not immediately apparent, and the study of prime numbers was, for a long time, the quintessential example of a purely aesthetic mathematical pursuit. Due to systems like RSA, however, prime numbers are now some of the most crucial mathematical objects from a practical standpoint, showing the unexpected ways in which mathematics find applications!

We briefly discuss how primes allow us modify KidRSA into a practical cryptographic system. Remember that in KidRSA we chose integers n, e, and d with $ed = 1 \pmod{n}$. We encrypted a message through multiplication, sending m to $e \cdot m \pmod{n}$. This is very insecure, as an attacker can deduce the secret d easily from the publicly available e and n. We discussed sending instead $m^e \pmod{n}$. We can compute this easily using fast modular exponentiation; unfortunately, it's not clear how to recover m from $m^e \pmod{n}$. Should we use the same d as before, or perhaps a new d?

Remember, it does us no good to be able to quickly send a message unless the recipient can quickly decode it and the attacker cannot. Fortunately, there is a small modification that allows for a rapid recovery of m *only* if we know a secret d related to e and n. The idea is to let $n = pq$, where p and q are distinct, large primes. In RSA we choose e and d so that $ed = 1 \pmod{(p-1)(q-1)}$.

Our purpose in this section is not to explain why these choices work, but to introduce the mathematics needed to understand the justifications later, and to implement RSA. In particular, we'll have more formal proofs here than we have had in other sections. The reason for this is we're using more advanced mathematics than before, and we need to be careful to only use true statements. We now turn to proving some needed properties about prime numbers.

In mathematics, common procedure is to first show something exists, then if possible show there are infinitely many, and finally if we can determine approximately how many there are. All three issues are important for RSA. We need to have large primes to have a reasonable level of security. Of course, we also have to be able to *find* these primes. Thus, if very few numbers are prime, it might be prohibitively expensive to locate them, making RSA impractical.

We first show there are infinitely many primes. The proof below is over 2000 years old and is one of the oldest proofs still taught. It uses a great technique, that of **proof by contradiction**. This means that we will assume that the theorem is false, and then arrive at a contradiction though logical steps. By arriving at a contradiction, we can conclude that the original assumption (that the theorem is false) must have been false; of course, this means that the theorem must be true!

The idea of proof by contradiction is illustrated by the following reasoning by someone who wants to know whether or not it recently rained.

```
Let's assume it recently rained.  Then there
must be puddles in the street.  But I can see
that there are no puddles in the street. This
is a contradiction.  Thus it must not have
recently rained.
```

The kind of reasoning above is different from a mathematical proof, in that the successive statements don't necessarily follow from one another. For

TABLE 7.2. Values of the polynomial $f(n) = n^2 - n + 41$.

n	f(n)	Primality of $f(n)$
1	41	prime
2	43	prime
3	47	prime
$\vdots$	$\vdots$	$\vdots$
37	1373	prime
38	1447	prime

example, it may be possible for it to rain without making puddles in some way that the speaker hadn't imagined (maybe the street is covered with a giant tent, for example). In a mathematical proof, deductions are not just "reasonable" like those from our discussion on the rain. They absolutely must follow without any additional assumptions.

Theorem (Euclid): *There are infinitely many primes.*

In other words, there is no "largest" prime number. We could try to satisfy ourselves of this fact by looking for really big prime numbers. Currently, $2^{43112609} - 1$ is one of the largest known, with 12,978,189 digits. Unfortunately, just because we keep finding more and more primes does not mean we have a proof that there are infinitely many. It's very possible to be misled by small data. As this is such an important lesson to learn that we give an example before returning to a proof of the infinitude of primes.

Let $f(n) = n^2 - n + 41$. Euler studied this polynomial in the 1770s. After looking at some of its values in Table 7.2, you'll see why he found this quadratic so interesting.

In the interest of space we only record a subset of the values of $f(n)$, but *all* n up to 38 give prime numbers. This is quite interesting; it seems like no matter where we evaluate this polynomial, we always get a prime! Based on the data above, it's natural to conjecture that $f(n)$ is *always* prime for any positive integer n.

Is this true? Perhaps if we go further we'll find a counter-example. Continuing, we see $f(39) = 1523$ is prime, $f(40) = 1601$ is prime, but $f(41) = 1681$ is composite. As we continue, we find more and more composite numbers. We're left with an interesting polynomial, and an important lesson. Just because the first 40 values generated prime numbers *does not mean* the polynomial always evaluates to a prime. Hopefully this example illustrates the importance and need for rigorous proofs. Just because something *appears* to be true does not mean it is. In particular, if we tried to use this polynomial to generate large primes for RSA, we would be trouble.

Now motivated and appreciative of the need for proofs, we prove the infinitude of primes.

Proof of Euclid's theorem. We prove the infinitude of primes by **contradiction**. This is another common proof technique which complements proof by induction (described in §3.7). The way proofs by contradiction work is we assume the negation of what we want to prove, and show that this leads to a contradiction. Thus our initial assumption is false, and the desired claim is true.

We return to Euclid's theorem. We assume there are only finitely many primes, and show this leads to a contradiction. Since we're assuming there are only finitely many primes, we can list them in principle (though in practice it might not be feasible to write them all down!) Let's write them in increasing order:

$$2,\ 3,\ 5,\ 7,\ 11,\ 13,\ 17,\ \dots,\ p_L,$$

where p_L is the "last" prime number. We aren't assuming anything about how many primes there are; just that it is a finite number. Our proof won't use any properties about p_L other than there are *no* primes greater than p_L.

What we're going to do now is create a new prime. Whatever we do, it's reasonable to expect the primes 2, 3, ..., p_L will play some role. The simplest thing we could do is multiply them all together; what works is only slightly more complicated. We multiply all of these prime numbers together and add 1 to the result:

$$s \ := \ 2 \cdot 3 \cdot 5 \cdot 7 \cdot 11 \cdot 13 \cdot 17 \cdots p_L + 1.$$

The most natural question to ask is whether or not s is prime. There are two possibilities: either s is prime, or it isn't. What we do now is break the argument up into two cases. We'll see that in each case we get a new prime not on our list. This contradicts our assumption that 2, 3, ..., p_L are all the primes, and it completes the proof.

- **Case 1: s is prime.** As $s > p_L$, s can't be one of the L primes on our list. Thus, in this case, we've found a new prime: s.

- **Case 2: s is not prime.** As $s > 2$, s is composite. As every composite number is divisible by a prime, we can ask, Which prime numbers divide s? Well, since s is 1 more than a multiple of 2 (we can view it as $(3 \cdot 5 \cdots p_L)2+1$), it can't be divisible by 2. Similarly, since it's 1 more than a multiple of 3, it can't be divisible by 3. We keep arguing like this, and find that *none* of the primes on our list divide s as each leaves a remainder of 1; however, s is composite so there must be *some* prime dividing it. We've therefore discovered a new prime that is not on our list. We don't know what that prime is, but we know there must be another prime.

We're done! There are only two possibilities. Either s is prime or it's composite, and in each case we find a new prime not on our list. Thus our assumption that there are only finitely many primes is false, and there must be infinitely many. □

If you were paying *very* close attention, you might have noticed one issue with the proof above. We accidentally assumed another fact in our proof! We said every composite number is divisible by a prime, but maybe that is what is false, not our assumption that there are only finitely many primes. We thus need to prove that every composite is divisible by a prime; we sketch the proof in Exercise 7.8.45. This illustrates how careful one must be; it's very easy to assume something accidentally.

Now that we know there are infinitely many primes, the next challenge is to count them. While we don't have useful formulas to generate the primes, interestingly we have pretty good formulas to count how many primes there are between 1 and x. We'll talk a bit more about this topic in the appendix to this section, as this material is not directly needed for our cryptography investigations.

Appendix to §7.5. More on Primes

We end this section with a few interesting observations about primes. These aren't needed for RSA, but we include them in the spirit of general inquisitiveness.

Mathematicians' fascination with prime numbers may stem in part from the contrast between their straightforward definition and the seemingly random "behavior". For example, in spite of the fact that prime numbers have a simple definition and satisfy many strict mathematical relations, there seems to be no obvious pattern to which numbers are prime, and, for example, there are no known formulas to simply generate the prime numbers one after another.

This duality between their deterministic and random-like behaviors was remarked upon by Don Zagier in a lecture in 1975:

> There are two facts about the distribution of prime numbers of which I hope to convince you so overwhelmingly that they will be permanently engraved in your hearts. The first is that, despite their simple definition and role as the building blocks of the natural numbers, the prime numbers grow like weeds among the natural numbers, seeming to obey no other law than that of chance, and nobody can predict where the next one will sprout. The second fact is even more astonishing, for it states just the opposite: that the prime numbers exhibit stunning regularity, that there are laws governing their behavior, and that they obey these laws with almost military precision.

Our first exploration involves Euclid's theorem. His proof leads to an interesting sequence: 2, 3, 7, 43, 13, 53, 5, 6221671, 38709183810571, 139, 2801, 11, 17, 5471, 52662739, 23003, 30693651606209, 37, 1741, 1313797957,

887, 71, 7127, 109, 23, 97, 159227, 643679794963466223081509857, 103, 1079990819, 9539, 3143065813, 29, 3847, 89, 19, 577, 223, 139703, 457, 9649, 61, 4357, ... This sequence is generated as follows. Let $a_1 = 2$, the first prime. We apply Euclid's argument and consider 2+1; this is the prime 3 so we set $a_2 = 3$. We apply Euclid's argument and now have $2 \cdot 3 + 1 = 7$, which is prime, and set $a_3 = 7$. We apply Euclid's argument again and have $2 \cdot 3 \cdot 7 + 1 = 43$, which is prime and set $a_4 = 43$. Now things get interesting: we apply Euclid's argument and obtain $2 \cdot 3 \cdot 7 \cdot 43 + 1 = 1807 = 13 \cdot 139$, and set $a_5 = 13$. Thus a_n is the smallest prime not on our list generated by Euclid's argument at the nth stage. There are a plethora of unknown questions about this sequence, the biggest of course being whether or not it contains every prime. This is a great sequence to think about, but it is a computational nightmare to enumerate! You can learn more at the Online Encyclopedia of Integer Sequences (its homepage is `http://oeis.org/`; the page for our sequence is `http://oeis.org/A000945`). You can enter the first few terms of an integer sequence, and it will list whatever sequences it knows that start this way, provide history, generating functions, connections to parts of mathematics, ... This is a *great* website to know if you want to continue in mathematics.

In the time since Euclid lived, mathematicians have proved many results about the prime numbers. Nevertheless, many things about the prime numbers remain unknown and mysterious. A great example is whether or not there are infinitely many **twin primes** (two integers $p < q$ are twin primes if both are prime and $q = p+2$). It is believed there are infinitely many twin primes; in fact, well-believed conjectures suggest there are about $.66x/\log^2 x$ twin prime pairs at most x. This question is not just of academic interest; it was how Nicely [**38, 39**] discovered the error in the Pentium chip!

We now turn to finding the correct rate of growth of the number of primes at most x. First, some notation.

Counting primes: $\pi(x)$: We let $\pi(x)$ (π is the Greek letter "p") denote the number of primes at most x:

$$\pi(x) \ := \ \#\{p \leq x : p \text{ is prime}\}.$$

- From Euclid we know $\pi(x) \to \infty$ as $x \to \infty$. The celebrated **prime number theorem (PNT)** says that for large x, $\pi(x)$ is approximately $x/\log x$ (here $\log x$ is the natural logarithm of x).
- The percentage of numbers at most x that are prime is $\pi(x)/x$. By the prime number theorem, this ratio is approximately $1/\log x$. This implies that the percentage of numbers that are prime tends to zero, but very slowly: when $x = 10^{400}$ we still have 0.1% of integers are prime.

A proof of the prime number theorem is well beyond the scope of this book, but we can at least comment on it. The fact that the percentage of numbers that are prime tends to zero slowly means we have plenty of primes to work with, and we need not worry about running out of primes. The situation would be very different if we looked at **perfect squares** (these are integers of the form k^2 for some integer k). If

$$S(x) \ := \ \#\{n \leq x : n \text{ is a perfect square}\},$$

then $S(x) \approx \sqrt{x}$. Why? The numbers which are perfect squares at most x are

$$1^2,\ 2^2,\ 3^2,\ 4^2,\ \ldots,\ \lfloor\sqrt{x}\rfloor^2.$$

Thus the percentage of numbers at most x that are perfect squares is $S(x)/x \approx \sqrt{x}/x = 1/\sqrt{x}$. If $x = 10^{400}$, the percentage of numbers that are perfect squares is about $1/10^{200}$. If we want to find a perfect square near a given x, there's a very good chance there are none close; however, it is very likely that there's a prime nearby.

While we can't prove the prime number theorem in this book, we can at least show it's reasonable. We show in Exercise 7.8.49 how to get a lower bound for $\pi(x)$ from Euclid's argument.

In the next section we prove Fermat's little Theorem, an old and fundamental result about prime numbers. This theorem is used both to see why RSA works *and* in to find the large prime numbers necessary for RSA's key-generation.

7.6. Fermat's little Theorem

We end this chapter with one last result, Fermat's little Theorem. We'll use it for two different purposes when we get to RSA: it can be used to help find the large primes we need, and in implementing RSA.

Fermat's little Theorem (FℓT): For any prime number p, and any number a which is not a multiple of p, a^{p-1} is always congruent to 1 modulo p:

$$a^{p-1} = 1 \pmod{p}.$$

Note Fermat's little Theorem immediately implies that if a is relatively prime to n and a^{n-1} is not congruent to 1 modulo n, then n must be composite.

This is, at first glance, a slightly mysterious fact. Let's try it out with a "random" example. Suppose we compute $3^{12} \pmod{13}$. This can be done quickly with fast modular exponentiation. We have $12 = 8 + 4$, so $3^{12} = 3^8 \cdot 3^4$. Modulo 13, we have 3^2 is 9, 3^4 is 3, and 3^8 is 9. Thus

$$3^{12} \pmod{13} \ = \ 3^8 \cdot 3^4 \pmod{13} \ = \ 9 \cdot 3 \pmod{13} \ = \ 1 \pmod{13},$$

which agrees with Fermat's little Theorem.

If the modulus is not prime, on the other hand, the conclusion of Fermat's little Theorem can fail. Consider $3^{11} \pmod{12}$. We again compute using fast modular exponentiation. We have $11 = 8+2+1$, so $3^{11} = 3^8 \cdot 3^2 \cdot 3^1$.

After a little bit of algebra, we find $3^{11} = 3 \pmod{12}$. This leads to a very important observation: if there is an a such that a^{n-1} is not equal to 1 modulo n, then n cannot be prime. Unfortunately, the other direction need not be true. It's possible for $a^{n-1} = 1 \pmod{n}$ and yet n is composite (for example, $37^{56} = 1 \pmod{57}$, but 57 is composite).

We now turn to a proof of Fermat's little Theorem. There are many different proofs of this important fact. Fortunately there's an elementary, self-contained proof, which we now give. It's based on the fact that if $a \cdot b = a \cdot c \pmod{n}$ and a and n are relatively prime (which means they do not have any common divisors other than 1), then $b = c \pmod{n}$. In other words, if a and n are relatively prime, then we may cancel the a's.

Proof of Fermat's little Theorem. Let $p > 2$ be a prime number (it's easy to check it's true for $p = 2$), and a an integer relatively prime to p. Consider the list

$$a, \ 2a, \ 3a, \ \ldots, \ (p-1)a$$

of the first $(p-1)$ multiples of a. *No two elements on this list can be congruent to each other modulo p!* To see why, imagine that two of them are actually congruent; let's say $ka = \ell a \pmod{p}$, where k and ℓ are distinct numbers between 1 and $p-1$. Since a is relatively prime to p, it has an inverse modulo p, and so we can multiply both sides of this congruence by it to give $k = \ell \pmod{p}$. This is a contradiction, since k and ℓ are supposed to be different, so our assumption that $ka = \ell a \pmod{p}$ is false. Thus, all of the multiples $a, 2a, 3a, \ldots, (p-1)a$ are different modulo p.

Since the numbers $a, 2a, 3a, \ldots, (p-1)a$ are all congruent to different things modulo p, they must actually be congruent to each of the numbers $1, 2, 3, \ldots, (p-1)$ modulo p exactly once, just not in that order. That means that if we multiply them all together, we have

$$a \cdot 2a \cdot 3a \cdots (p-1)a \ = \ 1 \cdot 2 \cdot 3 \cdots (p-1) \pmod{p}.$$

We write this more compactly as

$$a^{p-1} \cdot (p-1)! \ = \ (p-1)! \pmod{p}. \tag{7.1}$$

We're almost done. Observe that since $(p-1)!$ is not divisible by p, $(p-1)!$ is relatively prime to p. Thus we can multiply both sides of the congruence in line (7.1) by the inverse of $(p-1)!$ (modulo p) to get the congruence

$$a^{p-1} \ = \ 1 \pmod{p}. \tag{7.2}$$

Remember, this holds only under the assumption that a is not a multiple of p. If we multiply both sides of this congruence by a, we get the congruence from Fermat's little Theorem, completing the proof. □

Remark. For those loving mathematical induction, we sketch a more advanced proof. We need the **binomial theorem**, which states $(x+y)^n = \sum_{k=0}^{n} \binom{n}{k} x^k y^{n-k}$. Note that if p is prime, then the binomial coefficient $\binom{p}{k}$ is divisible by p for $1 \le k \le p-1$. Using this and the binomial theorem, one can show $a^p = (a-1)^p + 1 \bmod p$ by writing a as $a - 1 + 1$ and expanding $((a-1)+1)^p$. If we now write $a-1$ as $a-2+1$ we find

$(a-1)^p = (a-2)^p + 1 \bmod p$, so $a^p = (a-2)^p + 2 \bmod p$. Continuing by induction we eventually arrive at $a^p = a \bmod p$.

Fermat's little Theorem can be used to solve a major problem for us: finding the large primes necessary for RSA key-pair generation. Why is finding large primes such a problem? Suppose we have chosen a large number, such as

$$x = 421249166674228746791672110734681729275580381602196445017243909397,$$

and we want to know if it is prime. What can we do? We need to make sure that it isn't divisible by any (positive) numbers other than itself and 1. The simplest way to check this would be to just try dividing x by 2, 3, 4, 5, 6, $\dots$, $x-1$; unfortunately, that's $x-2$ divisions. As x is 66 digits long, all the computers in the world working together couldn't do that many divisions in a trillion years.

You might have noticed, though, that it's unnecessary to try dividing by numbers like 4 and 6 since it's enough to check whether s has any prime factors. This saves us a bit of time, as we only have to try primes at most x. Unfortunately, the prime number theorem tells us that there are about $x/\log x$ primes at most x, which means there are still far too many candidate divisors to check. Note how quickly $x/\log x$ grows; it's almost as fast as x. If we look at the ratio of the number of primes at most x to integers at most x, we see the percentage that are prime is about $\frac{x/\log x}{x} = 1/\log x$. This means that when x is a billion we still have around 4% of integers are prime; at the astronomically high 10^{200}, we have over a quarter of a percent of integers are prime. There are important practical implications of this; the primes are far too numerous for us to try to brute force this problem. We need a better approach for real world applications.

We do significantly better when we notice that if x is composite, then x must have a nontrivial divisor less than $\sqrt{x}$. For our 66-digit number above, this means we only need to check on the order of 10^{33} numbers (or a little less if we only check the primes at most 10^{33}). The fastest supercomputer today could maybe do this many divisions in a few hundred thousand years or so—still too long to wait for Bob to make his RSA keys! And this was just for a 66-digit number. The numbers used for RSA can be hundreds of digits long.

The motivation for the **Fermat primality text** is that Fermat's little Theorem tells us a property that that all prime numbers have; namely, that they satisfy the congruence $a^{p-1} = 1 \pmod p$ for any values of a from 1 to $p-1$ (since p is prime, all these values of a are relatively prime to p). Maybe then, if we pick a random number q and check for lots of values of a that $a^{q-1} = 1 \pmod q$ (which we can do quickly using fast modular exponentiation), then q is likely to be prime. Remember, the prime number theorem tells us that primes are not too uncommon, so a computer would not have to make too many guesses before stumbling upon even really large prime numbers.

To formalize this notion, consider the following process.

> **Using Fermat's little Theorem (FℓT) to find primes:** The following tests whether or not q is prime. If the program says "not prime", then q is definitely composite.
>
> (1) Pick a random a between 1 and q.
> (2) Compute $a^{q-1} \pmod{q}$.
> (3) If $a^{q-1} \neq 1 \pmod{q}$, then q is `not prime`.
> (4) If $a^{q-1} = 1 \pmod{q}$, then q may be prime. Return to step 1 and choose a new a, or exit the program and return `possibly prime`.

Look at step 3. Since FℓT tells us that $a^{q-1} = 1 \pmod{q}$ if q is prime and a is between 1 and q, then if we find such an a for which this congruence does not hold, then q must not be prime! There are two ways this can happen: the first way is if a is not relatively prime to q. Since $1 < a < q$, this can't happen if q is prime, and so it means that q has to be composite. The second way this can happen is if a *is* relatively prime to q, but the Fermat congruence still doesn't hold. Since this violates Fermat's little Theorem, this also shows that q isn't prime. Such a value a is called a **Fermat witness**, because it uses Fermat's little Theorem to "witness" the fact that q is a composite number.

Step 4 is also interesting. In this case, we've tried an a and found that $a^{q-1} = 1 \pmod{q}$. Unfortunately, this doesn't tell us anything as q *may* be prime, but maybe not. For example, $8^{20} = 1 \pmod{2}1$. For example, for $q = 341$, if we tried $a = 2$ we would compute $2^{340} = 1 \pmod{341}$, and suspect that 341 is prime. Upon trying $a = 3$, however, we would compute $3^{341} = 56 \pmod{341}$ and discover that it is not (some trial and error reveals that $341 = 11 \cdot 31$).

The questions are: For how many a's do we need to compute $a^{q-1} = 1 \pmod{q}$ before we can decide that q is prime? Are there any numbers q which are not prime, but for which $a^q = a \pmod{q}$ holds for *all* a? In other words, are there any composite numbers that have no Fermat witnesses at all? It turns out that there are. The smallest such number is 561. These are called **Carmichael numbers**, and fortunately they are very rare. Until 1994 it was not even known if there were infinitely many of them. After 561, the next smallest is 1105, then 1729. As numbers get bigger, they get even rarer. In fact, there are so few large Carmichael numbers that if we randomly choose a hundred-digit number, there is less than a $\frac{1}{10^{32}}$ chance that it will be a Carmichael number.

What about composite numbers which are not Carmichael numbers? For those, we know that there is at least one Fermat witness—so, at least one

a between 1 and p for which $a^p \neq a \pmod{p}$. This is not that useful. If we have to do $p-2$ fast modular exponentiations to decide that p is probably prime, we might as well just do trial division (at least then we only have to check up to $\sqrt{p}$). It turns out, however, that any composite number which has at least one Fermat witness has lots and lots of them:

Theorem: *If a composite number q has at least one Fermat witness, then at least half the numbers relatively prime to q are Fermat witnesses for q.*

We give a proof in the appendix to this section. This theorem is very important for primality testing with Fermat's little Theorem, because it implies that if q is composite and we randomly choose an a between 1 and q, then if a is relatively prime to q we have at least a 50% chance that we will discover that q is composite by finding that $a^{q-1} \neq 1 \pmod{q}$. On the other hand, it turns out that if a is *not* relatively prime to q, we have a 100% chance of finding this out! That's because if some number d divides both a and q, then it will divide both a^{q-1} and q, and so a^{q-1} will be congruent to a multiple of d modulo q.

Thus if q is composite, then there is at least a 50% chance that we will find it out (i.e., that we end up in step 3 in the Fermat primality test). This doesn't seem good enough: after all, we want to be *really sure* that numbers are prime before using them with RSA. The solution to this problem is simply to *repeat the primality test multiple times so long as it returns "maybe prime"*. If we ever find a value which returns "not prime", we choose a different possible prime value q and start over. But if the algorithm keeps returning "maybe prime" for lots of different randomly chosen values of a, then this means there's an excellent chance that q is prime (unless it is a Carmichael number).

It's reasonable to assume that if q is composite, then there is at most a 25% chance that the algorithm returns "maybe prime" twice in a row with different randomly chosen a's. If we choose three random a's, then the probability is $.5^3 = .125 = 12.5\%$. After 10 times, the probability drops to $.0976\ldots\%$, and after 20 times, its down to $.0000953\ldots\%$. If the test was repeated a few hundred times, the probability that it would incorrectly identify a number as prime would be less than the probability of randomly choosing a particular particle in the known universe.

To apply the Fermat primality test in practice, then, we simply repeat the test enough times that we reach a desired level of confidence.

Appendix to 7.6. Proof of Number of Witnesses

Assume as stated in the theorem that q is a composite number and has a Fermat witness a that is relatively prime to q. Thus $a^{q-1} = s \pmod{a}$, where $s \neq 1 \pmod{q}$. Now choose any number b which is relatively prime to q, but is not a Fermat witness for q (so $b^{q-1} = 1 \pmod{q}$); if there is no such b, then well over half the candidates are witnesses. We have $a \cdot b$ is a Fermat witness to q.

To see why, first observe that since a and b are relatively prime to q, $a \cdot b$ is also relatively prime to q. Next we need to check that $(a \cdot b)^{q-1}$ is not congruent to 1 modulo q. We have

$$(a \cdot b)^{q-1} \;=\; a^{q-1}b^{q-1} \;=\; s \cdot 1 \;=\; s \pmod q,$$

and so $a \cdot b$ is a Fermat witness for q.

Assume now that we have many different numbers $b_1, b_2, b_3, \ldots, b_k$ between 1 and q, all of which are *not* Fermat witnesses to q. What we said above implies that

$$a \cdot b_1,\ a \cdot b_2,\ a \cdot b_3,\ \ldots,\ a \cdot b_k$$

are all Fermat witnesses. Moreover, since a is relatively prime to q, these must all be different modulo q: if two were the same, we could multiply both by the inverse of a and find that to of the b_i's were the same, which would be a contradiction. This means that there are at least as many numbers relatively prime to q which *are* Fermat witnesses as there are which are not, completing the proof. □

7.7. Summary

In this chapter we defined possible public-channel cryptography systems: one used graph theory to define a perfect code, another defined and analyzed KidRSA. We saw yet again one of the fundamental dangers of cryptography: just because something looks hard, that does not mean it is hard. We saw how KidRSA can be easily cracked by the extended Euclidean algorithm. Fortunately, all was not lost. We shift from multiplication to exponentiation, which we can do very rapidly (like the extended Euclidean algorithm). We continued building up the mathematical prerequisites which we need to tackle RSA in Chapter 8.

7.8. Problems

• From §7.1: The Perfect Code Cryptography System

EXERCISE 7.8.1. *What are the degrees of the vertices in the graph in Figure 7.1?*

EXERCISE 7.8.2. *Assign the labels A through I to the vertices in the graph from Figure 7.3 so that it matches the labeling of the graph in Figure 7.1. Note for example that since A and E are adjacent in the drawing of the graph in Figure 7.1, A and E must be adjacent in your labeling of the graph. On the other hand, since C and E are not adjacent in the earlier drawings of the graph, they cannot be adjacent in your labeling of this graph.* Hint: *Note that in Figure 7.1, A was the only vertex of degree 1. Thus it is easy to tell which vertex must be A, and now it should be easy to figure out which vertex is E.*

EXERCISE 7.8.3. *Give an example of a graph with three vertices that does not have a perfect code. Given any integer n, can you find an example of a graph with n vertices that does not have a perfect code? If yes, find one.*

EXERCISE 7.8.4. *Consider a graph with n vertices. One way to find a perfect code is to look at all sets with just one vertex, and see if any of these are a perfect code. If none work, try all sets with exactly two vertices, and see if any are a perfect code. Show that the number of candidate sets to check is approximately 2^n. This grows exponentially with n, and thus such a brute force approach is impractical.*

EXERCISE 7.8.5. *You want to send Robert a message: the number* 30. *Assume Robert sends you the graph from Figure* 7.8 *to use for the perfect-code system. Give two labelings of this graph: the labeling produced by "spreading out" your message on the graph, and the labeling you send to Robert. Check that Robert recovers the correct message.*

EXERCISE 7.8.6. *An eavesdropper who doesn't know the perfect code can deduce the "spread-out" labeling of graph from the "clumped" labeling by solving a system of linear equations. Imagine Eve intercepts the graph in Figure* 7.12 *and labels the vertices as we did in Figure* 7.1. *Let $x_A, x_B, \dots, x_I$ denote the values placed at vertices A, B, ..., I. We do not know these values, but we do know the values on the intercepted graph; let's denote these observed values by $y_A, y_B, \dots, y_I$. Further, we know the rule used to construct the y-values from the x-values. For example, $y_A = x_A + x_E$. Write down the system of equations and solve for the original values $x_A, \dots, x_I$. Note that this allows you to decipher the message* without *having to find the perfect code! In general, if the graph has n vertices one has to solve a system of n linear equations in n unknowns, which can be done through linear algebra.*

The next few problems involve graphs; some of these require linear algebra. We always assume our graphs are **simple** (so we do not have multiple edges connecting two vertices) without **self-loops** (so no vertex is connected to itself). We define d-regular in Exercise 7.8.7, connected and disconnected in Exercise 7.8.12, and adjacency matrix in Exercise 7.8.13.

EXERCISE 7.8.7. *A simple graph without self-loops is d-**regular** if each vertex has exactly d edges going from it to d distinct vertices. What is the fewest vertices we need to have a d-regular graph? Give an example of a d-regular graph with that number of vertices.*

EXERCISE 7.8.8. *Let $n > 2$ be a positive integer. Draw a picture of a 2-regular graph for $n = 5$, $n = 10$, and $n = 20$.*

EXERCISE 7.8.9. *Draw a 3-regular graph for some $n > 10$, or prove no such graph exists.*

EXERCISE 7.8.10. *Find a perfect code on a 2-regular graph with* 20 *vertices, and transmit the message* 16.

EXERCISE 7.8.11. *If you were able to find a 3-regular graph in Exercise* 7.8.9, *can you find a perfect code on it? If yes, transmit the message* 20.

EXERCISE 7.8.12. *A graph is* ***connected*** *if there is a path of edges connecting any two vertices; otherwise the graph is* ***disconnected.*** *Draw a connected graph with at least ten vertices, and a disconnected graph with at least ten vertices.*

EXERCISE 7.8.13. *The* ***adjacency matrix*** *of a simple graph G without self-loops on n vertices is the $n \times n$ matrix where the entry in the ith row and j^{th} column is* 1 *if vertex i and j are connected by an edge, and* 0 *otherwise. Note that the adjacency matrix depends on the labeling of the vertices. Write down the adjacency matrix for the graph from Figure* 7.1 *(assign the label* 1 *to vertex A,* 2 *to vertex B, and so on).*

EXERCISE 7.8.14. *Write down the adjacency matrix for a* 2*-regular graph on ten vertices.*

EXERCISE 7.8.15. *(For those who know linear algebra). Let G be a d-regular graph with n vertices. Prove G is connected if and only if d is an eigenvalue with multiplicity* 1 *of the adjacency matrix of G. This provides a* very *nice way to test if a graph is connected!*

- **From §7.2: KidRSA**

EXERCISE 7.8.16. *The first step of KidRSA creates an integer M from integers a, b by setting $M = ab - 1$. Prove that you can find infinitely many pairs of integers (a, b) where each of these is at least* 2 *and M is a prime.* Hint: You may use the fact that there are infinitely many primes.

EXERCISE 7.8.17. *The first step of KidRSA creates an integer M from integers a, b by setting $M = ab - 1$. Prove that you can find infinitely many pairs of integers (a, b) where each of these is at least* 2 *and M is the square of an integer.*

EXERCISE 7.8.18. *The first step of KidRSA creates an integer M from integers a, b by setting $M = ab - 1$. Can you find infinitely many pairs of integers (a, b) where each is at least* 2 *and M is a multiple of* 1701?

EXERCISE 7.8.19. *Given an integer M, find the best bound you can on the number of pairs (a, b) of integers with $a \leq b$ and $ab - 1 = M$.*

EXERCISE 7.8.20. *Carry out the KidRSA key generation procedure with the choices $a = 5, b = 4, a' = 3, b' = 6$. Indicate which key you publish, and which you keep secret.*

EXERCISE 7.8.21. *Using the KidRSA values from Exercise* 7.8.20, *someone uses your public key to send you the encrypted message* 27. *What is the decrypted message?*

EXERCISE 7.8.22. *Your friend is worried that KidRSA is not secure, and comes up with what he believes is an improved version. He chooses eight numbers* $a, b, a', b', A, B, A', B'$ *such that* $ab - 1 = AB - 1 = M$. *Similar to KidRSA, he now finds pairs* $(e, n), (d, n)$ *and* $(E, n), (D, n)$. *To encrypt a message, he first multiplies by* e *and reduces modulo* n, *and then multiplies by* E *and reduces modulo* n. *To decrypt, you first multiply by* d *and reduce modulo* n, *and then multiply by* D *and reduce modulo* n. *Does this provide any additional security? Why or why not? Would there be additional security if the encryption/decryption pairs had different values of* n *(so now* $ab - 1 \neq AB - 1$*)?*

- **From §7.3: The Euclidean Algorithm**

EXERCISE 7.8.23. *Show that if* $d'|x$ *and* $d'|y - x$, *then* $d'|y$.

EXERCISE 7.8.24. *Find the following gcd's.*

(a) $\gcd(573, 1042)$
(b) $\gcd(690, 1219)$
(c) $\gcd(695, 842)$
(d) $\gcd(335, 1407)$

EXERCISE 7.8.25. *Find the following modular inverses—or, when there is none, give the greatest common divisor of the two numbers.*

(a) $269^{-1} \pmod{552}$.
(b) $321^{-1} \pmod{529}$.
(c) $641^{-1} \pmod{1000}$.
(d) $105^{-1} \pmod{196}$.

EXERCISE 7.8.26. *We only need the extended Euclidean algorithm in the special case when the greatest common divisor is* 1. *Show, however, that we obtain a similar relation even if the greatest common divisor is* 2 *or more. In particular, if* $x \leq y$ *are nonnegative integers, it always returns integers* a *and* b *such that* $ax + by = \gcd(x, y)$.

EXERCISE 7.8.27. *Consider three adjacent rows in the extended Euclidean algorithm:*

	y	x
$\vdots$	$\vdots$	$\vdots$
p	b_p	a_p
q	b_q	a_q
r	b_r	a_r

Fill in the details to prove that $r \leq p/2$.

- Case 1: *If* $q \leq p/2$, *then* $r \leq p/2$.
- Case 2: *If* $q > p/2$, *then* $p - q < p/2$, *which implies* $r < p/2$.

Thus every two rows is at least a factor of 2 *smaller than two rows above. As* $2^{\log_2 x} = x$, *we see we need about* $\log_2 x$ *sets of two steps to reduce* x *down to* 0 *or* 1 *and terminate the algorithm, so we need about* $2\log_2 x$ *steps.*

EXERCISE 7.8.28. *We found a great upper bound for the Euclidean algorithm in Exercise 7.8.27. While there was a danger it could be slow (as it is possible for a step to replace the pair (x, y) with $(x-1, x)$), we saw that every* two *steps sends the pair (x, y) to a pair (u, v) with $u < v$ and $u \leq x/2$. This implies that the Euclidean algorithm is fast—it runs in at most $2\log_2 x$ steps. A natural question is whether or not this can be improved; perhaps if we look at every four steps we will see an even greater savings. Unfortunately, we can't do better than $c\log_2 x$ steps for some $c > 1$ in general, as there are pairs (x, y) that take a long time. Investigate* worst case lower bounds *for the Euclidean algorithm by letting x and y be two adjacent **Fibonacci** numbers. (Remember the Fibonacci sequence is given by the recurrence relation $F_{n+2} = F_{n+1} + F_n$, with initial conditions $F_0 = 0$ and $F_1 = 1$.)*

EXERCISE 7.8.29. *Use the Euclidean algorithm to find the greatest common divisor of* 1701 *and* 24601. *How many steps are needed?*

EXERCISE 7.8.30. *Discuss how you would find the greatest common divisor of three numbers. What is the fewest number of steps you would need for three general numbers x, y, and z?*

For an additional problem involving the Euclidean algorithm, see Exercise 11.7.27 (which is the start of a series of problems related to prime numbers).

- **From §7.4: Binary Expansion and Fast Modular Exponentiation**

EXERCISE 7.8.31. *Find 10011_2, 1001001_2, and $10011_2 + 1001001_2$; for the sum, write the answer in both binary and decimal form.*

EXERCISE 7.8.32. *Find 11001_2, 101010101_2, and $110011_2 - 1001011_2$; for the difference, write the answer in both binary and decimal form.*

EXERCISE 7.8.33. *Show $25_{10} = 31_8$. (This is the punchline to the joke: Why does the computer scientist confuse Christmas with Halloween?)*

EXERCISE 7.8.34. *Let B_1 and B_2 be two different bases. Find all positive integers x such that the digits of x's base B_1 expansion are the same as the digits of x's base B_2 expansion. For example, if $B_1 = 5, B_2 = 10$, and $x = 23$, then the base 5 expansion is 43_5, while the base 2 expansion is 10111_2; as 43 is not the same as 10111, this x does not work.* Hint: Choose some values for B_1 and B_2 to build up your intuition, and then try to generalize.

EXERCISE 7.8.35. *Describe how to add two numbers quickly in binary, where we write the output in binary.*

EXERCISE 7.8.36. *Prove each positive integer x has one and only one base B expansion.* Hint: First show an expansion exists by subtracting off multiples of the highest power of B less than x, then repeat this procedure with the reduced number. Once you have existence, to show uniqueness imagine that there are two decompositions. Set them equal and find a contradiction.

EXERCISE 7.8.37. *Show*

$$25 \cdot 47 \pmod{14} = (25 \pmod{14}) \cdot (47 \pmod{14}) \pmod{14}.$$

EXERCISE 7.8.38. *Prove that if x, y are integers and n is a positive integer that $x \cdot y \pmod{n} = (x \pmod{n}) \cdot (y \pmod{n}) \pmod{n}$.*

EXERCISE 7.8.39. *For each of the following problems, solve using fast modular exponentiation.*

(a) $11^{80} \pmod{27}$
(b) $15^{43} \pmod{25}$
(c) $17^{165} \pmod{29}$
(d) $8^{250} \pmod{31}$.

EXERCISE 7.8.40. *Using fast modular exponentiation, how many steps are needed to compute* $17823941^{82521821} \pmod{748283710527193}$.

EXERCISE 7.8.41. *Approximately how many steps would you expect it to take to compute*

$$3^{564568493762517129347456462819918172636476381919273764361819} \pmod{2385712349}$$

using fast modular exponentiation?

EXERCISE 7.8.42. *The purpose of this section was to show you that sometimes, if we're sufficiently clever, we can solve problems significantly faster than you'd expect. If you know some linear algebra, think about matrix multiplication. If A is an $n \times n$ matrix, you might expect it will take on the order of n^3 operations to find A^2, as we need to find n^2 entries, and each entry requires n multiplications. It turns out we can do this much faster. Look up the* ***Strassen algorithm****, and describe how it works. Compare how many steps it needs for a 128×128 matrix versus the number of steps needed from standard matrix multiplication.*

EXERCISE 7.8.43. *Read about the mathematician Frank Cole and his famous factorization of the Mersenne number $M_{67} = 2^{67} - 1$. Would you rather do the multiplication of the suggested factors in base 10 or base 2? Why? Do the multiplication in whichever base you prefer* without *using a computer (as Cole did, to thunderous applause).*

• From §7.5: Prime Numbers

EXERCISE 7.8.44. *Determine which of the following are prime:* 91, 103, 113.

EXERCISE 7.8.45. *Show if $n \geq 2$ is composite, then some prime p must divide it.* Hint: This problem can be a bit tricky. Try for a proof by contradiction. Imagine there is a smallest composite number n such that no prime divides it. As n is composite, we can write it as xy for some $x, y \geq 2$. Show $x, y < n$. As we assumed n is the smallest composite not divisible by a prime, there must be a prime dividing x. Show that prime also divides n.

EXERCISE 7.8.46. *Using a table of prime numbers, check what percentage of the prime numbers at most* 100 *occur in "twins" (i.e., are part of a twin prime pair). Also check the percentages for primes at most* 200 *and at most* 300.

EXERCISE 7.8.47. *In Euclid's proof we multiply the first L primes together. Can we prove there are infinitely many primes by adding rather than multiplying? For example, maybe we consider* $2 + 3 + \cdots + p_L$, *or* $2 + 3 \cdots + p_L + 1$.

EXERCISE 7.8.48. *Euclid proved that there are infinitely many primes by considering* $p_1 \cdots p_L + 1$. *A similar argument proves there are infinitely many primes congruent to* 3 *modulo* 4. *Prove this claim.* Hint: Assume not, and consider $4p_1 \cdots p_L + 3$ where $p_1, \dots, p_L$ are now all the primes congruent to 3 modulo 4.

EXERCISE 7.8.49. Hard: *Use Euclid's method to prove that there is a constant* C *such that the number of primes at most* x *is at least* $C \log \log x$. *While this is far smaller than the true value, it at least tends to infinity.*

EXERCISE 7.8.50. *Let* $\mathcal{S}$ *be a set of numbers such that there is a* C *so that the number of elements of* $\mathcal{S}$ *at most* x *is at most* $C \log \log x$. *Prove this* does not *force the sum of the reciprocals of elements of* $\mathcal{S}$ *to diverge. In particular, deduce that the lower bound on the number of primes at most* x *from the previous problem is not sufficient to prove the sum of the reciprocals of the primes diverge. They do; see* **[13]** *for a proof of this and many other facts about the distribution of the primes.*

EXERCISE 7.8.51. *The volume of a sphere of radius* r *is* $\frac{4}{3}\pi r^3$. *The sun's radius is about* 432,450 *miles, and the Earth's radius is around* 3,958 *miles. How many Earths can fit inside the sun?*

EXERCISE 7.8.52. *Exercise* 7.8.51 *illustrates how easy it is to unknowingly make assumptions. When we're asked how many Earths fit inside the sun, there are two interpretations. The easier is to assume we can break the Earths into pieces, and thus we just need to divide the volume of the sun by the volume of the Earth. The other alternative is to assume we can't break the Earths into smaller pieces, and thus we have to pack spheres inside a giant sphere. This is a hard problem. Let's try a simpler one. How many circles of radius* 1 *can we fit inside a circle of radius* 4*? What if the larger radius is* 10*?*

EXERCISE 7.8.53. *Read Nicely's papers* **[38, 39]** *where he finds the Pentium bug. Discuss the mathematics of the issue and the financial impact (Intel lost hundreds of millions of dollars).*

- **From §7.6: Fermat's little Theorem**

EXERCISE 7.8.54. *Use Fermat's little Theorem to test whether or not* 7 *is prime.*

EXERCISE 7.8.55. *Use Fermat's little Theorem to test whether or not* 10 *is prime.*

EXERCISE 7.8.56. *$F\ell T$ only tells us that $a^{p-1} = 1 \pmod p$ for prime values of p, and for other values, it is not always true (for example, for $p = 8$, the theorem would be false since $3^7 = 3 \pmod 8$). Where in the proof did we use that p is a prime number? Identify the step in the proof that depended on p being a prime number.*

EXERCISE 7.8.57. *Show that if n is composite, then it has a factor at most $\sqrt{n}$. This implies that we need only check for factors up to $\sqrt{n}$ to see if a number is prime.*

EXERCISE 7.8.58. *As one only needs to divide by primes up to $\sqrt{n}$ to check for primality, estimate how many divisions it could take to determine whether a* 100*-digit number is prime.*

EXERCISE 7.8.59. *Choose any number a between* 2 *and* 560*, and use fast exponentiation to check that $a^{560} = 1 \pmod{561}$.*

EXERCISE 7.8.60. *Find nonzero integers a, b, c, and n such that $a \cdot b = a \cdot c \pmod n$ but b is not equal to c modulo n.*

EXERCISE 7.8.61. *Test the number* 1089 *for primality using the Fermat primality test (show all your fast modular exponentiation calculations). Test until the algorithm returns "not prime", or until it returns "maybe prime" for three different values of a (whichever occurs first). For a number of this size not divisible by* 2, 3, *or* 5, *testing three times would be enough to impart a minimum confidence of roughly* 80% *that it is prime—not enough for real-life RSA implementations, but enough for this exercise.*

EXERCISE 7.8.62. *We use the notation $F\ell T$ for Fermat's little Theorem as **FLT** is already taken; that's used for **Fermat's Last Theorem**. Fermat's Last Theorem is in a sense a generalization of the Pythagorean theorem. The Pythagorean theorem says that if a right triangle has sides of length a and b and a hypotenuse of length c, then $a^2 + b^2 = c^2$. There are infinitely many triples of integers satisfying this, such as $3^2 + 4^2 = 5^2$ and $5^2 + 12^2 = 13^2$. FLT says that if $x^n + y^n = z^n$ with x, y, z integers and $n \geq 3$ an integer, then the only solutions have at least one of x, y, z equal to* 0. *This problem has been a terrific source of motivation for centuries, and was the impetus behind many mathematical discoveries. Look up some of the history of this problem, and some of the mathematical fields it helped initiate.*

EXERCISE 7.8.63. *The previous exercise discusses Fermat's Last Theorem. Using it, investigate what happens if n is a* negative *integer. Specifically, when can you find an integer solution to $x^{-m} + y^{-m} = z^{-m}$, where $xyz \neq -$ and m is a positive integer?*

EXERCISE 7.8.64. *Fermat's little Theorem can be generalized to* ***Euler's theorem****: if $\varphi(n)$ is the number of integers in $\{1, \cdots n\}$ that are relatively prime to n, then for all a relatively prime to n we have $a^{\varphi(n)} = 1 \pmod{n}$. For us, the most important case of this is when n is the product of two distinct primes p and q. Prove Euler's theorem in this case. You're asked to prove the general case in Exercise* 8.8.54.

Chapter 8

Public-Channel Cryptography

As in Chapter 7, here we concentrate on developing a two-key system, one key for encrypting a message and one for decrypting the message. The message itself can be numeric or alphabetic, including a conversion of text to numbers. In Chapter 7 we discussed several attempts at constructing a public key system. Both the perfect code and the KidRSA methods are based on taking a problem which is easy to solve with some extra information, but believed to be hard without it. For example, it is easy to construct a graph with a perfect code, but it is not easy to find a perfect code in a graph. Unfortunately, it turns out to be possible to decipher the message *without* finding the perfect code. For KidRSA, encryption is simple as it's just one modular multiplication. At first it seemed difficult to undo without knowing the multiplicative inverse, but we then saw that we could quickly compute this inverse using the extended Euclidean algorithm.

Though neither method is practical, the basic idea is sound. The main thrust of this chapter is to describe RSA, and some of the issues that arise in its implementation. Initially we consider numeric messages, but since we want to encrypt long messages, we concentrate on the development of the encryption system RSA for encrypting the keys only. We again try to build a system on a mathematical problem which is easy in one direction but believed to be hard to invert without special, additional secret information. This time, the pairing is multiplication and factorization. If we take two large primes p and q it is very easy to multiply them together; however, we currently do not have efficient algorithms to factor large numbers into primes unless we know some additional information. We therefore hope a cryptosystem based on this problem will be secure.

It's important to note, however, that this is just a hope; we do *not* have a proof of the security of RSA. All we can do is show that it's resistant to a large number of varied attacks and that to date no one has published a successful assault. Of course, if someone does learn how to factor large numbers, they may deliberately keep that information secret, and thus be able to read RSA encrypted messages without the senders or recipients being any wiser. In Chapter 11 we discuss some factoring algorithms and primality tests. Interestingly, there are ways to quickly determine if a number is prime or composite; however, for composite numbers we learn that it is factorable *without* knowing the factor—it's a pure existence argument!

Again, the remarkable fact about RSA and other public-key encryption systems is that they allow two people to securely communicate without having previously met to agree upon a secret key. This is a monumental achievement, and the culmination of our search for a fast, practical and believed-to-be-secure encryption method.

We start the chapter with a description of the mechanics of how to implement RSA, and wait until the final section to prove why it works. After we understand how RSA works, we discuss how it is used in practice. Frequently it takes too long to encrypt and decrypt a long message with RSA, so in practice RSA is used to transmit a secret to allow people to use less secure means of communication. We then turn to an issue which we have mostly ignored so far: just because we receive a message that claims to be from Alice does not mean it *is* from Alice. This leads to the concept of a digital signature. Without the ability to easily "sign" messages, e-commerce would break down. It's not enough to just attach our signature to the end of a message, though; we must convince the recipient that we meant to sign the transmitted message. For example, imagine we are tricked and accidentally sign a blank piece of paper; a nefarious person could then write any message above it and it would appear that we signed it. It is thus imperative that our signing process not only establish our identity, but also that we intended to sign the received message. One solution to this problem is with hash functions, our next topic. Finally, though RSA was the gold standard of encryption for years, it was not the first procedure to publicly transmit a secret. That honor goes to the Diffie–Hellman protocol, and we end with a brief tour of this revolutionary idea. Before its discovery, many practitioners doubted the possibility of exchanging a secret in public.

8.1. RSA

In this section, we introduce the RSA public-key encryption algorithm. RSA was first publicly described in 1977 by Ron Rivest, Adi Shamir, and Leonard Adleman, and it was named for these founders (Clifford Cocks found something similar in 1973 while working for the British Government Communications Headquarters, which classified the work). It was the first publicly known example of a secure, practical algorithm that could be used to encrypt secret messages without somehow establishing a secret key beforehand. Before the 1970s, nobody even had any reason to think that such a remarkable thing could be possible.

As a public-key cryptography system, RSA calls on the intended recipient of an encrypted message to create a public and private key. He publishes the public key, which anyone can then use to encrypt messages to him, and keeps the private key secret. In the case of the perfect code system, the ability to create a public/private key-pair depended on the ease with which one can create graphs with known perfect codes, in spite of the difficulty associated with finding them. In the case of KidRSA, the ability to create a key-pair depended on a method to create triples of numbers e, d, n such that $d = e^{-1} \pmod{n}$, in spite of the difficulty (we thought!) associated with finding modular inverses.

The RSA public/private key-pair generation is based on the difficulty of factoring large integers. Given a number like 15, it's easy enough to determine that it factors as $15 = 3 \cdot 5$. When we try to factor a larger number, like 1,010,021, it becomes apparent that factoring the number 15 depended largely on our ability to simply try all possible divisors; we could do this since 15 is quite small. We don't know any good way to factor a larger number like 1,010,021; certainly, trying all possible factors by hand is impractical! In real-life computer implementations of RSA, the numbers which have to be factored to break the system are hundreds of digits long, and would take even the world's fastest computers thousands or millions of years to factor. We have already seen in the perfect code and KidRSA systems that the "hard problem" that these public-key systems are based on tend to have an asymmetry to them (easy to do one way, hard to do the other). In the case of factoring, this asymmetry comes from the fact that it is very easy to generate numbers for which we know the factors: simply multiply some prime numbers together! Multiplication is very fast. In fact, even for hundred-digit numbers, this would in principle be possible to do by hand (although it might take you a couple days!). A computer can do such long multiplications in microseconds.

KidRSA is not secure because finding modular inverses is easy after all, once we know the extended Euclidean algorithm. Although no one knows a way to easily find perfect codes in graphs, the perfect code system is insecure because an attacker can circumvent the system by solving a system of equations, decrypting the message without ever discovering the perfect code. To date no one has figured out a way to break the underlying problem of RSA (by finding a way to quickly factor integers), and no one has found a way to circumvent the system by breaking it without factoring integers. That is not to say that they won't, however.

We'll discuss some of these issues later, but first let's go over the mechanics of RSA. See §2.5 for a review of some needed number theory notation. We additionally need the **Euler totient function**, φ. It's defined on positive integers n by setting $\varphi(n)$ equal to the number of positive integers at most n that are relatively prime to n; thus

$$\varphi(n) \;=\; \#\{m : 1 \leq m \leq n \text{ and } \gcd(m, n) = 1\}.$$

While this function has a rich history and many nice properties (some of which we develop in Exercises 8.8.7 through 8.8.11), for RSA all we need is Exercise 8.8.7: $\varphi(pq) = (p-1)(q-1)$ when p and q are distinct primes.

Suppose Bob wants to receive RSA messages. He needs to generate a public/private key-pair. He does this as follows.

RSA encryption: The following is the **RSA key generation** for Bob.

(1) Choose distinct prime numbers p and q.
(2) Set $n = pq$, and compute $\varphi(n) = (p-1)(q-1)$.
(3) Choose any e (for encrypt) relatively prime to $\varphi(n)$, and set d (for decrypt) equal to $e^{-1} \pmod{\varphi(n)}$.
(4) Bob's **public key** is (e, n), while (d, n) is his **private key**.

The messages (keys) are positive integers m that are less than n and relatively prime to n. Alice encrypts her message m by sending the ciphertext $c = m^e \pmod{n}$. Bob decrypts the received message c by computing $c^d \pmod{n}$, which equals m.

Unlike KidRSA, where encryption and decryption were done with multiplication, in RSA encryption and decryption are done with exponentiation: given a message m, Alice sends the ciphertext $c = m^e \pmod{n}$. Bob decrypts the ciphertext by computing $c^d \pmod{n}$. Fast modular exponentiation shows that we can find c and c^d extremely quickly; the hope is that exponentiation mixes things up more than just multiplication, leading to greater security. *It should not be obvious at this point why* $c^d \pmod{n} = m$ *and thus recovers the original message.* It does, though. The proof properly belongs in a higher level mathematics course, but fortunately we've already developed just the bits we need and can give a quick, self-contained proof in §8.6.

For now, let's do an example to get a feel for how RSA works. Suppose Bob wants to create a private/public key-pair and chooses $p = 3$ and $q = 11$. Then $n = 33$ and $\varphi(n) = (3-1)(11-1) = 20$. If he chooses $e = 3$, then d is $3^{-1} \pmod{20}$, which is 7 (we check, and do indeed find $3 \cdot 7 = 21$, which is 1 modulo 20). Thus Bob publishes the public key $(3, 33)$. His private key is $(7, 33)$.

If now Alice wants to encrypt the message "5" using Bob's public key, she computes $5^3 = 26 \pmod{33}$. When Bob receives the message 26, he decrypts by calculating $26^7 \pmod{33}$, which will recover the plaintext 5.

Something seems quite strange about this, however. How can this be secure? Knowing that $n = 33$ and $e = 3$, can't Alice (or an attacker Eve) figure out that $\varphi(n) = 20$ and then compute the inverse of $3 \pmod{20}$ to find d, just like Bob did? At first glance, this seems reasonable, but figuring out what $\varphi(n)$ is from n requires *factoring* n (while Bob knows that $\varphi(33) = 20$ since he knows p and q, the hope is that without knowing the factorization of n there is no easy way to deduce $\varphi(n)$, and this requires an attacker to find the modular inverse of 3 without knowing the problem's modulus!) For large numbers, factoring isn't feasible. Note that in addition to keeping d secret, Bob must keep $\varphi(n)$ secret as well, and, by extension, the primes factors p and q of n, since these can easily be used to compute $\varphi(n)$.

One final point to consider is whether RSA can actually be carried out efficiently. We've already said that it should be infeasible for someone to break RSA by factoring large integers, but maybe even generating RSA keys and encrypting messages is already too cumbersome to be practical. Fortunately, this is not the case. Let's examine the steps of RSA key generation.

The first step is choosing prime numbers p and q. In practice, the numbers that are used for RSA are hundreds of digits long; in particular, they're much too large to factor. It turns out that, in spite of the fact that we can't factor really large numbers, there are nevertheless good ways of finding really large prime numbers. We discussed this a bit when we talked about Fermat's little Theorem in §7.6, and will explore these issues in more detail in Chapter 11.

The second step of the process involves only subtraction and multiplication, which is very easy to do.

In the third step, we need to be able to choose e relatively prime to $\varphi(n)$, and then compute the inverse of e modulo $\varphi(n)$. Both of these tasks can be done with the extended Euclidean algorithm: we can choose an e we hope will be relatively prime to $\varphi(n)$, and carry out the extended Euclidean algorithm with e and $\varphi(n)$. If the algorithm returns a gcd of 1, then the numbers are in fact relatively prime, and it also gives the inverse! If it gives a gcd greater than 1, we try a different value for e and run the algorithm again. We discuss some related issues in Exercise 8.8.12.

Finally, the actual encryption and decryption operations, consisting of modular exponentiations, can be done very efficiently using the streamlined method of fast modular exponentiation. We need to make sure the message m is relatively prime to $n = pq$, but this is fairly easy to do. It's very rare for m to share a nontrivial divisor with n, as the only way that could happen is if p or q divides m; see Exercise 8.8.5 for some results on how rare this is. We can use the extended Euclidean algorithm to quickly determine whether or not m and n share a factor (if their gcd is 1, then they are relatively prime).

Thus the steps of RSA are easily implemented. "All" that remains is proving why $c^d \pmod{n} = m$, where $c = m^e \pmod{n}$. We hold off on proving this result until the end of the chapter (see §8.6). Instead, we next investigate some of the implementation issues of RSA. These range from whether or not RSA is fast enough to be practically useful, to the ability to authenticate messages. These topics are some of the most important in cryptography, and provide an excellent introduction to many modern problems.

8.2. RSA and Symmetric Encryption

Throughout the book we've harped on the need for speed. It isn't enough that something is mathematically possible; to be practical we need to be able to do it relatively quickly, ideally in real time. RSA meets these requirements if we are only sending a short message, just a few numbers. Unfortunately, it is too slow for long messages, and there are lots of occasions where lengthy amounts of information need to be sent.

Let's examine what's happening. Messages encrypted by RSA must be numbers, and they must be at most n, where n is the publicly displayed modulus. It's easy to convert text to a number (we can let `A` be 01, `B` be 02 and so on, and just run the digits together). If n is say 400 digits, this leads to an unacceptably small restriction on our message size. We can get around this by breaking our message into small parts, each encrypted separately and then sent in a stream. Unfortunately, if the message is large (perhaps it is a website, which may include images) this would require lots of encryptions. Although the operations used for RSA (such as the extended Euclidean algorithm and fast modular exponentiation) can be done efficiently, they still take long enough that we don't want to have to perform them thousands or millions of times. For example, a book can have between half a million and a million alphanumeric characters. We need a better way than applying RSA to each bit.

The solution to this problem is to encrypt messages using a symmetric encryption method (such as AES) and use RSA simply to transmit the keys used for the symmetric method. Thus while the message itself (which may be quite large) is encrypted with a block or stream cipher (both of which are *much* faster than RSA encryption), only the key (which is potentially much smaller than the message) need be encrypted using RSA.

As an example, suppose we have published the RSA public key $(3, 55)$ and have a corresponding private key of $(27, 55)$. We might receive the following message:

> Dear [our name here]. The secret message is
> `QZODK BFUZS WQKEU EQMEK`
> Encryption is with the Caesar cipher. The key, encrypted with your RSA key, is 23.

Notwithstanding the fact that the Caesar cipher is very insecure, this message could be transmitted without any security at all—via email, or even on a billboard. That's because determining the key used for the encryption requires knowing our RSA private key (or breaking RSA, for example by factoring 55, which with such small numbers is also quite feasible). In this case, since our key is $(27, 55)$ we recover the symmetric encryption key by calculating $23^{27} \pmod{55}$, which gives 12. Subtracting 12 from the Caesar cipher text recovers the message `ENCRY PTING KEYSI SEASY` ("encrypting keys is easy").

Of course, in real life applications the symmetric cipher used is not the Caesar cipher, but something secure like AES. Let's see how one could use BabyBlock (see §6.7) in tandem with RSA.

Suppose we want to send the message `HURRY` to someone with whom we have not exchanged any secret keys. We can nevertheless encrypt the message using BabyBlock with a key of our choosing—say, 1010. The message `HURRY` becomes 0011110100100011000111000 in binary, and is padded with three 0's to produce 0011110100100011000111000000, so that the number of digits is a multiple of 4. BabyBlock encryption with the key 1010 results in a binary message of 1101001111001101111100100000, which converts to `.PG_FY`. We can transmit an encrypted version of this message using BabyBlock, and then send the key encrypted with RSA. Note that if we view the key 1010 as a binary number, it corresponds to the number 18. If the recipient has published an RSA public key of $(5, 91)$ we would calculate the encryption $18^5 = 44 \pmod{91}$, and transmit the following to the agent:

`.PG_FY` Encryption is with BabyBlock. The key, viewed as a binary number and encrypted with your RSA key, is 44.

If BabyBlock was substituted for a secure cipher like AES, and RSA was being done with suitably large numbers, such a message could be transmitted in the open. To recover the message, the recipient uses their private key, which is $(29, 91)$ in this case, to decrypt the key. They calculate $44^{29} \pmod{91}$ and recover 18. By writing 18 in binary as 1010, they find the BabyBlock key which can be used to decrypt the message.

8.3. Digital Signatures

Public-key cryptography is a game-changer for modern communications. It means that people can securely conduct commerce over insecure networks like the internet. It means that people separated by large distances who never meet can still communicate in privacy. Nevertheless, there is a separate problem concerning communication at large distances: the problem of authentication.

To introduce the problem, suppose the instructor of a cryptology course sends out homework assignments by email. There is probably no reason to encrypt such emails, since they aren't secret. Suppose, however, that a student in the class—Albert, say—receives an email with a ridiculous assignment ("Do all the problems in the book", for example). Assuming this is out of character for the instructor, Albert may doubt that the message was really sent by the instructor. Actually, it's not too difficult to "spoof" emails so that they appear to have originated from someone other than the true sender. Thus Albert may suspect that a prankster has sent the email containing the bogus assignment, but has made it seem to originate from the instructor's address. Even if the email is encrypted, this offers no reassurance to Albert that the email was sent by his instructor, as this requires only Albert's public key, which may be public knowledge.

How can the instructor convince Albert that they are the sender of the email? In this case, since the two have met before, they probably have some shared information, so that the instructor could convince Albert by sending an email along the lines of:

```
I am really your instructor; I remember 2 weeks
ago that you were 20 minutes late and told me
your dog ate your homework.  And I really sent
you that message containing the hard assignment.
```

But what if they didn't have any shared information? For example, what if the two had never actually met before? Or, more likely, perhaps there is no time to send messages back and forth. The professor may be teaching the course in the United States, but flying to say Bristol, England, to meet colleagues and give a talk, and might be unreachable for a day or two due to travel.

Instead of being a student in a cryptology class, it's far more likely that Albert is a consumer purchasing something from a website, and instead of worrying about whether the assignments he is receiving are authentic, he is worried about whether the website he is about to give his credit card number to is authentic. This authentication problem is dealt with by modern web browsers. They must verify that a website is being run by who it claims to be run by. How can this be accomplished?

It turns out, amazingly, that RSA provides a very simple solution to these authentication problems. In general, the setting in which **digital signatures** are used concerns the situation in which a sender (Alice) wants to convince a receiver (Bob) that she is the true sender of a message (or a website, for example).

When Alice wanted to *encrypt* a message to Bob, Bob must generate an RSA key-pair; however, when Alice wants to *sign* a message (regardless of who it is sent to), it is Alice who must generate a key-pair. Once she has generated a public key (e, n) and a private key (d, n), Alice uses her *private* key to sign a message. Given a message m (some number $1 \leq m < n$), Alice computes the signature as $s = m^d \pmod{n}$. She would transmit this signature along with the original message (which may be encrypted). To verify the message's signature, Bob uses Alice's public key by computing $m = s^e \pmod{n}$. Note that this recovers m for the same reason that RSA decryption recovered an encrypted message (because $m^{de} = m^{ed} = m \pmod{n}$). The fact that the signature checks out against the original message using Alice's public key should convince Bob that it was signed with Alice's private key, which only she should know. Further, Alice is able to send all of this to Bob in just one email; there is no need to have a chain of emails in an attempt to establish identity.

In practice, digital signatures would be completely unworkable without the use of **hash functions**, covered in the next section. Apart from the fact that hash functions are a practical necessity when using digital signatures, as we shall see, hash functions are fascinating in their own right: their security properties turn out to be startlingly counterintuitive and seemingly paradoxical.

8.4. Hash Functions

We've discussed in the previous section that RSA is too slow to encrypt large files in real time. We got around this problem by encrypting messages using a symmetric cipher (AES is a good choice in real life), and just using RSA to transmit the key. A major issue is that we want to make sure the sender can verify our identity. This led to digital signatures, but we have the same problem when we want to sign a message. Suppose, for example, Alice tells Bob that she'll drive him to the airport if he buys her lunch. Being very untrustworthy, and worried that Alice might back out of their deal, Bob types up the following "contract":

```
I Alice agree to drive Bob to the airport this
Saturday in exchange for the lunch he is buying
me today, Wednesday.
```

Bob wants Alice to sign the message with her private RSA key and tell him the signature. That way, if she backs out on the deal, Bob can go to the "authorities", show them that Alice's public key (which is public knowledge) recovers the message from the signature, which means that the contract must have in fact been signed by Alice with her public key.

The problem with this scenario is that it requires turning this message into a number whose size is at most the modulus of Alice's public key. Even with a message this short, there are more than 10^{140} different messages of this length, even if we ignore case, spacing, and punctuation. Thus turning this message into a number will result in a number with at least 140 digits; this is small enough (though not by much) to be encrypted with the giant RSA keys used in computer implementations of RSA, but certainly too long a number for us to work with here. In general, most messages we want to sign are far too big even for computer-based implementations of RSA.

How can Alice "sign" a reasonably sized number that will nevertheless authenticate the entire message as a genuine agreement by Alice? Recall that for encryption we dealt with the size problem by using RSA to encrypt just the key for a symmetric cipher.... Does that work in this case?

Suppose for example that Alice encrypts the message with the Caesar cipher with a shift of 2, producing the following:

```
KCNKE GCITG GVQFT KXGDQ DVQVJ GCKTR QTVVJ KUUCV WTFCA KPGZE
JCPIG HQTVJ GNWPE JJGKU DWAKP IOGVQ FCAYG FPGUF CA
```

She could now sign the number 2 with her RSA private key, and tell the signature, the key, and the encrypted contract to Bob.

Bob, however, can now make it seem like Alice signed a totally different contract. For example, Bob could encrypt the message "Alice must take Bob to the airport and buy his plane ticket" with the Caesar cipher using the key 2, and the signature would match the resulting encrypted message just as well as it matches the real one. Alice could play the same game: if she did back out on their deal and refused to take Bob to the airport, Alice could make her own bogus contract (such as "Bob has to buy Alice lunch every day for a month, and she doesn't have to take him to the airport") and claim that this is the contract that she *really* meant to sign.

The lesson to learn here is that it's not enough to just sign a message, since a key can be used to encrypt any message. We need something more involved. A logical candidate is the message itself. Somehow, the signature should incorporate information about the message, so we know *which* message Alice truly meant to sign.

This is where **hash functions** come in. The purpose of a hash function is to take a (possibly quite long) message and return a "hash" of the message, which is a sort of "fingerprint" of the message. Alice, wishing to authenticate a message, signs this hash value rather than the whole message.

Consider the following toy example of a hash function:

The ToyHash algorithm: Given a message, let v be the number of vowels (defined for ToyHash as `a`, `e`, `i`, `o`, or `u`), c the number of consonants, and w the number of words. The ToyHash value of the message is $v^2 + cv + w$ MOD 19.

Let's return to the example of Alice's promise to Bob:

```
I Alice agree to drive Bob to the airport this
Saturday in exchange for the lunch he is buying
me today, Wednesday.
```

This message has 37 vowels (not counting any y's), 55 consonants, and 22 words. Thus the ToyHash hash value of this message is

$$37^2 + 55 \cdot 37 + 22 \text{ MOD } 19 \;=\; 6.$$

Alice now signs this hash value (6) with her RSA key, and gives Bob the result as a digital signature. If Bob tried to claim that Alice had actually signed a different contract, the different contract would likely have a different hash value and thus not match Alice's signature. Even with the small RSA keys we've been using to do the calculations by hand, Alice can sign the entire message with just one RSA operation. At first, this looks promising, as the signature involves the entire message.

On the other hand, ToyHash never gives a hash value larger than 18 (when we reduce modulo 19, the possible values are 0, 1, ..., 19). This is a significant problem, and it turns out that the ToyHash function is remarkably insecure. What does it mean to say that ToyHash is insecure? Roughly speaking, the problem is that it is easy to make it seem like Alice signed a different contract than the one she intended to sign. Returning to the earlier example, recall that ToyHash produces a hash value of 6 on the message "I Alice agree to drive Bob to the airport this Saturday in exchange for the lunch he is buying me today, Wednesday." We suppose that Alice has signed this hash value using RSA and told Bob the result as a digital signature of the original message.

If Bob is clever, however, he might try to claim that Alice actually signed the following contract:

```
I Alice promise to buy Bob an airplane ticket to
Hawaii this Saturday in exchange for buying me
lunch today, Wednesday.
```

This contract has 40 vowels, 57 consonants, and 21 words, giving a ToyHash value of $40^2 + 57 \cdot 40 + 21$ MOD $19 = 6$, the same as the original contract! Bob can make an untrue but apparently convincing claim that Alice signed this bogus contract, since her signature matches the hash value of this message.

The situation above, where two messages have the same hash value, is called a **collision**. Collisions are a problem for hash functions, because each collision represents multiple messages which the hash function cannot distinguish between; for such pairs of messages, neither party can prove which message was actually signed. Finding collisions can involve a bit of luck, but for some hash functions (such as ToyHash) simple strategies suffice. For example, if Bob wanted to go to Cuba instead of Hawaii, he might have found that the bogus contract

```
I Alice promise to buy Bob an airplane ticket to
Havana this Saturday in exchange for buying me
lunch today, Wednesday.
```

has a ToyHash value of 5 (it has 39 vowels, 59 consonants, and 21 words). By studying the formula $v^2 + cv \cdot 40 + w$ MOD $19 = 6$, he might notice that he can increase the hash value by 1 if he can increase the number of words by 1 without changing the number of vowels or consonants. He can do this by inserting a space by "air" and "plane":

```
I Alice promise to buy Bob an air plane ticket
to Havana this Saturday in exchange for buying me
lunch today, Wednesday.
```

The result has a ToyHash value of 6, and so forms a collision with a genuine contract. Although it is not really common or even correct to write "air plane" as two separate words, the meaning is clear anyway, and contracts are not typically invalidated because of grammar mistakes!

Above we saw a major weakness of ToyHash; collisions are common. Moreover, they can be found without too much work as it's easy to predict the impact on the hash value of certain kinds of changes to the message (for example, increasing the number of words). Let's begin by examining the cause of the first problem.

Collisions must be common simply because ToyHash can only return 19 different hash values (0 through 18). This means that out of any 20 messages, we can be guaranteed that at least two share a hash value; if not, then each of the 20 possible hash values has at most one message, and that can account for at most 19 messages. This turns out to be a powerful observation, and is a great use in a variety of mathematical problems. It's known as the **pigeon-hole principle**, **Dirichlet's pigeon-hole principle**, or the **box principle**; typically whenever something has many names it's important! We give a few examples of its utility in Exercises 8.8.30 to 8.8.34.

Thus it seems that one obvious way to improve ToyHash is to simply increase the number of possible of hash values so that collisions don't occur (or at least occur infrequently). Remember, though, that the very purpose of the hash function is to reduce the size of the value we need sign with RSA; accomplishing this *requires* collisions!

To see why this is the case, let's imagine that we typically deal with binary messages of around 10,000 digits (approximately 10kb) and use a hash function which produces hash values 100 digits long. The number of possible messages in this situation is roughly $2^{10,000}$, while the number of possible hash values is roughly 2^{100}. There are far more possible messages than hash values, meaning that lots and lots of messages get the same value. There are roughly $2^{10,000}/2^{100} = 2^{9,900}$ times more messages than hash values, meaning that a "typical" hash value corresponds to $2^{9,900}$ different messages! This same kind of reasoning shows that there will always be collisions if the hash function produces values which are smaller than the sizes of the messages.

We thus have a thorny problem with two competing issues. On the one hand, for a hash function to do its job, it must produce a hash value which will typically be small compared to the size of a message. On the other hand, accomplishing this very thing means that the hash function has lots of collisions, which might be exploited by someone wishing to find bogus contracts which match a signature value they know for another contract.

In spite of this, there are secure hash functions. As already discussed, it is impossible to have a hash function which does its job without producing lots of collisions; the property of secure hash functions is not that such collisions are particularly rare, but simply that they are *hard to find.*

To see how this could be possible, let's again assume that typical messages in some situation consist of 10,000 binary digits, and we are using a hash function that produces hash values 100 digits long. We already pointed out that in such a situation, each hash value corresponds to $2^{9,900}$ messages. It's possible to imagine many hash functions for this situation where it's easy to find a collision. Suppose, for example, that the hash function simply considered the message as a binary number, and reduced it modulo 2^{100}. The result would be an integer which could be written as a binary number of at most 100 digits. Collisions are very easy to find in this case. In fact, given *any* message, a collision can be found by considering it as a binary number, and producing another message by adding to it 2^{100}. The problem with this hash function (a problem shared by ToyHash), is that it is easy to predict how the hash value changes with changes to the message.

Our goal is to find a complicated hash function where it is nearly impossible to figure out how one could produce a message with a certain hash value without simply trying random messages and calculating their hash values (as the number of messages to try is so astronomically large as to be impractical). In our example there are 2^{100} hash values; it could take $2^{100}+1$ guesses before we found two messages with the same hash value. (In this scenario, there is actually a not-too-tricky technique to probably find a collision with just 2^{50} steps, but that's still a lot, exceeding 10^{15}!) A good hash function is designed to try to make this kind of wild guessing essentially the best strategy. It is perhaps slightly astonishing just how possible this is. **SHA-1 (Secure Hash Algorithm 1)** is one hash function in widespread use in everyday internet communications; it outputs a 160-bit hash value. In spite of the fact that it has been studied extensively since its release in 1995, and in spite of the fact that for very small messages, "gazillions" of collisions are guaranteed to exist by the very nature of a hash function, **as of 2012, no one knows even a single instance of a collision for the hash function SHA-1**. SHA-1 is an old hash function, and is being replaced by newer hash functions that produce bigger hash values and are expected to be even more secure. This is the amazing thing about hash function security: in spite of the fact that all hash functions have lots (and lots!) of collisions, it's possible to design a hash function for which we don't expect anyone to ever be able to find even a single one!

8.5. Diffie–Hellman Key Exchange

RSA is an outstanding achievement: it provided the first practical, secure encryption method allowing people to encrypt and decrypt messages without having any shared secret information ahead of time. As we saw in Section 8.2, however, practical considerations mean that when we use RSA, we typically use it not to send our message, but just to send a secret key to use to encrypt our message with another (symmetric) cipher. If this is all we're trying to do, there's another way of agreeing on a secret key: the **Diffie–Hellman key exchange protocol**, named after Whitfield Diffie and Martin Hellman, who discovered the method in 1976.

While this section is not needed for RSA, their protocol was one of the most important milestones in the history of cryptography: the discovery of a method to allow two people to exchange a secret by yelling in a crowded room! Of course, this is just a colorful way of saying that we have a way to allow two people to reach a common secret where every step they do is under the watchful glare of attackers.

Here's a fun way of describing the problem and its solution. We first give a toy example illustrating the basic concept, and then give the rigorous mathematical version. Imagine that Alice is going to send Bob a message by painting the message on a wall. The wall is out in the open, and lots of people can see it. To keep eavesdroppers from knowing which message is the actual message intended for Bob, she paints many different messages on the wall in lots of different colors. If she and Bob can agree on a special color ahead of time, then Alice will paint the true message in that color, and Bob can tell which message is intended for him. If you don't like the fact that the message is out in the open, instead of painting the message it's enough for Alice and Bob to be able to agree upon the same secret color, which can be used for future messages (the shade of the secret color plays the role of the shared secret for an encryption scheme).

Let's consider the following protocol, which should allow Alice and Bob to agree on a secret paint color even if all of their communications are subject to eavesdropping.

The secret paint protocol: Bob and Alice each go to the paint store and buy some paints in primary colors, which allows them to mix different colors of paint. Then Alice and Bob each mix two gallons of their own secret color, which they don't share with anybody (including each other). Bob also mixes up several gallons of a third "base" color. He gives two gallons to Alice, and keeps one for himself.
Bob mixes together one gallon each of his secret paint and the base paint, and sends this to Alice. Alice mixes together one gallon each of her secret paint and the base paint, and sends this to Bob. Finally, Bob adds one gallon of his secret paint to the paint he receives from Alice, and Alice adds one gallon of her secret paint to the paint she receives from Bob. At the end, each has three gallons of paint mixed in a $1:1:1$ ratio from the base paint, Bob's paint, and Alice's paint; this is the secret color to be used for their secret message.

Why does this procedure work to agree on a color through public communication channels? We assume that eavesdroppers can see all the paint Alice and Bob exchange. (Maybe someone at the post office is opening their packages, for example.)

An eavesdropper has therefore seen three paints transferred in this protocol:

(1) The base paint.
(2) The paint mixed in a 1:1 ratio with Bob's paint and the base paint.
(3) The paint mixed in a 1:1 ratio with Alice's paint and the base paint.

The point is that there seems to be no way for the eavesdropper to use these paints to make paint which is in a 1:1:1 ratio with the base paint,

Alice's paint, and Bob's paint. For example, if the eavesdropper mixes together paints (2) and (3), she'll get paint in a 2:1:1 ratio of base paint, Alice's paint, and Bob's paint. If she uses any of paint (1), the situation just gets worse.

This really does seem to be a workable way to publicly agree upon a paint color unknown to any eavesdroppers. Interestingly, Alice and Bob can't control what paint they're going to end up with at the end; what they can do is make sure they have a **shared secret**, which they can then use to send other messages. Although based on modular arithmetic, the Diffie–Hellman key protocol is conceptually very similar to the secret paint protocol, and shares this basic characteristic: it can't be used to send a message directly, but at the end of the protocol, both Alice and Bob share a common secret number which they can then use as a key for another encryption method.

Diffie–Hellman key exchange: The following procedure allows Alice and Bob to agree upon a secret key.

(1) Alice chooses a prime number p and a base g modulo p, and tells both to Bob. Since all communications are over public channels, g and p are now public knowledge.
(2) Alice chooses a large exponent $a < p$ and keeps it private; Bob chooses a large exponent $b < p$ and keeps it private.
(3) Alice computes $g^a \pmod p$ and sends it to Bob, while Bob computes $g^b \pmod p$ and sends that to Alice. We must assume an attacker has intercepted these messages, that is: $p, g, g^a \pmod p$ and $g^b \pmod p$ are all public knowledge (though a and b are private).
(4) Alice now computes $(g^a \pmod p)^b \pmod p$, which equals $g^{ab} \pmod p$, and sends this to Bob. Similarly, Bob computes $(g^b \pmod p)^a \pmod p$, which equals $g^{ab} \pmod p$.

Alice and Bob have reached a secret known only to the two of them: $g^{ab} \pmod p$.

Like the other (practical and impractical) approaches to public-channel cryptography we've considered, the secret paint protocol was based on the notion of a problem which is easy to do one way, but seems hard to reverse. In this case, that problem is mixing paint: although it is easy to mix two paints together, it seems like it might be hard to precisely "subtract" one paint color from another, which is what would allow an eavesdropper to recover the secret paint shared by Alice and Bob.

On the other hand, the Diffie–Hellman key exchange protocol is based on the problem of modular exponentiation. We saw in Section 7.4 that it is easy to quickly compute $g^a \pmod p$ given g, a, and p; this is used in steps

(3) and (4) above. It seems like the most obvious way for an eavesdropper to break the Diffie–Hellman key exchange protocol is to find a given the publicly transmitted g, p, and $g^a \pmod{p}$, or, alternatively, find b given g, p, and $g^b \pmod{p}$. Since the problem amounts to finding the exponent which gives a certain result for a given base modulo the given number p, it is the modular arithmetic analog to taking the logarithm, and is known as the **discrete logarithm problem**.

The security of Diffie–Hellman thus depends on the fact that the discrete logarithm problem is believed to be hard; that is, although we can quickly carry out modular exponentiation, we don't know a good way to discover in general what exponent was used to arrive at a certain result from a certain base and modulus. Note the presence of a modulus is essential to the difficulty of the method; if we consider the related problem without reductions modulo n, then we can easily find the exponent. For example, if we know the integers g^a and g, then

$$\frac{\log(g^a)}{\log(g)} = \frac{a\log(g)}{\log(g)} = a.$$

The problem is fundamentally different with a modulus. Now we get

$$\frac{\log(g^a \pmod{n})}{\log(g)},$$

but we can't simplify the numerator (or at least no easy way beckons) because of the reduction modulo n. The difficulty is that reducing modulo n masks the power of g.

The discrete logarithm problem is not hard for all g and p, however, so in practice p and g must be chosen carefully to achieve good security. For example, there is no guarantee that $g^x \pmod{p}$ takes on all values modulo p for different values of x, and if it takes on very few values, then an eavesdropper won't need too many guesses to find $g^{ab} \pmod{p}$. Further, there are algorithms that efficiently compute the discrete logarithm for any base g when the modulus p has the property that $p-1$ has only small prime factors.

8.6. Why RSA Works

It's now time to prove why RSA works. The only difficulty is showing that if $c = m^e \pmod{n}$ is the ciphertext, then $c^d \pmod{n}$ equals the original message m. The proof is significantly more involved than the corresponding decryption proofs for the perfect code cryptosystem and KidRSA. While this makes the reading more intense, the hope is that this means the system is more secure. We've already developed all the mathematics for the arguments below.

The RSA theorem: Let $n = pq$ be the product of two primes, let e and d be two positive integers such that $ed = 1 \pmod{(p-1)(q-1)}$, let m (for message) be a positive integer less than n and relatively prime to n, and set c (for ciphertext) equal to $m^e \pmod n$. Then $c^d \pmod n = m$, the message.

Before proving this theorem, which explains why RSA works, we briefly comment on the conditions. We are assuming m is relatively prime to n. As n is the product of two very large primes, a vanishingly small percentage of the numbers at most n share a factor with n (see Exercise 8.8.47 for a guided discussion on just how small), so this is a reasonable assumption. Exercises 8.8.48 and 8.8.57 ask you to think about the consequences of Alice finding a message that shares a factor with n.

We'll go through the proof slowly, as we have two goals. While obviously we want to prove the result so we know we can use RSA, we also want to highlight how to find proofs of deep mathematical statements.

Proof of the RSA Theorem. We're trying to show $c^d \pmod n$ equals m, where $c = m^e \pmod n$. Because we can freely reduce modulo n before or after exponentiation without changing the answer modulo n, this is equivalent to showing $m^{ed} = m \pmod n$. Our proof uses Fermat's little Theorem from §7.6, which says that if a is relatively prime to a prime r, then $a^{r-1} = 1 \pmod r$. After some thought, this should look like a reasonable strategy. After all, we need to exponentiate, and the only fact we know about exponentiation is Fermat's little Theorem.

Let's re-examine our set-up, in the hopes that this will provide a clue as to how to proceed. We want to show $m^{ed} = m \pmod n$, where $ed = 1 \pmod{(p-1)(q-1)}$. The second statement means there is an integer j such that

$$ed \;=\; 1 + j(p-1)(q-1),$$

and thus we are reduced to showing $m^{1+j(p-1)(q-1)} = m \pmod n$.

Now we're making progress! Using the laws of exponents, we have

$$m^{1+j(p-1)(q-1)} \;=\; m \cdot m^{j(p-1)(q-1)} \;=\; m \cdot \left(m^{(p-1)(q-1)}\right)^j.$$

This is great, as we can see a factor of m by itself. To finish the proof, all we have to do is show that $m^{(p-1)(q-1)} = 1 \pmod n$, as if this equals 1, so too does the jth power.

It should also be clearer now how Fermat's little Theorem (FℓT) enters. Our two primes are p and q, and we're raising a quantity to the $(p-1)(q-1)$ power. We're thus done if we can show $m^{(p-1)(q-1)} = 1 \pmod n$. Unfortunately, F$\ell$T only gives us results about congruences modulo a prime; n is not a prime but instead is a product of two primes. This isn't a fatal obstruction; all we need to do is apply FℓT *twice*! First we show $m^{(p-1)(q-1)}$ equals 1 modulo p, then we show it equals 1 modulo q. These two facts imply that $m^{(p-1)(q-1)} = 1 \pmod{pq}$. We now do these three steps.

- We have
$$m^{(p-1)(q-1)} = \left(m^{q-1}\right)^{p-1}.$$
This is a natural regrouping to do. Remember we need to have something to the $p-1$ power, which is why we group this way. Of course, before we use *any* theorem we always must make sure the conditions are satisfied. Is m^{q-1} relatively prime to p? Yes, because m is relatively prime to p (you are asked to prove this in Exercise 8.8.49). Thus, by Fermat's little Theorem,
$$m^{(p-1)(q-1)} = \left(m^{q-1}\right)^{p-1} = 1 \pmod{p}.$$
In particular, this means there is a k such that
$$m^{(p-1)(q-1)} = 1 + kp.$$
- A similar argument as above shows
$$m^{(p-1)(q-1)} = \left(m^{p-1}\right)^{q-1} = 1 \pmod{q},$$
and thus there is an ℓ such that
$$m^{(p-1)(q-1)} = 1 + \ell q.$$
All you need to do is go through the above proof, switching the roles of p and q. We can do this as we aren't using *any* special properties of p or q (for example, if we assumed $p < q$ and then used in the proof that p was the smaller factor of n, then more work would be needed).
- Since $m^{(p-1)(q-1)} = 1 + kp$ and $m^{(p-1)(q-1)} = 1 + \ell q$, we have $m^{(p-1)(q-1)} - 1$ is a multiple of p *and also* a multiple of q. This means there is some b such that $m^{(p-1)(q-1)} - 1 = bpq$, or (at last!)
$$m^{(p-1)(q-1)} = 1 \pmod{pq}.$$
This completes the proof, as $n = pq$.

8.7. Summary

This chapter was the most mathematical to date. To discuss why RSA is believed to be secure, however, requires far more powerful mathematics, and is an active area of research.

We talked a lot about how RSA is felt to be secure as we believe factoring is difficult. It's possible, however, that someone might be able to crack RSA *without* being able to factor large numbers. We know that the ability to factor quickly means RSA is secure, but this is not an if-and-only-if matter. Remember in the perfect code cryptosystem that we saw it's possible to crack a message without having to find the perfect code. Perhaps something similar is true here; perhaps one can crack RSA without having to factor n. This is called the **RSA problem**.

We continued with a discussion of digital signatures. It does not matter how much you trust a friend; it is absolutely essential that you verify their identity. Unscrupulous people try hard to get you to divulge personal information, and you must be alert. We don't want banks or credit card companies releasing our funds to anyone claiming to have our permission; we want a good confirmation procedure. This led to a quick tour through hash functions, though there is a lot more to the story.

Towards the end we had a quick discussion of the Diffie–Hellman method for key exchange. Interestingly, in the British Government Communications Headquarters, RSA was discovered first and Diffie–Hellman second, while in the public land of academia, the orders were reversed. Using just elementary mathematics, two people can share enough information so they can reach the same answer in a way that is *believed* to be impractical for someone else to mimic, even if they hear all the information our two people publicly exchange.

Finally, without using the language of more advanced courses, we develop just enough mathematics to prove why RSA works.

8.8. Problems

• From §8.1: RSA

EXERCISE 8.8.1. *Generate a public/private RSA key-pair using the primes* $p = 11$, $q = 23$. *You should use the extended Euclidean algorithm. Since you get to choose* e, *there is more than one possible answer to this problem!*

EXERCISE 8.8.2. *Generate a public/private RSA key-pair using the primes* $p = 13$, $q = 19$. *You should use the extended Euclidean algorithm. Since you get to choose* e, *there is more than one possible answer to this problem!*

EXERCISE 8.8.3. *What would happen if you tried to implement RSA and* $p = q$*? Would it still work? Would there be security issues?*

EXERCISE 8.8.4. *If you receive the message* 124, *encrypted using the public key you produced in the previous problem, what does the message decrypt to?*

EXERCISE 8.8.5. *If* $n = pq$ *is the product of two distinct odd primes, show that at least one out of every three consecutive integers is relatively prime to* n. *Thus if the last digit of a message is kept free, it is always possible to choose a final digit so that the message is relatively prime to* m. Hint: Proceed by contradiction. Assume not, thus either p or q divides at least two of the three consecutive integers. Show this implies p or q divides 2.

EXERCISE 8.8.6. *Finding* $\varphi(n)$ *is equivalent to factoring* n*; there is no computational shortcut to factoring. Clearly, if one knows the factors of*

$n = pq$, one knows $\varphi(n) = (p-1)(q-1)$. Show that if you know $\varphi(n)$ and n, then you can recover the primes p and q. Hint: Set $k = n + 1 - \varphi(n)$ and show that the two prime factors of n are the integers $(k \pm \sqrt{k^2 - 4n})/2$.

EXERCISE 8.8.7. *Show that $\varphi(p) = p - 1$ if p is prime and $\varphi(pq) = (p-1)(q-1)$ if p and q are distinct primes.*

EXERCISE 8.8.8. *If p is prime, show that $\varphi(p^k) = p^k - p^{k-1}$.*

EXERCISE 8.8.9. *If* $\gcd(m, n) = 1$, *show* $\varphi(mn) = \varphi(m)\varphi(n)$.

EXERCISE 8.8.10. *Using Exercises* 8.8.8 *and* 8.8.9, *show that if $n = p_1^{r_1} p_2^{r_2} \cdots p_k^{r_k}$ is the factorization of n into distinct prime powers (so $p_1 < p_2 < \cdots < p_r$), then*

$$\varphi(n) \;=\; \left(p_1^{r_1} - p_1^{r_1-1}\right)\left(p_2^{r_2} - p_2^{r_2-1}\right)\cdots\left(p_k^{r_k} - p_k^{r_k-1}\right).$$

EXERCISE 8.8.11. *It's interesting to compare the size of $\varphi(n)$ to n. The largest the ratio $\varphi(n)/n$ can be is when n is a prime. What n have a low ratio?* Hint: It might help to think of the definition of φ, namely $\varphi(n)$ counts the number of integers at most n that are relatively prime to n.

EXERCISE 8.8.12. *For RSA, we must choose an e relatively prime to $\varphi(n) = (p-1)(q-1)$. One way to find such an e is to randomly choose a candidate, and then keep incrementing by 1 until we reach a number relatively prime to $\varphi(n)$. The more prime factors $\varphi(n)$ has, the harder this is to do. This* suggests *that some choices of p and q may be better than others. If $p \geq 5$ is a prime, then $p-1$ cannot be prime as it must be divisible by 2. If there are no other divisors of $p-1$, then we say $p-1$ is a* ***Germain prime*** *(in other words, $\frac{p-1}{2}$ is prime). If p and q are Germain primes, show that one out of every six consecutive integers is relatively prime to $\varphi(n)$.* Hint: This problem is very similar to Exercise 8.8.5.

EXERCISE 8.8.13. *What would happen if we tried to do RSA and instead of choosing two primes p and q, we chose two composite numbers?*

EXERCISE 8.8.14. *Research the life of Clifford Cocks, and write a short report about his role in the history of cryptography.*

- **From §8.2: RSA and Symmetric Encryption**

EXERCISE 8.8.15. *Discuss a way to convert a substitution alphabet cipher into a number, and then describe how one can use RSA to securely transmit such a key.*

EXERCISE 8.8.16. *Can you use RSA to transmit RSA public and private keys? If so, describe how.*

EXERCISE 8.8.17. *A big theme of this chapter is the need for speed. Look up* ***Horner's algorithm*** *and write a short report about it. One nice application is in quickly drawing pictures in fractal geometry.*

EXERCISE 8.8.18. *Continuing our introduction to efficient algorithms, write a short report on the* ***Strassen algorithm*** *for matrix multiplication.*

EXERCISE 8.8.19. *The famous* ***Fibonacci numbers****, given by the recurrence* $F_{n+1} = F_n + F_{n-1}$ *with* $F_0 = 0$ *and* $F_1 = 1$*, occur throughout mathematics. While it is possible to use the recurrence relation to compute any term, such an approach is very inefficient and time consuming, as it requires us to find all the terms less than the desired one.* ***Binet's formula*** *allows us to jump immediately to any term (and can be generalized to other recurrences). Write a short report on Binet's formula.*

• From §8.3: Digital Signatures

EXERCISE 8.8.20. *Assume you have created the key-pair* $(3, 55)$ *(the public key) and* $(27, 55)$ *(the private key). You wish to produce a signature for the message "9". Which key do you use? What is the signature?*

EXERCISE 8.8.21. *Assume now that you are the recipient of the message and signature from the previous exercise. Carry out the calculation used to verify the signature.*

EXERCISE 8.8.22. *Suppose Alice is sending Bob an important message. She wants Bob to be convinced the message originates from her, so she uses her public/private keys as stated in the section. Bob now knows, as this is publicly available, that* $(e_{\text{Alice}}, n_{\text{Alice}})$ *and a message* $m^{d_{\text{Alice}}} \pmod{n_{\text{Alice}}}$. *Is this enough information for Bob to deduce* d_{Alice}*? What would be the consequences if he could determine* d_{Alice}*?*

EXERCISE 8.8.23. *Two of the key players in cryptography in general, and digital signatures in particular, are Whitfield Diffie and Martin Hellman. Write a short report about their roles.*

EXERCISE 8.8.24. *There are many digital signature schemes, including Lamport signatures, Merkle signatures, and Rabin signatures. Write a short description of one of these.*

• From §8.4: Hash Functions

EXERCISE 8.8.25. *Assume you have created the key-pair* $(3, 55)$ *(the public key) and* $(27, 55)$ *(the private key). You wish to produce a signature for the message:*

`Your instructions are to bombard the enemy with water balloons.`

Compute the ToyHash value of this message and sign using RSA. (Which of your keys do you use for this purpose?)

EXERCISE 8.8.26. *Briefly explain why it is necessary to use a hash function for the purpose of creating digital signatures, rather than using RSA to sign the message directly.*

EXERCISE 8.8.27. *Assume Bob wants to go anywhere in Africa. Find a bogus contract that involves Alice paying his way to a city or country in*

Africa, whose ToyHash value matches that of the original genuine contract Alice actually signed.

EXERCISE 8.8.28. *If we ignore spaces, case, and punctuation in text messages, there are* 26^n *possible messages that are* n *letters long (although most of them won't make much sense). On average, approximately how many different* 100*-letter-long messages does a ToyHash value correspond to?*

EXERCISE 8.8.29. *Are all ToyHash values approximately equally likely for English messages? Explore this by choosing some book, and calculating the ToyHash value for the first sentence on each of the first* 100 *pages. Is this a large enough sample set to trust the answer?*

EXERCISE 8.8.30. *As remarked earlier, collisions in hash functions are a consequence of* ***Dirichlet's pigeon-hole principle*** *(also known as the* ***box principle****): if* $n+1$ *pigeons must be placed in* n *boxes, at least one box must have at least two pigeons. This principle is used to solve a variety of problems. For example, show that if* $a_1, \dots, a_{n+1}$ *are distinct integers in* $\{1, \dots, 2n\}$*, then two of them add to a number divisible by* $2n$. Note: these two numbers need not be distinct. If instead we required the two numbers to be distinct, how many of the $2n$ numbers must we choose?

EXERCISE 8.8.31. *Use the pigeon-hole principle to show any rational number has a finite or repeating decimal expansion.*

EXERCISE 8.8.32. *Let* S *be any set of* $n+1$ *distinct elements chosen from* $\{1, 2, \dots, 2n\}$*. Show* S *contains at least two elements* a, b *with* $a|b$. Hint: Use the pigeon-hole principle from the previous problem. Also, look at how many times 2 divides each number on your list.

EXERCISE 8.8.33. *We can also use the pigeon-hole principle to find good rational approximations to irrational numbers. Fix an irrational number* $\alpha \in [0,1]$ *and an integer* Q*. Consider* $\alpha \pmod 1, 2\alpha \pmod 1, \dots, (Q+1)\alpha \pmod 1$*, which are all in* $[0,1]$*. As*

$$[0,1] \;=\; [0,1/Q] \cup [1/Q, 2/Q] \cup \cdots \cup [(Q-1)/Q, 1]$$

is the union of Q *intervals, since we have* $Q+1$ *numbers at least two must be in the same interval, which is of length* $1/Q$*. Use this to show that we can find integers* p, q *with* $1 \le q \le Q$ *such that* $|\alpha - p/q| \le \frac{1}{qQ}$*, which is at most* $1/q^2$*. This means we can approximate a number very well; the error is on the order of the* square *of the denominator, which is significantly better than truncating the decimal expansion (where the error is on the order of the denominator).*

EXERCISE 8.8.34. *Building on the previous problem, let* $\alpha = \pi - 3$ *and* $Q = 10$*. Show this leads to* $|\alpha - 1/7| \le 1/7^2$*, or* $|\pi - 22/7| \le 1/49$ *(the approximation is actually closer than* $1/49$*). This is a little better than truncating* π *after one decimal and approximating it as* 3.1 *or* 31/10*. To really* *see an improvement, you need to take larger Q. Explore the approximation from $Q = 100$.*

EXERCISE 8.8.35. *Read about SHA-0, SHA-1, and SHA-2, and write a short note about how they work and what is known about their collisions.*

EXERCISE 8.8.36. The pigeon-hole principle *(see Exercise* 8.8.30*) says that if there are $n+1$ pigeons in n boxes, at least one box has two pigeons. In practice, it often happens much earlier. Consider birthdays. Assume all days are equally likely to be someone's birthday, that everyone's birthday in the class is independent of everyone else's birthday, and no one is born on February 29th (so there are 365 candidate days). The* **birthday problem** *asks how many people we need to have in a room before there is at least a 50% chance that two share a birthday. Show that we need about 23 people to have a 50% chance of a shared birthday. If you have a class with more than 30 people, what's the probability of at least one shared birthdate? If your teacher is game, try this in class. The next exercise continues the discussion about this surprising result, that far fewer people are needed than one might expect.*

EXERCISE 8.8.37. *Generalize the previous problem to the case with n days in a year. Show that one now needs about $\sqrt{2n \log 2}$ people to have a 50% chance. What matters here is not the constant, but the n-dependence. It's not nearly as bad as being on the order of n; square-root of n suffices with high probability. The warning, however, should be heeded: it is easier to find a collision than one might expect, and this is the basis behind* **birthday attacks**.

- **From §8.5: Diffie–Hellman Key Exchange**

EXERCISE 8.8.38. *Look up the history of the Diffie–Hellman key exchange as well as RSA. Are you surprised which came first? Read about the British efforts and what they had, and when. Write a short note about your findings.*

EXERCISE 8.8.39. *Generalize the pirate example so three people can share a secret (in this terminology, so that three people on three different islands can see the ring). What is the smallest number of moves you need to accomplish this? Here a move is the pirate ship taking a box from one island to another.*

EXERCISE 8.8.40. *Generalize the previous problem to four people. Is it simpler if the number of people is a power of 2? Is there a nice way to generalize from four to eight people?*

EXERCISE 8.8.41. *Justify the steps in creating the secret in the* agreeing on a secret through shouting *method.*

EXERCISE 8.8.42. *Let $n = 69$, $g = 17$, $a = 13$, and $b = 19$. Find g^{ab}* $\pmod{n}$.

EXERCISE 8.8.43. *Let $n = 1967$, $g = 1030$, $a = 1001$, and $b = 843$. Find* $g^{ab} \pmod{n}$.

EXERCISE 8.8.44. *Using notation as in the previous problem, compute* $\log(g^a)/\log(g)$ *and* $\log(g^a \pmod{n})/\log(g)$. *The presence of the logarithm makes* $g^a \pmod{n}$ *of the same size as* g^a.

EXERCISE 8.8.45. *Read about the discrete logarithm problem, and write a short report. In particular, research the Pohlig-Hellman algorithm for attacking the discrete logarithm problem. See also Exercise* 11.7.16.

• **From §8.6: Why RSA Works**

EXERCISE 8.8.46. *Let $p = 7$ and $q = 13$. Find a pair (e, d) for RSA, and encrypt and decrypt the message* 55.

EXERCISE 8.8.47. *Let $n = pq$ be the product of two primes $p < q$. Show the number of integers at most n that are not relatively prime to n is $p+q-1$. Thus the percentage of numbers at most n that are not relatively prime to n is $\frac{p+q-1}{pq}$, which is less than $1/p + 1/q$, which itself is less than $2/p$. If p has* 200 *digits (which is often the case in real world applications), this leads to a probability of choosing a bad message for RSA of around* 1 *in* 10^{200}. *This probability is so vanishingly small we may safely ignore it.*

EXERCISE 8.8.48. *Suppose Alice wants to send a message to Bob. She looks at his public key (e, n) and chooses a message m. Amazingly, even though Exercise* 8.8.47 *says it is almost impossible to find a message m that shares a factor with n, Alice got lucky and beat the odds: her message* does *share a factor with n! There are massive implications: what can Alice now do?*

EXERCISE 8.8.49. *Prove that if m is relatively prime to p, then m^{p-1} is also relatively prime to p.*

EXERCISE 8.8.50. *Prove that if i and j are positive integers, then $x^{ij} = (x^i)^j$. We have to be more careful with exponents. Find values of positive integers x, a and b such that $x^{(a^b)} \neq (x^a)^b$.*

EXERCISE 8.8.51. *Where in the proof of the RSA theorem did we use that m is relatively prime to p and q? Would the result still hold if m shared a factor with p or q (i.e., is $m^{ed} = m \pmod{n}$ even if m is not relatively prime to n)?*

EXERCISE 8.8.52. *Imagine $n = pqr$ is the product of three distinct primes, and $cde = 1 \pmod{n}$. If m is relatively prime to n, is $m^{cde} = 1 \pmod{n}$? Note that this would be a generalization to the RSA theorem.*

EXERCISE 8.8.53. *Often in RSA we wrote $\varphi(n)$ for $(p-1)(q-1)$, with $n = pq$ the product of two primes; φ is called the* **Euler totient function**, *and it gives the number of integers in $\{1, 2, \ldots, n\}$ that are relatively prime to n. Show the following are true.*

(1) *If m is a prime, then $\varphi(m) = m - 1$.*
(2) *If $m = p^2$ is the square of a prime, then $\varphi(m) = p^2 - p$.*
(3) *If $m = p^k$ for k a positive integer, then $\varphi(m) = p^k - p^{k-1}$.*
(4) *If $m = pq$ is the product of two distinct primes, then $\varphi(m) = (p-1)(q-1)$.*
(5) *If $m = ab$ with a and b relatively prime, then $\varphi(m) = \varphi(a)\varphi(b)$.*
(6) *If $m = p_1^{r_1} \cdots p_k^{r_k}$ (where the primes are distinct), then $\varphi(m) = \prod_{i=1}^{k}(p^{r_i} - p^{r_i - 1})$.*

It's worth noting that we may re-interpret Fermat's little Theorem as $a^{\varphi(p)} = 1 \pmod p$, while for RSA we have $m^{\varphi(pq)} = 1 \pmod{pq}$.

EXERCISE 8.8.54. *Let $m \geq 2$, and let a be a positive integer relatively prime to a. Using the previous problem, show that $a^{\varphi(m)} = 1 \pmod m$. This is **Euler's theorem**, and it generalizes Fermat's little Theorem.*

EXERCISE 8.8.55. *Find a lower bound for $\varphi(m)$, where φ is the Euler totient function of the previous problem. In particular, can you prove that if m is large, then $\varphi(m) > \sqrt{m}$? Can you do better? What about an upper bound?*

The last few problems introduce the **Chinese Remainder Theorem**, and discuss its application to RSA. The Chinese Remainder Theorem states that if n_1, n_2 are two relatively prime integers, then given any residues $a_1 \pmod n)_1$ and $a_2 \pmod n)_2$ there is an x such that $x = a_1 \pmod n)_1$ and $x = a_2 \pmod n)_2$.

EXERCISE 8.8.56. *Prove the Chinese Remainder Theorem.* Hint: Use the Euclidean algorithm.

EXERCISE 8.8.57. *In RSA we needed to assume the message m was relatively prime to n. Use the Chinese Remainder Theorem to show that the RSA Theorem also holds if m is not relatively prime to n; i.e., if $n = pq$ is the product of two distinct primes and $ed = 1 \pmod \varphi)(n)$, then $m^{ed} = m \bmod n$.*

EXERCISE 8.8.58. *The Chinese Remainder Theorem also allows us to speed up the implementation of RSA. Consider a message m modulo n, with ciphertext $c = m^e \bmod n$. If we can solve $x = c^d \pmod p$ and $x = c^d \pmod q$, then by the Chinese Remainder Theorem we know $x \pmod p)q$. We can compute these two equivalences significantly faster than it takes to find $c^d \pmod n$. We may write $d = k(p-1) + s$ for some integer k and $s \in \{0, 1, \ldots, p-2\}$. Thus $c^d \pmod p$ is the same as $c^s \pmod p$, and s is typically of size p; note the modulus p is around $\sqrt{n}$. We have similar results for computing $c^d \pmod q$. Show that while we have two calculations each half as long as the original, they involve moduli that have roughly half as many digits as n. This means the fast exponentiation runs faster, and typically we are able to speed up RSA by nearly a factor of 4.*

Chapter 9

Error Detecting and Correcting Codes

We've discussed at great lengths why it's important to encrypt messages. If we have some information that we only want certain individuals to get, it would be foolish to shout it out, or write it in English and leave it lying on the table for our friends to read. Clearly there are times when encryption is necessary. We need to transmit encoded messages to our friend, who then decodes it and acts accordingly; however, what if the message is garbled in the transmission? We've all had experiences where a computer hangs up for no reason or displays strange text. What happens if there's an error in transmitting our message? For example, we often use binary for our messages, and send strings of 0's and 1's which are then converted to letters. What happens if one of the 1's is accidentally transmitted as a 0 (or vice-versa)? This leads to the subject of this chapter: error detection and, if possible, error correction.

Though this book is about encryption, the mathematics of error detection and correction is used in other settings as well. Great examples include streaming video on public sites or bar codes on products. There the information is meant to be publicly available, and there is a real need to ensure that the correct data is received.

In this chapter we explore these issues and some of their resolutions. We begin with a more detailed introduction to the problem, which we then follow with two interesting riddles. The resolutions of these puzzles are related to key concepts in the field; we encourage the reader to try to solve these problems first before reading the solution. We then build on these discussions and give various examples of error detecting and correcting codes, and end with a few applications.

9.1. Introduction

When communicating directly in a language (such as English), error detection and correction happen automatically, to some extent. For example, imagine you receive the message

> Hi,
> I will meet you at the restautant for lunch.

This message contains an error. Nevertheless, without contacting the sender for any additional information, you know that the intended message was

> Hi,
> I will meet you at the restaurant for lunch.

We are very confident of our correction as "restautant" is not an English word, and, moreover, the only English word which is a one-letter change away from it is "restaurant". Note that carrying out this correction doesn't even require us to understand the message—a simple computer program which knows the list of valid English words could automatically replace "restautant" with "restaurant". On the other hand, if "meet" had been replaced with "mete" in the message, a human who understands the meaning of the message would be able to correct the mistake, but a computer which only knows the list of valid English words (which would contain "mete" as a separate entry) would not recognize that there is an error.

When we send an encrypted message, on the other hand, there seems to be no possibility for automatic error correction/detection at all. A small error in transmission may turn most if not all of the message's decryption into gibberish, and before decrypting there is typically no restriction on the content of the message. For example, when computers send encrypted messages, they send strings of 1's and 0's—effectively, 1's and 0's are the "words" allowed by the language. If any digit is altered by a transmission error, there is no way to detect it, as the result will always be an allowed word.

While this motivating example deals with an encoded message, the same issues arise in other situations too. Imagine we want to go to a website to watch some streaming video, maybe our favorite baseball team is playing or we're looking at the highlights from yesterday's games. It's quite likely that the video isn't encrypted, and a string of 0's and 1's are sent to our computer, which then makes the movie for us to see. If we're thousands of miles away and there are lots of intermediate sites, there are many chances for digit errors; it would be terrible if there were so many of these that we couldn't reconstruct the picture. If a 1 is switched to a 0, how can we detect (and correct) the problem?

The problem of error detecting and correcting codes is to define languages (called **codes**) of **codewords** which can be used (usually by a computer) to transmit arbitrary data (encrypted messages, video, websites, etc.)

By restricting a transmission to allowable codewords, we enable small mistakes in transmission to be detected and/or corrected automatically, just on the basis of which codewords are considered valid, in analogy to how we corrected the error in the word "restaurant" in the example above.

For another example, we turn to the seemingly mundane task of purchasing a pack of gum (or almost any product) from almost any store. Gone are the days of a cashier staring at the product, looking for the price tag. Typically, the item is scanned; a laser runs over the UPC symbol and the correct price is displayed. This, however, is only the start of the story. If the product is correctly scanned, the store can update its inventory in real time; no longer must hundreds of man-hours be spent walking the shelves to keep track of purchases. Further, many stores "force" customers to sign up for rewards cards to get lower prices. This allows them to track purchases over time, providing a wealth of useful information.

The purpose of this chapter is to describe some of the issues of error detection and correction. The importance of both are clear. We just scratch the surface of these topics, but we go into enough detail to see some of the truly wonderful, amazing methods. To do the subject justice requires a little abstract algebra (group theory), linear algebra, and probability; however, we can go a long way with just some elementary observations. It's frequently the case that it's very hard for someone to have the flash of insight that leads to the discovery of one of these methods; however, it's often not that difficult to follow in their footsteps and learn their method.

9.2. Error Detection and Correction Riddles

Math riddles are often fun stepping stones to advanced theories with interesting applications. This is particularly true for the two riddles below. The solution to the first introduces a concept which is helpful in designing an error detection code, while the second riddle leads to an idea that can actually correct errors!

Both of these riddles are well known throughout the mathematics community. We've had a little fun with the phrasing. Try and solve these before reading the answers. For the first, it's possible to get a less than optimal solution that is still better than trivial. Both riddles involve hats that are either white or black; the connection to transmitting data is clear when we replace white with 1 and black with 0.

9.2.1. Error Detection Riddle

Imagine 100 *mathematicians are standing in a line. They close their eyes and you place a black or a white hat on each person. You can do this any way you wish; you can give everyone a white hat, or just throw them on randomly. Each mathematician can* **only** *see the color of the hats of the people in front of them. Thus the first person sees no hats, the second sees* *just the hat of the first person, and so on until the last person, who sees the colors of the* 99 *hats in front of her.*

When you say go, the last person says either "white" or "black" but not both; immediately after she speaks the second to last person says either "white" or "black" but not both. This continues until the first person speaks, saying either "white" or "black" (but not both). After all 100 *have spoken, we count how many said the color of their hat, and how many said the opposite color. For each person who said their color correctly, you give everyone in line one dollar; however, for each person who was wrong, everyone in line gives you one dollar.*

Remember, these are not 100 *ordinary people. These are* 100 *mathematicians, and they are allowed to talk to each other and decide on a strategy to maximize their expected winnings. They are thus extremely intelligent, and if there is a good idea they will find it! You get to put the hats on any way you wish, they get to look, and the* k^{th} *person sees the hat colors of the* $k-1$ *people in front of them. What is the minimum amount they can guarantee themselves earning? Or, to put it another way, what is the smallest* N *such that, no matter what you do, at least* N *of the* 100 *mathematicians correctly say their hat color!*

We want to find out how many people we can guarantee will say their hat color correctly. Thus, if ever someone is guessing, we have to assume they guess wrong. One possible strategy is for everyone to just say "white"; however, you know their strategy. If everyone is going to just say "white", then all you have to do is give everyone a black hat. In this case $N = 0$, which is really bad. If the hats were to be placed randomly, then yes, this strategy should lead to about half the people saying the correct color, but that is not this problem. In this problem, you can be malicious as you place the hats!

So our mathematicians need to be a bit more clever. Each person cannot be entirely devoted to finding their hat color—they have to somehow help each other. After a little thought, they come up with the following plan. All the even-numbered people say the hat color of the person in front of them. By doing this, all the odd-numbered people know their hat color! For example, if person 99 has a white hat, then person 100 says "white"; while she may be wrong, person 99 now says "white" and is correct. If this is their strategy, you'll of course make sure all the even people have the opposite hat color as the person in front of them; however, there is nothing you can do about the odd people. They *always* get their hat color right, and thus with this strategy $N = 50$ (in other words, at least half the people are always right).

So, here's the question: Can you do better than $N = 50$ (i.e., better than 50%)? Amazingly, yes! See how well you can do before you read the solution below. Can you do $N = 66$? Or $N = 75$? Clearly, the best possible is $N = 99$, as there is no way to ever help the last person. Is that possible? Can we find a way so that 99 out of 100 are correct?

Solution: There are several strategies, all of which do far better than just half are correct. We give just two. Let's first review the strategy that ensures that at least half are correct. We had all the even people say the hat color of the person in front of them. We used **one** piece of information to get **one** piece of information. Can we do better?

Let's examine what person 3 sees. In front of him are two people. There are four, and only four, possibilities: she sees WW, WB, BW, BB (where of course W means a white hat and B a black hat; the first letter denotes the hat of person 1 and the second letter the hat of person 2). Further, person 2 gets to see person 1's hat; there are only two possibilities here: W or B. What we're going to do is have the third person say something which, when combined with what the second person sees, allows first the second person and then the first person to deduce their hat color. Let's have the third person say "white" if the two hats have the same color, and "black" if the two hats have opposite colors. As soon as the third person says this, the first two people know whether or not they have the same or opposite colors. If they have the same, then the second person just says the color of the first person, and then the first person says that color as well; if their colors are opposite, then the second person says the opposite color of what he sees in front of him, and then the first person says the opposite color of the second person.

For example, if the third person sees BW, he says "black", as the hats are different colors. Seeing person 1 wearing a black hat, person two says "white", and then person 1 says "black". In this way we can make sure two out of every three people are correct. This means that we can take $N = 66$, or almost two-thirds of the people are guaranteed to say the correct color with this strategy.

Before reading the next solution (which is the best possible), try Exercises 9.9.1 and 9.9.2 to see how far you can push the above construction.

We now jump to the best strategy. It's absolutely amazing, but we can make sure that 99 out of 100 people are correct! How? The last person counts up how many white hats she sees, and how many black hats. The number of white hats plus the number of black hats must add up to 99, which is an odd number. Thus there has to be an odd number of white hats or an odd number of black hats, but not both. Here's the strategy: the last person says "white" if there is an odd number of white hats, and "black" if there is an odd number of black hats.

Why does this strategy work? Let's say the last person sees 73 white hats and 26 black hats. She therefore says "white" as there is an odd number of white hats. What should the second to last person do? The only difference between what he sees and what the last person sees is that he cannot see his hat color. There are only two possibilities: he sees 72 white hats and 26 black hats, or he sees 73 white hats and 25 black hats. He knows that there is an odd number of white hats. If he sees 72 white hats, he knows that he must be wearing a white hat, as otherwise the last person would not have said white. Similarly, if he sees 73 white hats, then he must be wearing a black hat, as otherwise the last person wouldn't have said there are an odd number of white hats.

The process continues. Each person keeps track of what has been said, and whether or not initially there was an odd number of white or black hats. Before speaking each person can see whether or not there are an odd number of white or black hats in front of them, and speak accordingly. Let's continue our example. We'll say there are 73 white hats, and for definiteness let's assume the 99th person has a white hat. Thus the last person says "white" as there is an odd number of white hats. The second to last person now immediately says "white", because he sees an even number of white hats (if he didn't have a white hat, the last person would have said "black"). Now let's look at the 98th person. She knows that the last person saw an odd number of white hats. The second to last person said "white". This means that there must be an even number of white hats on the first 98 people. Why? If there were an odd number of white hats on the first 98 people, then the last person would see an even number of white hats (because the odd number from the first 98 plus the one white hat from the 99th person would add up to an even number, which contradicts the last person seeing an odd number). So, the 98th person knows there are an even number of white hats on the first 98 people. If she sees 71 white hats, then she says "white", otherwise she says "black".

The key concept in this strategy is that of **parity**. All we care about is whether or not there is an even or an odd number of white hats on the first 99 people. The last person transmits this information, and the other people use it wisely. Later in this chapter we'll expand on the notion of parity and use it to create an error detecting code.

9.2.2. Error Correction Riddle

This riddle is famous in not just mathematics, but also economics. It too is a hat problem, but unlike the last one (where you could put the hats down however you want), this time the hats are randomly assigned. This randomness has profound consequences, which we'll see later in the chapter.

Three mathematicians enter a room, and a white or black hat is placed on each person's head. Each person is equally likely to have a white or a black hat, and the hat assignment of one person has no effect on the hat assignment to anyone else. Each person can see the other people's hats but not their own.

No communication of any sort is allowed, except for an initial strategy session before the game begins. Once they have had a chance to look at the other hats, each person has to decide whether or not to say "white", "black", or remain silent. Everyone must speak (or remain silent) at the same time. If everyone who speaks says the color of their hat, then everyone gets one million dollars; however, if even one person who speaks says the wrong color for their hat, then everyone loses one million dollars. If no one speaks, then no money is won or lost.

What is the best strategy for the mathematicians? In other words, what is the largest value of p so that, if this game is played many, many times, then the mathematicians are expected to win p percent of the time, and lose $1 - p$ percent of the time.

Unlike the first riddle, here each hat is randomly assigned; each person gets a white hat half the time and a black hat half the time. There is a very simple strategy that ensures that the mathematicians never lose: no one ever speaks! Unfortunately, with this strategy they also never win.

There's an easy way to make sure they win half the time. One person is told to always say "white" and the other two are told to always remain silent. Half the time the person will be correct in saying white, and half the time they will be wrong. Thus, we can easily get $p = 1/2$.

Is it possible to do better? It seems absurd to think about getting a p greater than $1/2$. After all, each person who speaks says either white or black, and they are equally likely to have a white or a black hat. Doesn't this mean that anyone who speaks will be right half the time and wrong half the time? Further, if more people speak, it's even worse, as they only win if everyone who speaks is right. It therefore seems impossible to come up with a better strategy, yet there is one, and this strategy leads to error correcting codes.

Solution: There's actually a strategy that works 75% of the time when the hats are randomly placed: each person looks at the other two people. If you see two hats of the same color, you say the opposite color; if you see two hats of opposite colors, you stay silent.

That's it! It's simple to state, but does it work, and if so, why? Let's tackle whether or not it works first. Imagine the three hat colors are WBW. Then the first person sees BW, the second sees WW and the third sees WB. Only the second person sees two hats of the same color. So only the second person speaks, saying "black" (the opposite color); the other two people are silent. What if instead it was WWW? In this case, everyone sees two hats of the same color, so everyone speaks and says "black", and everyone is wrong.

Table 9.1 looks at who speaks and whether they are correct or incorrect.

There are several remarkable facts that we can glean from this table. The first and most important, of course, is that the strategy is successful exactly three-fourths of the time. This is truly amazing. Each person has an equal chance of having a white or a black hat, yet somehow we manage to do better than 50%. How can this be?

The answer lies in the three columns saying whether or not each person is correct or incorrect. While the outcome column is very nice for our three people (saying they win six out of eight times), it is the individual right/wrong columns that reveal what is really going on. Note each person is correct twice, incorrect twice, and silent four times. Thus, each person is only correctly saying their hat color half the time they speak (and only a

TABLE 9.1. The various outcomes for our hat strategy. The first three columns are the hat colors of the three people; the next three columns are what each person says (if they remain silent, we leave it blank); the next three columns are whether or not a speaker is correct; and the final column is whether or not the players win or lose.

			#1	#2	#3	#1	#2	#3	Outcome
W	W	W	black	black	black	wrong	wrong	wrong	lose
W	W	B			black			right	win
W	B	W		black			right		win
B	W	W	black			right			win
W	B	B	white			right			win
B	W	B		white			right		win
B	B	W			white			right	win
B	B	B	white	white	white	wrong	wrong	wrong	lose

quarter of the time overall). Notice, however, the sideways M-shaped pattern in who is correct and who is incorrect. We've somehow arranged it so that the wrong answers are piled up together, and the correct answers are widely separated. In other words, when we are wrong, boy are we wrong! All three people err. However, when we are right, only one person is speaking. We thus take six correct and six incorrect answers and concentrate the incorrect answers together and spread out the correct answers.

The arguments above explain how it works, but it doesn't really say why it works. To understand the "why", we have to delve a bit deeper, and it is this explanation that plays such a central role in error correcting codes. We have a space of eight possible hat assignments:

$$\{\text{WWW, WWB, WBW, BWW, WBB, BWB, BBW, BBB}\}.$$

Each assignment is equally likely; thus one-eighth of the time we have WWW, one-eighth of the time we have WWB, and so on. We partition our space into two disjoint subsets:

$$\{\textbf{WWW}, \text{WWB, WBW, BWW}\}, \quad \{\text{WBB, BWB, BBW, } \textbf{BBB}\}.$$

What is so special about this partition is how the elements are related. In the first set, the second, third and fourth elements differ from the first element, which is WWW, in only one place. To put it another way, if we start with WWW we get any of the next three elements by changing *one and only one hat color*. Further—and this is the key point—the only way we can get something in the second set from WWW is to change *at least two hat colors*. We have a similar result for the second set. The first, second and third elements can be obtained from BBB by switching exactly one of their colors; further, nothing in the first set can be switched into BBB unless we change at least two hats.

This partitioning of the outcome space is at the heart of our solution. We have split the eight elements into two sets of four. The first set is either WWW or anything that can be made into WWW by switching exactly one hat. The second set is either BBB or anything that can be made into BBB by switching exactly one hat. Further, these two sets are disjoint—they have no elements in common, and thus they split our space into two equal groups. Later in this chapter we'll see how this partition can be used to build an error correcting code.

9.3. Definitions and Setup

It's time to switch from our informal discussion to a more rigorous description of error detection and correction. To do so, we need a few definitions. We've repeatedly talked about codes in this chapter (error correcting codes, error detecting codes). What do we mean by this? A **code** is a collection of strings formed from a given **alphabet**. The alphabet might be the standard one for the English language, or it might be the binary set $\{0, 1\}$. If the alphabet has just two elements we refer to it as a **binary code**. The elements of the code are called the **codewords**. If every codeword has the same length, we have a **fixed length code**.

In practice, here's what happens. We have a message or some data that we want to transmit. We choose a code C and convert our message to codewords from C. We transmit that string of codewords to the intended recipient, who converts the string back into an ordinary message. The key question is, How can we construct C to allow us to detect some errors in transmission? What if we want to be able to automatically *correct* errors as they happen? Can we do all this without increasing the size of the message too much?

Let's look at some examples.

EXAMPLE 9.3.1. *Take $A = \{a, b, c, \dots, x, y, z\}$ as our alphabet, and let C be the set of all finite strings of elements in A. Clearly, C is a code; moreover, every English word is in C, as are nonsense words like qwerty and noitadnuof. It isn't a fixed length code, as "the" and "bulldog" are both valid codewords, and they have different lengths.*

EXAMPLE 9.3.2. *Take $A = \{0, 1\}$ as our alphabet, and let C be the set of all strings of elements in A of length 4. We easily see that C is a code; further, it is a fixed length code, as each codeword has length 4.*

In this chapter we concentrate of fixed-length codes. This has the advantage that we don't need to do any extra work to know when one codeword stops and the next begins. For example, with the code from Example 9.3.2, if we receive the message

$$001010011010,$$

then we know that the codewords sent were 0010, 1001, and 1010.

Since in practice error correcting codes are used for transmission by computers, from here on we concentrate on codes over the alphabet $\{0, 1\}$. Thus all words are made up of only 0's and 1's. We need one last definition. The **Hamming distance** between two binary strings (both of the same length, say n) is the number of places where the two strings differ. We write $d(s, t)$ to mean the Hamming distance between two strings s and t.

For example, imagine $n = 10$ and our two strings are 0011100101 and 0011100001. These two strings are almost identical; the only difference between them is in the eighth position. Thus

$$d(0011100101, 0011100001) \;=\; 1.$$

A very important example is the binary code

$$\{111,\ 110,\ 101,\ 011,\ 100,\ 010,\ 001,\ 000\}.$$

We have

$$\begin{aligned} d(111, 111) &= 0, \\ d(111, 110) &= d(111, 101) \;=\; d(111, 011) \;=\; 1, \\ d(111, 100) &= d(111, 010) \;=\; d(111, 001) \;=\; 2, \\ d(111, 000) &= 3. \end{aligned}$$

If you view the two codewords as points in n-dimensional space, the Hamming distance measures how far apart they are, given that we can only walk parallel to the coordinate axes. Thus $d((0, 0), (1, 1)) = 2$; we can either take the path

$$(0, 0) \longrightarrow (0, 1) \longrightarrow (1, 1)$$

or the path

$$(0, 0) \longrightarrow (1, 0) \longrightarrow (1, 1).$$

This is different than the normal distance between the two points, which is found using the Pythagorean theorem. In this problem, the normal distance is just $\sqrt{(1-0)^2 + (1-0)^2} = \sqrt{2}$.

The **minimum distance of a code** C is the smallest Hamming distance between two distinct codewords in C. We denote this by $d(C)$. If we didn't force the two codewords to be distinct, then the minimum distance of any code would be zero, as $d(w, w) = 0$; thus this is a necessary constraint. The **maximum distance of a code** C is the maximum distance between two distinct codewords.

We end this section with another definition that we need. We say a code C is k **error detecting** if the following holds: no matter what codeword of C we choose, if we change at most k digits, then the resulting string is not a codeword of C. The motivation for this definition is simple: if a message is transmitted with a 3 error detecting code, then transmission errors can be recognized as long as you make at most three errors in a single codeword (if you make more than that, you may accidentally land on a different valid codeword).

If C is k error detecting but not $(k+1)$ error detecting, then we say C is **exactly k error detecting**. Note that an exactly 0 error detecting code is not useful for detecting errors at all: it means that making a single digit error can still leave you with a valid codeword (different from the one you intended).

EXAMPLE 9.3.3. *Consider the binary code*

$$\{00,\ 01,\ 10,\ 11\}.$$

This code unfortunately cannot detect any errors. Why? We have a fixed length code (of length 2*). There are four possible codewords of length* 2, *and all of them are in our code. If we change any digit of any codeword, we end up with another codeword.*

The above example shows us that, in order to be k error detecting (for some $k \geq 1$), it's necessary that our code is only a subset of all possible words. If we have a binary code of fixed length n, then there are 2^n possible codewords. Clearly, if every word is in our code, then we cannot detect any errors. The question becomes how many words can we include and have a 1 error detecting code, and which words can these be? We can then of course ask the same question for a 2 error detecting code, and a 3 error detecting code, and so on.

EXAMPLE 9.3.4. *Consider the binary code*

$$\{000,\ 111\}.$$

This code is exactly 2 *error detecting. Even though it can detect one error and it can detect two errors, it cannot tell whether one or two errors were made!*

9.4. Examples of Error Detecting Codes

The purpose of this chapter is to resolve two important issues in information theory. The first is to construct an error detecting code, and the second is to construct an error correcting code. In this section we tackle the first problem, and give a variety of error detecting codes. Remember, our goal here is *not* to figure out what message was sent, but rather to determine whether or not there was a transmission error. This is a much easier problem. Most of the codes we'll see here are exactly 1 error detecting. It's not unreasonable to emphasize these codes. Hopefully the probability of making an error is small; if that is the case, it's unlikely we'll have two or more errors in a single codeword.

In this section we briefly describe several different types of error detecting codes, and then discuss them in greater detail in the following section, both historically and mathematically.

EXAMPLE 9.4.1. *Imagine our code is*

$$C = \{1000,\ 0100,\ 0010,\ 0010,\ 1110,\ 1101,\ 1011,\ 0111\}; \tag{9.1}$$

this is a fixed length binary code (the length is 4*), made up of words with an odd number of* 1*'s. The minimum distance of this code is* 2*. While this can be seen by an unenlightening brute force computation, there's a much better way. If we take any two words with exactly one* 1*, then they differ in two places, and similarly if we take two words with exactly three* 1*'s. What happens if we take one of each? If the* 1 *from the codeword with exactly one* 1 *aligns with one of the* 1*'s from the word with exactly three* 1*'s, then the* 0 *of the word with three* 1*'s aligns with the* 0*, but the other two* 1*'s do not, and thus the separation is* 2; *a similar argument gives* 2 *when the* 1 *does not align.*

Imagine we receive the message 0101*. We were expecting a codeword from* C*; it is impossible that this was the message. Why? Our message has to be one of the eight words in* (9.1)*; however, all of the codewords have an odd number of* 1*'s, and* 0101 *has an even number of* 1*'s. There* must *have been an error!*

This is terrific—we know there was a mistake, and we can ask for the message to be resent. We unfortunately cannot tell where the error is; perhaps the message was supposed to be 0100, or perhaps it was meant to be 1101. We can't tell, but we do know an error was made.

There are many other error detecting codes. Here's another, which was inspired by our first riddle.

EXAMPLE 9.4.2 **(Parity code)**. *Let* C *be the binary code of all words of length* 4 *such that the sum of the digits of a codeword is divisible by* 2 *(alternatively, we say the sum of the digits is congruent to* 0 *modulo* 2*). There are thus* $2^4 = 16$ *binary codewords of length* 4; *half of these have digits summing to zero modulo* 2*. Why? We can choose the first three digits any way we wish, and then the last digit is forced on us. For example, if we have* 101*, then the final digit must be a* 0*, while if we have* 010 *the final digit must be a* 1*. There are* $2^3 = 8$ *ways to choose the first three digits, so there are eight codewords. Thus*

$$C = \{0000,\ 0011,\ 0101,\ 1001,\ 1100,\ 1010,\ 0110,\ 1111\}. \tag{9.2}$$

If we receive the message 1011*, we know there was a transmission error as the digits do not sum to zero modulo* 2 *(i.e., there is not an even number of* 1*'s). Going through all the cases, we see if we take any codeword and change exactly one digit, then it is no longer a codeword. Thus, the code is* 1 *error detecting. It is not* 2 *error detecting. To see this, we can change* 0011 *to* 0101 *by changing the middle two digits, and this is a valid codeword in* C*.*

These two examples give us two error detecting codes; however, they seem very similar. Both involve codes with eight words, and both are exactly 1 error detecting. Let's look at a really different example.

EXAMPLE 9.4.3 **(Fixed number code)**. *We take our binary code C to be all binary words of length 5 with exactly two 1's:*

$$C = \{00011,\ 00101,\ 01001,\ 10001,\ 00110,\\ 01010,\ 10010,\ 01100,\ 10100,\ 11000\}.$$

Note the minimum distance of this code, like the other two examples, is 2. If there's exactly one error, then we'll detect it, as there will then be either three 1's or just one 1. If there are exactly two errors, we may or may not detect it. If the message was meant to be 00011 but we receive 01111, we know there was an error; however, if we receive 00101, then we would not realize an error occurred. This code was used by Bell Telephone Laboratories in the 1940s (see page 8 of [**48**]*), as the ten different codewords could be set in correspondence with the ten different digits, and thus this gave a way of programming in the decimal system.*

We have two candidates for an exactly 1 error detecting code: the code from Example 9.4.2 and the code from Example 9.4.3. Which is better? It's a little hard to compare the two codes. The first has codewords of length 4, the second has codewords of length 5. Both can detect exactly one error, but is it better to be able to detect one error in four digits or one error in five? We're comparing apples and oranges, and that's always a bit messy. We need to find a way to judge how good each code is.

One very good metric is to look at how much information each code allows us to convey. Let's look first at the code from Example 9.4.2. The codewords there have length 4, with the fourth digit determined by the first three. Thus we have three free digits, and can effectively send any of eight words: 000, 001, 010, 100, 101, 110, 011, 111. If we were to look at a string of length 20, we could send five codewords. As each codeword can be one of eight possibilities, this means we can send $8^5 = 32,768$ messages. What about the code from Example 9.4.3? There we have ten possible codewords of length 5. So, if we were to send a message of length 20, we could send four codewords, giving us a total of $10^4 = 10,000$ messages.

It should now be clear why we looked at messages of length 20. This is the least common multiple of 4 (the length of the codewords of the first code) and 5 (the length of the codewords of the second code). The first code allows us to transmit almost three times as much information as the second. From this perspective, the first code is far superior. So, if all we care about is the amount of information transmitted, the first code is better; however, clearly this is not the only thing we care about. We are also concerned with detecting errors. In blocks of length 4, the first code can detect one error, while the second code can detect one error in blocks of length 5.

Comparing these two codes turned out to be easier than we thought—in each of the two natural metrics, the first code is clearly superior.

We end with one final example of an error detecting code.

EXAMPLE 9.4.4 **(Repetition code)**. *This is the "tell me twice" code, where we take binary code of fixed length and replace each word with its double. For example, if we started off with the code*

$$\{00,\ 01,\ 10,\ 11\},$$

the new code becomes

$$\{0000,\ 0101,\ 1010,\ 1111\}.$$

It's very easy to detect exactly one error; if there is just one error, then the first two digits is not the same pair as the last two digits. Thus, this is a 1 *error detecting code. It isn't a* 2 *error detecting code as* 1010 *could be the intended message, or it could have been* 0000 *corrupted in two places.*

In Example 9.4.4, we have a four digit code with exactly four codewords. Note this is much worse than Example 9.4.2. There we also had a four digit code, but we had eight codewords. Both codes can detect one error in a block of length 4, but the code from Example 9.4.2 can transmit twice as many messages per block of four.

9.5. Error Correcting Codes

We now move to the exciting, long-awaited topic of error correction. We describe some simple and easily implemented codes that not only detect errors, but actually correct them as well!

The simplest is an expanded form of the repetition code of Example 9.4.4. Instead of repeating the message once, let's repeat it twice. In other words, if we want to send one of 00, 01, 10, 11, then our code is

$$\{000000,\ 010101,\ 101010,\ 111111\}.$$

The idea behind this code is that, if mistakes are rare, the probability of two mistakes is very unlikely, and the majority opinion is probably correct. For example, imagine we receive the codeword 101110. This isn't in our code, and thus we know there's been a transmission error. If there can be only a one-digit error, then the original message must have been 101010; it's the only codeword whose distance is at most 1 unit from 101110. All three of the blocks say the first digit of the message is 1; however, two of the blocks say the second digit is a 0 while one block says it's a 1. Going by the majority, we declare the fourth received digit to be in error, and the intended message to be 101010.

We've done it—we have a code that not only allows us to detect but also correct errors. Unfortunately, it's quite expensive. How costly is it? Well, in our fixed code of length 6 in the above example, only two digits were used for the message, and the other four were used for detection and correction. That means only one-third of the message is actually conveying information. Can we do better?

Let's revisit the second riddle from §9.2. The key observation in our solution of the hat problem is that of the eight binary words of length 3, each word is either 000, 111, or differs in exactly one digit from one of 000 and 111. Let's consider a code based on this. We transmit either 000 or 111. Our received message is either one of these, or differs from one of these in exactly one digit. If our received message is not 000 or 111, we take whichever of these two is exactly one digit away. Note that this method is the same as repeat it twice.

While this actual code turns out to be a disappointment (it's the same as just the first digit is free, or again an efficiency of one-third), the idea of using the Hamming distance to partition the set of possible codewords into disjoint subsets is very useful. There are entire courses devoted to this subject; we cannot begin to do it justice in a section or two, but we can at least give a flavor of what can be done. In this section we content ourselves with describing a very powerful error correcting code, the **Hamming** $(7, 4)$ **code**. Our exposition will be entirely unmotivated—we'll simply state what it is and see why it works. In the next section, we discuss how Hamming and others found this code.

The general idea is easy to state. Imagine we have a binary fixed length code C (of length n) which is k error correcting. This means our codewords all have the same length (namely n), and are just strings of 0's and 1's. Further, any two distinct codewords differ in at least $k+1$ places. It turns out C has the following wonderful property: each of the 2^n binary strings of length n is within distance $\lfloor \frac{k}{2} \rfloor$ of at most one element of C. Thus our code C can detect and correct $\lfloor \frac{k}{2} \rfloor$ errors!

Why is there at most one codeword within distance $\lfloor \frac{k}{2} \rfloor$ of any binary string of length n? Suppose there were two such codewords, w_1 and w_2, within distance $\lfloor \frac{k}{2} \rfloor$ of a string s. Since w_1 differs in at most $\lfloor \frac{k}{2} \rfloor$ places from s, and w_2 differs in at most $\lfloor \frac{k}{2} \rfloor$ places from s, w_1 and w_2 differ in most

$$\left\lfloor \frac{k}{2} \right\rfloor + \left\lfloor \frac{k}{2} \right\rfloor \leq k$$

places, which is impossible: since the code is k error detecting, any two codewords must differ in $k+1$ places!

We can rephrase this argument as pointing out that the Hamming distance d satisfies the **triangle inequality** $d(x, z) \leq d(x, y) + d(y, z)$. Letting $x = w_1$, $y = s$, and $z = w_2$, we have

$$d(w_1, w_2) \ \leq \ d(w_1, s) + d(s, w_2) \ \leq \ 2 \left\lfloor \frac{d-1}{2} \right\rfloor \ \leq \ k,$$

which gives the same contradiction.

We have shown the following result.

THEOREM 9.1. *Any k error detecting code is $\lfloor \frac{k}{2} \rfloor$ error correcting.*

Thus the code from Example 9.3.4 is a 1 error correcting code! Notice that really this is again a triple-repetition code, this time based just on 0 and 1. So once again, it takes three times as much data to send a message with this code as to send it directly.

The Hamming (7,4) code does much better. The code has seven binary digits, four are free message digits and three are check digits. Thus, using the code only increases the message length by a factor of $\frac{7}{4}$; the message gets 75% longer.

The message digits are the third, fifth, sixth, and seventh, and the check digits are the remaining three. (There are good reasons to have these as the message digits and not the first four; one could of course re-order the discussion below and make the message the first four bits.) The code C is the following 16 words:

$$1111111, 0010110, 1010101, 0111100, 0110011, 1011010, 0011001, 1110000,$$
$$0001111, 1100110, 0100101, 1001100, 1000011, 0101010, 1101001, 0000000.$$

A tedious calculation (there are better ways) shows that the minimum and the maximum distance of the code is 4; equivalently, this means each codeword differs from any other codeword in exactly four places!

The Hamming $(7,4)$ code has 16 codewords. Each codeword is exactly four units from any other codeword, thus the code is 3 error correcting. Since $\lfloor \frac{3}{2} \rfloor = 1$, this is a 1 error correcting code. Another way of stating this is that each of the $2^7 = 128$ binary strings of length 7 is at distance 1 from at most one codeword.

We have therefore shown that the Hamming $(7,4)$ code can correct one error. How efficient is it? How much information does it transmit? As we have four message digits and three check digits out of seven, four-sevenths or about 57% of the digits convey information. This is far superior to the repetition code, where only one-third of the digits were transmitting information. In fact, over half of our digits convey information, something that is of course impossible with a repetition code.

The Hamming $(7,4)$ code has a lot of nice features. Our message is digits 3, 5, 6, and 7. We discuss in the next section why we take the check digits 1, 2, and 4 to be the values they are; for now, let's just accept these (though we encourage you to try and find a nice pattern among these three numbers). There are three parts to using an error detecting code. First we encode our message, then we transmit it, and finally the recipient tries to decode the message. Our code tells us how each message is encoded. For example, while 1111 becomes 1111111, we have 1011 is sent to 0110011. We then transmit our message, and any single error can be detected and corrected. To decode our message, is it enough to just drop the first, second, and fourth digits? Unfortunately, no; this works if there are no transmission errors, but of course the entire point is in correcting errors! Fortunately the decoding isn't too bad. So long as there's at most one error, only one of our 16 codewords is within 1 unit of the received message; whatever codeword is this close is the decoding. Thus the Hamming (7, 4) code has an extremely simple decoding routine; it's so easy we don't have to worry at all about any difficulties in implementing it.

Our purpose is to just show that efficient, easily implementable error correcting codes exist. There's an extensive literature on this important subject; see for example [**7, 10, 11, 27, 48**] for introductions at various levels. Stepping back, what we have just shown is already quite stunning: we can construct an easy to implement code that can detect and correct an error while still having more than half of its digits devoted to our message!

9.6. More on the Hamming (7, 4) Code

Before describing how Hamming arrived at this code, it's interesting to note why he was looking for it. In the 1940s Hamming worked at Bell Laboratories. He had access to computer time, but only on the weekends. The way it worked was his program was fed into the computer, which ran until an error was found. At the first sign of an error, the program halted and the computer moved on to the next person's task. For two straight weekends, errors developed and Hamming was left with nothing. Frustrated, he wondered why couldn't the computer be taught to not just detect errors, but also to correct them.

EXAMPLE 9.6.1 **(Square code)**. *Hamming had a series of papers developing error correcting techniques. One fun example is a* (9, 4) *binary block code. Consider a four digit number in binary, say* $b_1b_2b_3b_4$. *We write the number in a square*

$$\begin{matrix} b_1 & b_2 \\ b_3 & b_4 \end{matrix}$$

and extend it to

$$\begin{matrix} b_1 & b_2 & b_5 \\ b_3 & b_4 & b_6 \\ b_7 & b_8 & b_9 \end{matrix}$$

as follows: b_5 *equals* $b_1 + b_2$ *modulo* 2, b_6 *equals* $b_3 + b_4$ *modulo* 2, b_7 *equals* $b_1 + b_3$ *modulo* 2, b_8 *equals* $b_2 + b_4$ *modulo* 2, *and finally* b_9 *is* $b_1 + b_2 + b_3 + b_4$ *modulo* 2. *We thus have four message digits and five check digits. It's a nice exercise to show that this can detect and correct any single error (in fact, we don't even need* b_9*). This code conveys more information than the repeating block code. While the "tell me three times" or the "majority rules" repeating code had one-third of its digits conveying information, here four-ninths or about* 44% *of the digits convey information (if we take the improved version where we remove the superfluous* b_9*, then half the digits are transmitting information). While this is worse than the Hamming* (7, 4) *code, it's a marked improvement over the repetition code. Note that this method is a natural outgrowth of parity checks.*

We described the square code for a codeword with four digits of information; of course, we could do a larger one (and we sketch some of the details in Exercise 9.9.34). If we had r message digits in each codeword, then we need to have that $r = m^2$ for some m as we have to be able to arrange the message in a square. How many check digits are there if our message is of size m^2? There are $2m + 1$ check digits: m from the m rows, another m from the m columns, and then the final check for the row and column checks; of course, we can similarly remove the last check and make do with just $2m$ check digits. Note that as m grows, almost all (percentage-wise) of the digits of our message are transmitting information, and almost none are being used for checking. Specifically, the size of the code is

$$m^2 + 2m + 1 = (m + 1)^2,$$

m^2 digits are used for transmitting information, and thus the percentage being used for information is

$$\frac{m^2}{(m+1)^2} = \left(\frac{m+1-1}{m+1}\right)^2 = \left(1 - \frac{1}{m+1}\right)^2.$$

As m tends to infinity, this can be made as close to 1 as we wish (though always, of course, a little bit less than 1).

What does this mean? Our previous codes had efficiencies like a third or four-ninths; now we can make a code as efficient as desired, with (percentage-wise) almost all digits devoted to transmitting information! There is, of course, a trade-off. We don't get anything for free. This code can still only detect one error per codeword. Correcting one error in a codeword of length $(m+1)^2$ is much worse than correcting one error in each codeword of length four or seven. We've returned to the standard trade-off; would we rather have a greater percentage of our message conveying information, or would we rather be able to detect a greater percentage of errors? Clearly, it's much better to detect one error out of every three bits than one error out of a thousand bits. The choice between correcting more errors and having a more space-efficient code is not always clear, and depends on the bandwidth and error rate of our transmission medium. Certainly, square codes are not the end of the story.

We now describe the derivation of the Hamming (7, 4) code. We have seven digits at our disposal, and the goal is to devote four of them to our message and three of them to checks. We use parity checks, which is very common when working with 0's and 1's. We're going to ignore the geometry and the distance metric, and concentrate on how the check digits are used. We have 16 possible messages, or equivalently 16 numbers.

We place our four digit message into slots 3, 5, 6 and 7 of our seven digit message. We now have to assign values to the first, second and fourth digits. Let's write our seven digit message as $d_1d_2d_3d_4d_5d_6d_7$, where we know d_3, d_5, d_6, and d_7, and we need to determine d_1, d_2, and d_4. We set

$$\begin{aligned} d_1 &= d_3 + d_5 + d_7 \bmod 2, \\ d_2 &= d_3 + d_6 + d_7 \bmod 2, \\ d_4 &= d_5 + d_6 + d_7 \bmod 2. \end{aligned} \tag{9.3}$$

Remember that if k is an integer, then k mod 2 is 0 if k is even and 1 if k is odd.

Why does Hamming's method work? We have three parity checks. The first involves (d_3, d_5, d_7), the second (d_3, d_6, d_7) and the last (d_5, d_6, d_7). We assume there is at most one error. If the message is transmitted correctly, then (9.3) is satisfied. If there is an error, then at least one of the three equations in (9.3) cannot hold. We now explore all the various possibilities.

Imagine that only one test fails. We'll assume it's the first test that fails; the argument is similar for the other two tests failing. If only the first fails, then d_1 (one of our check digits!) is wrong. To see this, note that if the second two equations hold, then $d_2, d_3, \ldots, d_7$ must all be correct, and therefore the cause of the error must be d_1.

We now assume that exactly two tests fail. Again, the argument is similar no matter which two tests fail, so let's assume it's the first two tests that fail. If the first two equations fail, then the fact that the last holds means d_4, d_5, d_6 and d_7 are correct. The only element in common between the first two equations is d_3, which therefore must be in error.

Finally, imagine all three parity checks fail. The only way this can happen is if d_7 was in error. This is because d_7 is the only variable to appear in all three equations; each other variable is in at most two, and thus cannot cause more than two parity checks to fail.

After a long discussion or proof, it's worthwhile stepping back and seeing what was the fundamental idea or observation. Many arguments in this chapter, including the one we just finished, are based on the notion of parity. While we introduced it to solve the first hat riddle, it turns out to be a fundamental and powerful concept. We'll see another example of its utility in the next section.

9.7. From Parity to UPC Symbols

We've just scratched the surface of the consequences and power of parity. It's worth mentioning in some detail one of the most common applications of parity checks, something you probably see every day: **UPC symbols**. UPC stands for **Universal Product Code**; it's the vertical lines you see on products (see Figure 9.1). Our goal here is to help you appreciate the applicability of math the next time you purchase a product.

The story of how UPC codes came about is a fascinating one. It was originally called the Uniform Grocery Product Code, as it arose from a consortium of food chains. The impetus for this was the need to track purchases. There are many reasons for this. An obvious one is to make sure

FIGURE 9.1. UPC-A barcode for the book *Book of Pure Logic*, by George F. Thomson. (Image posted on Wikipedia Commons by Maly LOLek, February 2007.)

the store is properly stocked. For decades, this was done by hiring people to go through the shelves and manually keep track of the number of items they saw. This is expensive, it's time intensive, and it's prone to errors. Many people realized the advantages of recording purchasing decisions at the point of sale, namely at the checkout. Each purchased item has to go through the checkout before leaving the store. If we can record the purchases here, we can easily update the store's inventory. The first item, a 10-pack of Wrigley's Juicy Fruit gum, was scanned in 1974 at a Marsh's supermarket in Troy, Ohio. Barcodes now appear everywhere, from the smallest products to the shipping labels used to mail them.

The system we currently have, with a laser scanning a sequence of vertical lines, is now fast and cheap, but of course there's always the possibility of an error in scanning. To overcome this, the UPC barcodes use a **check digit** scheme, which is a generalized parity test. If the test fails, then there must have been an error, and the barcode is scanned again. In practice, this just means the cashier hasn't heard a beep or seen a light flash, and just runs the product through again.

We won't describe the mechanics of the UPC barcodes, as there are a plethora of good, detailed webpages and articles on this, such as

Mathematics of the UPC barcodes:
- ◇ `http://en.wikipedia.org/wiki/Universal_Product_Code`.
- ◇ `http://electronics.howstuffworks.com/gadgets/high-tech-gadgets/upc.htm`.
- ◇ `http://www.adams1.com/upccode.html` (which also has some useful links at the end).

Further reading on the history of UPC barcodes:
- ◇ `http://www.nationalbarcode.com/History-of-Barcode-Scanners.htm`
- ◇ `http://en.wikipedia.org/wiki/Universal_Product_Code`.
- ◇ `http://barcodes.gs1us.org/DesktopModules/Bring2mind/DMX/Download.aspx?TabId=136&DMXModule=731&Command=Core_Download&EntryId=79&PortalId=0`. This is a review by PriceWaterhouseCooper on the 25th anniversary of UPC barcodes (and is directly linked on the Wikipedia page).

Instead, we quickly discuss check digit algorithms. Typically, the last digit is some function of the former. One of the most common is the last digit is simply the sum of the previous digits modulo 10 (or modulo 2, depending on what base we're using). If the scanned number doesn't have its final digit match the sum of the previous, then we know an error has occurred. Note that while this method is capable of *detecting* an error, it isn't capable of *correcting* it. In practice, if errors are rare and it's easy to rescan, this isn't a serious problem.

Of course, the check digit algorithm above only catches certain types of errors. If we want the number 1 1 9 8 8, the sum modulo 10 is 7, so we would write 1 1 9 8 8 7. Imagine we read 1 1 9 3 8 7 (so one of the 8's was mistaken for a 3). The check digit test fails, as $1+1+9+3+8 \neq 7 \bmod 10$. If instead we read 1 9 1 8 8, however, the check digit test is successful even though we've made a mistake, switching two digits. This has led to many people developing more complicated check digit algorithms capable of detecting larger classes of errors.

There are many other check digit algorithms (a particularly nice one was created by Jacobus Verhoueff, which was designed to detect many of the most common errors made by people). As we've seen time and time again, there are trade-offs. Here we need to balance the simplicity of the algorithm with the types of errors that can be detected. In practice, however, there are no issues with using the more involved algorithms. The reason is that these are all being done by a computer, and it's only marginally harder for a computer to do the weighted sums from Exercise 9.9.40 than just summing the digits. For more, see [**19**].

9.8. Summary and Further Topics

We've just scratched the surface of both what can be done with error correcting and detecting, as well as the mathematics needed. Right now we can correct one error, but what if we wanted to correct two errors? Earlier exercises asked you to create some methods. It's not too hard to find a method that will correct two (or even n) errors *if* we're willing to accept a very small information rate. We quickly describe one such method below and then mention some more advanced techniques, which require significantly more mathematical background than the rest of the chapter.

We saw that repeating a message three times sufficed to catch one error; if now we repeat the message five times, we can catch two errors. For example, imagine our message is either 0 or 1, and we repeat it five times. We therefore transmit either 00000 or 11111. If at most two digits are flipped, we can figure out what was intended. This is clearly an improvement over our original method of repeating three times; for example, if there are two transmission errors among the first three message bits, we can now detect that! The problem, of course, is that most of our message is devoted to error

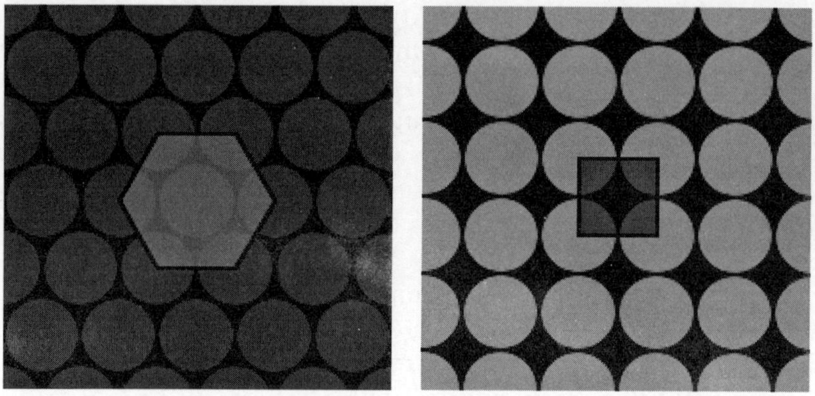

FIGURE 9.2. Packing the plane with circles in a hexagonal and a square pattern. (Images from Wikipedia Commons from user 99of9.)

detection and correction. Specifically, only 20% (or 1 in 5 bits) are now used to send information, much worse than the 33% (or 1 in 3 bits) from before.

While we could generalize this to a method capable of correcting n errors, the amount of the message devoted to information is painfully small. How bad is it? Using the above method, we repeat everything $2n + 1$ times and can detect and correct n errors, but only 1 out of every $2n + 1$ bits is conveying new information.

We clearly need a better way if we want to convey enormous amounts of information, such as downloading a large program from the web or watching streaming video. A fruitful approach involves the mathematics of **sphere packings** and **lattices**. We briefly describe these subjects.

Given some space, we can investigate how many unit spheres can be placed inside. In one dimension, a sphere is just a line segment, which means we can cover the entire real line without any gaps. In two dimensions our spheres are now circles, and it's been known for a long time that the best pattern is a hexagonal pattern, which covers more of the plane than a square pattern; see Figure 9.2 for these patterns. The optimal packing in three dimensions was only recently resolved.

Looking at the hexagonal and square packings in the plane in Figure 9.2, we see there's a lot of structure. In particular, the centers of the circles lie in a very ordered manner. In each case the centers form a **lattice**. Given independent vectors $\overrightarrow{v_1}, \dots, \overrightarrow{v_m}$ in n-dimensional space, they generate a lattice Λ where

$$\Lambda = \{a_1\overrightarrow{v_1} + \cdots + a_m\overrightarrow{v_m} : a_1, \dots, a_m \in \mathbb{Z}\}.$$

In other words, the lattice formed by the independent vectors is just the set of all linear combinations with integer coefficients.

While we haven't mentioned sphere packings and lattices explicitly in the chapter, they've been lurking in the background. It turns out that there's a connection between how tightly spheres can be packed together (not necessarily in three dimensional space!) and error correcting codes. Thinking back to the Hamming $(7,4)$ code, we can view this as having 16 unit spheres in a 7-dimensional space such that no two spheres intersect, and each of the 2^7 tuples is in or on exactly one sphere. This leads of course to the question of how does one optimally pack spheres, which we briefly discussed in one and two dimensions. It was only recently that Hales and his graduate student Ferguson solved the problem in three dimensions. Kepler conjectured that the optimal arrangement was to have alternating layers of spheres; this became known as the **Kepler conjecture**, and was one of the longest unsolved, important problems in mathematics (it's a little over 25 years older than Fermat's Last Theorem, which fell a few years before Kepler's conjecture did). In each layer, the centers form a hexagonal pattern, and each subsequent layer is placed so that its spheres go into the holes of the previous layer. The centers of the spheres form a lattice. In higher dimensions, packings where the centers of the spheres are on lattices often, *but not always*, lead to the densest packings. The lattices that arise are intimately connected with very complex groups. The name here says it all, as many of these are realized through the **Monster group**, a group with approximately $8 \cdot 10^{53}$ elements. All of a sudden, the numbers from the Enigma chapter don't look so large anymore!

Using the above mathematics, more powerful codes can be created. One such is the binary Golay code, where 12 of 24 bits are devoted to information, the code can detect up to seven errors and correct up to three errors. Golay described this code in [**20**]; at less than a page, many people say this paper has one of the greatest ratios of importance to page length! For an expanded version, see [**22, 48**] and the references therein, and for more on sphere packings and codes see [**11, 25**].

9.9. Problems

• From §9.2: Motivating Riddles

The following problems concern the hat riddle with n people in a line.

EXERCISE 9.9.1. *The above strategy can be improved with just a little more work to give 80%. Consider the fifth person. He sees four people in front of him. The first two people either have the same hat colors or different hat colors. Let's denote this by S_1 if they have the same color and D_1 if they have different colors; similarly, define S_2 and D_2 coming from the third and fourth people. There are thus four possible strings: S_1S_2, S_1D_2, D_1S_2, and D_1D_2. If we have two S's or two D's, the fifth person says "white"; if we have one S and one D, he says "black". Show how this allows the four other people to figure out their hat colors.*

EXERCISE 9.9.2. *Building on the idea in the previous exercise, can you design a strategy where the ninth person says something and then the previous eight can deduce their hat colors? If yes, how far can you push this idea?*

EXERCISE 9.9.3. *Generalize the hat problem to having three different hat colors. What is the optimal strategy now?*

EXERCISE 9.9.4. *Generalize the hat problem to having k different hat colors. What is the optimal strategy now?*

The following problems deal with three people seeing each other's hats (but not their own hat).

EXERCISE 9.9.5. *Generalize this hat problem to $4, 5, 6,$ or 7 people.*

EXERCISE 9.9.6. *Generalize this hat problem to having $2^k - 1$ people.*

EXERCISE 9.9.7. *What if we have three people and three hat colors, with each color being equally likely. What is the optimal strategy now?*

EXERCISE 9.9.8. *Building on the previous problem, find a number of people n and a strategy such that, with three hat colors, we have a better than one-third chance of winning.*

- **From §9.3: Definitions and Setup**

EXERCISE 9.9.9. *If we let the number of people increase, we see in the limit that the first strategy has 50% guessing correctly, while the second strategy has 100% (all but the last are correct). Are there other methods that give 100% in the limit? There are. If we have N people, prove the following method has at most $\lceil 1+\log_2(N-1)\rceil$ people saying the wrong color (here $\lceil x\rceil$ means the smallest integer at least x). Write the number of white hats the last person sees in binary; we need at most $k = \lceil 1+\log_2(N-1)\rceil$ digits. The last k people play the role of these digits, saying "white" if the corresponding binary digit is a one and "black" if it is a zero.*

EXERCISE 9.9.10. *Consider the two codes from Examples 9.3.1 and 9.3.2.*
(1) *For the code from Example 9.3.1, how many codewords are there of length exactly 2? Of length at most two?*
(2) *For the code from Example 9.3.2, how many codewords are there of length exactly 2? Of length at most two?*

EXERCISE 9.9.11. *Consider the binary code*

$$\{0,\ 1,\ 10,\ 11,\ 100,\ 101,\ 110,\ 111\},$$

and say you receive the message 1101. It's possible that this message is the codeword 110 followed by the codeword 1; it could also be the codeword 1 followed by the codeword 101, or 11 followed by 0 followed by 1. Could it be anything else, and if so, what?

EXERCISE 9.9.12. *Of course, a code can still have each message uniquely decipherable even if it isn't a fixed length code.*

(1) Imagine our code is the set

$$\{1,\ 1001,\ 111000111000\}.$$

Is this code uniquely decipherable?

(2) Imagine our code is the set

$$\{1,\ 10001,\ 111000111\}.$$

Is this code uniquely decipherable?

(3) Consider an r-ary code $C = \{c_1, c_2, \dots, c_n\}$ *with the length of* c_i *equal to* ℓ_i*. McMillan proved that if the code* C *is uniquely decipherable, then*

$$\sum_{i=1}^{n} \frac{1}{r^{\ell_i}} \leq 1;$$

unfortunately, this sum can be finite without the code being uniquely decipherable. Consider the code C *of all binary words of length at most* 10 *with an even number of* 1*'s. Thus* 100100011 *is in* C*, as is* 00100001*, but not* 1000000011*. Show* C *cannot be uniquely decipherable.* Hint: Remember that 0001 is a codeword of length 4, and is different from the codeword 1.

EXERCISE 9.9.13. *Let* C *be the binary code of words of length* 5 *with an odd number of* 1*'s. What is the minimum distance of the code?* Hint: The number of codewords is $\binom{5}{1} + \binom{5}{3} + \binom{5}{5} = 5 + 10 + 1 = 16$. This is a small enough number to be manageable; in other words, you could write down all the different codewords, but then you would have to look at all pairs of distinct codewords, and there are $\binom{16}{2} = 120$ different pairs! Fortunately you don't have to investigate all of these pairs; you just have to find the minimum. Try proving that the minimum cannot equal 1.

EXERCISE 9.9.14. *Let* C *be the binary code of words of length* 10 *with exactly eight* 1*'s. What is the minimum distance of the code? What is the maximum distance?* Hint: There are $\binom{10}{8} = 10!/8!2! = 45$ codewords in C; you clearly want a faster way than enumerating all of these!

EXERCISE 9.9.15. *Let* C *be the binary code of words of length* 10*. What is the minimum distance of the code?*

EXERCISE 9.9.16. *Let's consider the binary code*

$$\{00,\ 11\}.$$

Show this is a 1 *error detecting code but not a* 2 *error detecting code, and thus is exactly a* 1 *error detecting code.*

EXERCISE 9.9.17. *We defined the Hamming distance earlier. A distance function* d *is supposed to satisfy three properties:*

(1) *$d(x, y) \geq 0$ and equals zero if and only if $x = y$;*
(2) *$d(x, y) = d(y, x)$;*
(3) *for all z, $d(x, y) \leq d(x, z) + d(z, y)$.*

Prove the Hamming distance satisfies these three properties.

EXERCISE 9.9.18. *In the definition of a distance function above, imagine that instead of property (1) we had (1′) $d(x, y) = 0$ if and only if $x = y$. Prove that (1′), (2) and (3) imply (1).*

- **From §9.4: Examples of Error Detecting Codes**

EXERCISE 9.9.19. *Consider the code from (9.1). Imagine you receive the message*

$$111011010100101 1.$$

Could this have been the intended message? What if you receive

$$111010110011010 0.$$

Could that have been the intended message?

EXERCISE 9.9.20. *Consider the code from (9.2). Imagine you receive*

$$0011011011110111.$$

Could this have been the intended message? What about

$$1100101010101010 0011?$$

EXERCISE 9.9.21. *Using the code from Example 9.4.3, how many different messages of length 5 can be sent? If there is exactly one error, how many messages could be received?*

EXERCISE 9.9.22. *Construct a fixed length binary code that can detect two errors. What percentage of your code is devoted to information, and what percentage to check digits? Compare these percentages with the values from the parity code and the "tell me twice" code.*

EXERCISE 9.9.23. *Construct a fixed length ternary code that can detect two errors. What percentage of your code is devoted to information, and what percentage to check digits?*

EXERCISE 9.9.24. *Construct a fixed length binary code that can detect three errors. What percentage of your code is devoted to information, and what percentage to check digits?*

- **From §9.5: Error Correcting Codes**

EXERCISE 9.9.25. *Check that the Hamming distance is transitive by comparing $d(100110, 101101)$ and $d(100110, 100111) + d(100111, 101101)$. Of course this isn't a proof of transitivity, but it is a good check.*

EXERCISE 9.9.26. *Complete the argument above by showing the Hamming distance is transitive.* Hint: First show it is enough to prove this in the special case when our words have length 1!

EXERCISE 9.9.27. *There are* 120 *distances to check, as there are* $\binom{16}{2} = 120$ *ways to choose two out of* 16 *words when order does not count. Show* $d(0110011, 0100101) = 3$ *and* $d(1000011, 1101001) = 3$. *If you want to write a program to check all of these, the pseudo-code looks like:*

Let $\text{Ham}(i)$ *denote the* i^{th} *of the* 16 *codewords.*
$\min = 7; \max = 0.$
For $i = 2$ *to* 16,
For $j = 1$ *to* $i - 1$,
Let $d = d(\text{Ham}(i), \text{Ham}(j))$;
If $d < \min$, *then* $\min = d$;
If $d > \max$, *then* $\max = d$.
Print min. *Print* max.

EXERCISE 9.9.28. *Design a binary code that detects and corrects* two *errors. Calculate its efficiency; in other words, what percent of the message is devoted to transmitting information, and what percent to error correction/detection.*

EXERCISE 9.9.29. *Generalizing the previous problem, can you design a binary code that detects and corrects* three *errors? What about an arbitrary number of errors? Of course, your efficiency is probably going to go down as you increase the number of errors that can be corrected.*

EXERCISE 9.9.30. *Find some sequences that start* 1, 2, 4, *and guess which one is relevant for the Hamming codes. The* On-Line Encyclopedia of Integer Sequences *is an excellent source to investigate patterns. It not only suggests possible sequences, but often provides information on them (where they occur, generating functions, limiting behavior, and so on). It's online at* `https://oeis.org/`. *To use it, simply input a partial set of terms of a sequence, and it'll list all the sequences it knows containing it. For the pattern* 1, 2, 4 *there are many answers; a particularly fun one is sequence A000124; see* `http://oeis.org/A000124`. *This site is a terrific resource. The easiest way to appreciate its use is to think back to proofs by induction (described in §3.7), which are significantly easier if you have a conjectured form for the answer!*

- **From §9.6: More on the Hamming (7, 4) Code**

EXERCISE 9.9.31. *Given the message* 1011, *encode it using the algorithm above.*

EXERCISE 9.9.32. *List the 16 different messages in the code from Example* 9.6.1.

EXERCISE 9.9.33. *Assume you and your friend are using the method from Example* 9.6.1, *and you receive the message* 101111000. *You quickly realize this is not what your friend meant to send; assuming there was only one error in the transmission, what was the intended message?*

EXERCISE 9.9.34. *Generalize the square code and consider the square* $(16, 9)$ *code. Show that it can detect one error, and has nine out of* 16 *digits (or* 56.25%*), devoted to our message. Of course, similar to the original square code we don't need the final entry, and could construct a* $(15, 9)$ *code, which would lead to* 60% *of our message devoted to information! We could continue to push this method further, and look at square* $((n+1)^2, n^2)$ *codes. Almost all of the message is now devoted to the information we wish to transmit; unfortunately, we are still stuck at being able to detect only one error.*

EXERCISE 9.9.35. *Try your hand at further generalizations of the square code. What if you looked at a cube? For instance, if you took a* $3 \times 3 \times 3$ *cube with a* $2 \times 2 \times 2$ *subcube, you would have eight message digits. How many errors would you be able to detect? What if you looked at larger cubes? What if you looked at higher dimensional analogues, such as hypercubes?*

EXERCISE 9.9.36. *Instead of increasing the dimension and going up to a cube, what if we decreased the number of points and went from a square to a triangle: is it possible to create a code that corrects and detects one error? Consider three points in an equilateral triangle, say* x, y, *and* $z \in \{0, 1\}$. *We create check digits* $a = x + y$, $b = y + z$, *and* $c = x + z$ *(all sums are modulo* 2*), and thus the message* (x, y, z) *becomes* (x, y, z, a, b, c). *Half of the message is information, half parity checks. Does this give a code capable of detecting and correcting one error?* Hint: If it did, either Hamming missed it, we deliberately didn't mention it, or Hamming had some reason to prefer a square code.

EXERCISE 9.9.37. *Assume we're using the Hamming* $(7, 4)$ *code, and we receive the message* 0011001. *What message was meant?*

EXERCISE 9.9.38. *It's not a coincidence that the nonmessage digits are* 1, 2, *and* 4. *Let's try and do better. We use four parity bits instead of three. Find a construction that allows us to transmit a message of length* 15 *where* 11 *bits are the message and four bits are the parity checks; this is the Hamming* $(15, 11)$ *code, where now over* 70% *of the message is devoted to information. If we instead use five parity bits, how many bits of information can we transmit and still detect and correct one error? Of course, percentage-wise it's much worse to catch one error in* 15 *than one error in seven.*

EXERCISE 9.9.39. *Freedonia security caught ten spies from the Kingdom of Sylvania who attempted to poison the wine of Freedonia's king, Rufus T. Firefly the Great. The king keeps* 1000 *bottles of wine in his cellar. The spies managed to poison exactly one bottle, but were caught before they could poison any more. The poison is a very special one that is deadly even at one-trillionth the dilution of the poisoned wine bottle and takes* exactly 24 *hours to kill the victim, producing no symptoms before death. The trouble is that the king doesn't know which bottle has been poisoned and the wine is needed for the Royal Ball in exactly 24 hour's time! Since the punishment for attempted regicide is death, the king decides to force the spies to drink the wine. The king informs his wine steward that if he mixes wine from appropriate bottles for each spy, he will be able to identify the poisoned bottle by the time the ball starts and kill at most ten of the spies. Further, each spy drinks only once, though each spy's cup is potentially a mixture of wine from many bottles.* How should the wine be mixed? Is it possible to do it in such a way that at most nine spies die?

• From §9.7: From Parity to UPC Symbols

EXERCISE 9.9.40. *Consider the following check digit scheme, which is part of the UPC codes. Given an* n *digit number* $a_1 a_2 \cdots a_n$*, the check digit number is* $3a_1 + 1a_2 + 3a_3 + 1a_4 + \cdots \bmod 10$. *In other words, we add all the digits with an odd index and multiply that sum by* 3*, we add all the digits with an even index and multiply that sum by* 1*, and then take the sum of these two numbers modulo* 10*; this value is our check digit. (For* 19731870*, the check digit is* $3 \cdot (1+7+1+7) + 1 \cdot (9+3+8+0) \bmod 10$*, or* 2*.) Show this method is able to detect any single digit error. Can it detect two* adjacent *digits being switched?*

EXERCISE 9.9.41. *Consider the check digit scheme of the previous exercise. Which numbers below are correctly encoded? For the ones incorrectly encoded, what type of error was made?*

(1) 24610746567;
(2) 24611746567;
(3) 24601746567.

EXERCISE 9.9.42. *Write a short report on UPC symbols.*

EXERCISE 9.9.43. *Research the life of Jacobus Verhoeff, and write a short report about him and his check digit algorithm.*

• From §9.8: Summary and Further Topics

EXERCISE 9.9.44. *Calculate the percentage of the plane covered by circles in the hexagonal pattern, and show it's greater than the percentage in the square pattern. While we should really take a large portion of the plane and take the limit, we'll argue informally. Consider unit circles in each case, and see what percentage of the corresponding hexagon and what percentage of the corresponding square are covered, and then note that the plane can be perfectly tiled by hexagons or circles (this means we can cover the plane with translated copies of these figures without any overlap).*

EXERCISE 9.9.45. *Describe or sketch the lattices formed from the following:*

(1) $\overrightarrow{v_1} = (1,0)$, $\overrightarrow{v_2} = (0,1)$;
(2) $\overrightarrow{v_1} = (1,0)$, $\overrightarrow{v_2} = (1/2, \sqrt{3}/2)$;
(3) $\overrightarrow{v_1} = (1,0,0)$, $\overrightarrow{v_2} = (0,1,0)$, $\overrightarrow{v_3} = (0,0,1)$.

EXERCISE 9.9.46. *Look up the Kissing number problem, and write a short note about it. In particular, discuss the controversy between Newton and Gregory.*

EXERCISE 9.9.47. *Describe Hales' proof of the Kepler conjecture, highlighting the key ideas.*

Chapter 10

Modern Cryptography

In this chapter we look at encrypting and decrypting today. In particular we consider the traditional cryptographic tools in the context of available technology, as well as new tools. First, we revisit steganography, talked about briefly in Chapter 1, and consider how messages can be hidden in digital text, audio, and visual media. Then we consider the complicated field of quantum cryptography from an elementary perspective. It is hoped and feared that someday quantum cryptography will advance to the point that there are truly unbreakable encryption schemes. Lastly, we look at how cryptography, especially decryption, plays an important role in not only wars, but in securing the homeland against attacks such as those that occurred on September 11, 2001.

10.1. Steganography—Messages You Don't Know Exist

10.1.1. Introduction

Invisible ink dates back to Roman times and was used extensively and successfully as recently as during World War II. German spies used invisible ink to print very small dots on letters, blocks of text or images scaled down to the size of a regular dot, now called microdots. Even as late as 2008, al Qaeda used the invisible ink form of steganography. A British man, Rangzieb Ahmed, was alleged to have a contact book with al Qaeda telephone numbers, written in invisible ink. He was subsequently convicted of terrorism and is currently in jail.

Steganography is the technique of passing a message in a way that even the existence of the message is unknown. The term steganography is derived from the Greek *steganos*, which means covered, and *graphein*, to write. In the past, it was often used interchangeably with cryptography, but since 1967 it has been used exclusively to describe processes that conceal the presence of a secret message, which may or may not be additionally protected by a cipher or code. The content of the message is not altered through the process of disguising it. There are two types of steganography: technical or physical steganography, and linguistic.

Invisible ink is an example of technical steganography, as are hollowed out books, umbrella handles, and other things that occur frequently in spy novels. The use of technical steganography goes back to at least the fifth century B.C.E. The Greek historian Herodotus described the revolt against Persian rule that succeeded with the use of technical steganography. Two leaders of the revolt communicated secretly by shaving the head of a slave and tattooing a secret message on it. After the slave's hair grew back, he was sent to coconspirators who read the message by shaving his head. The Greeks were able to overthrow the Persians using the information in the concealed message.

Fifty years earlier, a second steganographic method was used by the Greeks against the Persians to turn back an invasion of Xerxes and his men. Demartus used a makeshift device created by scraping wax from two wooden tablets to alert the Spartans. He inscribed what he knew of the intentions of the Persians on the tablets and then replaced the wax covering. The seemingly plain tablets were passed to the Spartans untouched, who in turn scraped off the wax to read the message.

Why is encryption insufficient and steganography needed? The obvious answer is to hide the mere existence of files. Often no one can even prove such files exist, whereas courts can demand access to encrypted files, or a person can be tortured to give the password to encrypted data. With steganography we have the possibility that people are unaware of the existence of the message.

10.1.2. The Processes of Steganography

Let's start with some terminology. Steganography has various component parts.

- The **carrier** or **cover image** is the original file or image in which the secret message is to be embedded: text, image, audio, or video.
- The **payload** is the data (the image or message) to be hidden.
- The **package** or **stego-object** is the information embedded in the carrier.
- The **key** unlocks the message so it can be viewed by the receiver.

We depict the process in Figure 10.1. In functional notation, we can describe a **stego-function** as

$$f(C, P, K) \;=\; C + P + K \;=\; S,$$

where C is the carrier, P is the secret image, K is the key, and S is the result called the stego-object.

The **encoding density of a stego-image** is the percentage of bits in the payload to bits in the carrier. In other words the lower the density, the more hidden the file.

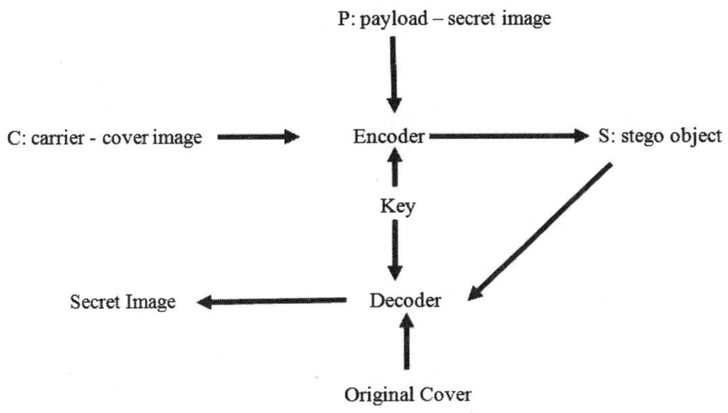

FIGURE 10.1. Schematic of steganography process.

Activity: Equipment required: deck of cards, fine point pen.

- Step 1: Choose two students.
- Step 2: These two students take the deck of cards and decide on an order to the cards.
- Step 3: Off to the side of the room they arrange the pack in that order.
- Step 4: They then write a message (to do something) on the side of the deck with a fine point pen, making sure no one sees them do it.
- Step 5: They shuffle the deck and pass the deck to another group of students.
- Step 6: Have the group do what the message indicated.

Before reading on, identify the carrier, payload, key, and stego-object in the activity.

In this activity, the **carrier** is the deck of cards, the **payload** is the message, the **stego-object** is the deck of cards with message on it, and the **key** is knowledge of the "right" order of the cards. Note that the encoding density of the stego-object is quite high, since the only thing added was the writing.

What things can be used today as carriers? Audio, video, still imagery, plain text, microdots, covert channels using Internet Control Message Protocol can all be used as carriers. Let's look at examples for each possible carrier. First we consider examples using plaintext carriers.

Using **plaintext**, one option is for the sender to send a series of numbers along with the message. There is prior agreement (which becomes the key) that the numbers correspond to the letter in the word in that place. A 0 corresponds to skipping a word. For example, consider the message "The men protest including raw meat. Some protest at least weekly. Beef undercooked is leading soldiers' illness complaints to grow", with number sequence 3111121415211114701 which is delivered separately. The receiver takes the message and highlights the third letter of the first word, the first letter of the second word, the first letter of the third word and so on as follows:

> Th**e** **m**en **p**rotest **i**ncluding **r**aw m**e**at. **S**ome pro**t**est **a**t leas**t** w**e**ekly. **B**eef **u**ndercooked **i**s **l**eading sol**d**iers' **i**llness complai**n**ts to **g**row.

Using only the highlighted letters, the message becomes "EmpireStateBuilding", indicating a potential terrorist target. Note that the 0 indicates that the word "to" is not considered.

Null ciphers (unencrypted messages) don't require a key. The letter in each word is uniform and is known ahead of time, for example the second letter of each word. The message is innocent sounding to anyone but the recipient. Null ciphers were used extensively during WWII, but have not been used much since.

Horizontal line (word) shift coding is a form of hiding in plain text, which uses horizontal spacing between each line (word). The lines (words) are shifted slightly left or right, the decoding is given in a codebook. An extreme example is shown in the following using the sentences of a paragraph:

> Some kids used vinegar or milk to write a message to friends
> that others couldn't read; an iron would reveal it. Some used
> invisible ink (such as milk, vinegar, juice or other liquids
> that are clear on paper, but darken when heated).

Assume our codebook is 1 left = tomorrow, 2 left = today, 3 left = next week, 1 right = bomb, 2 right = DC and 3 right = al Qaeda. The paragraph above is 3 right, 1 left, 1 right, 2 right, which yields: `al Qaeda tomorrow bomb DC`. Note that the last line is flush so encodes nothing.

Vertical line (word) shift coding uses vertical shifting of lines(words) so only part of the message is seriously read.

Feature coding uses extra white space characters, between consecutive words for example. The key is a lookup table available to each of the communicants, which matches the numbers of spaces to words or letters. Obvious short words are filled in by the receiver. For example, take the message:

```
My  wife    and   I are having troubles with our children.
```

The number of extra spaces are 2,4,3,1,1,1, and the key is 1 = tomorrow, 2 = al Qaeda, 3 = attack, and 4 = NYC. The decoding is "al Qaeda in NYC to attack tomorrow". Tomorrow is repeated in order to make the spacing less obvious, and "in" is filled in by the recipient.

10.2. Steganography in the Computer Age

Steganography processes in the computer age make use of the newest technologies to hide messages in plain sight. We saw some examples of tools that could be used with or without a computer. Now we look at examples that are particularly designed for computer tools. Indeed, the most used steganography today consists of digital images. These visual digital images come in many formats.

10.2.1. Visual Images

The most used are **still images**. These are messages hidden in digital imagery utilizing the limits of human vision. Human vision cannot detect differences in luminance (intensity) of color at the high frequency end of the color spectrum. A digital picture (or video) is stored as an image file using an array of colored dots, called pixels. An image file is merely a file containing a binary representation of the color or light intensity of each picture element (pixel) that makes up the image. Each pixel typically has three numbers associated with it, one for the red intensity, one for the green intensity, and one for the blue intensity. Images typically use 8-bit or 24-bit color. An 8-bit color scheme means that eight bits are used per pixel. These numbers or values range from 0 to 255 (for 256 or 2^8 values). Each number is stored as eight bits of 0's and 1's. An 8-bit bitmap representing the three colors, or intensities, at each pixel will have 2^8 different values at blue, 2^8 different values at red, and 2^8 different values at green. A difference of 1 in intensity is imperceptible. For example, if 11111111, which represents blue, is changed to 11111110, then this is undetectable to the human eye.

The 24-bits per pixel provide a much better set of colors. Using 24-bits per pixel for color means that changes are even more minimal and indiscernible to the human eye. Thus, in a standard 24-bit bitmap you have three color components per pixel: red, green, and blue. Each component has 2^8 or 256 values. When using a computer printer to construct the image, color intensity is given as a number value for each of red, green, and blue for each pixel, and then converted by the computer to 24-bit representations. The value designation is shown in Figure 10.2.

Using the color coding described in Figure 10.2, 191 converts to 10111111 for red's intensity, 29 is represented by 00011101 and 152 is represented in binary by 10011000, which produces a magenta color as shown for a color pixel.

Using 8-bit color a 640×480 pixel image using 256 colors requires a 307 KB file, whereas one twice as large using 24-bit color needs a 2.36 MB file. These are large files and need some form of compression to be workable.

The simplest way of hiding a message within an image file is to use what is called the **least significant bit insertion method**. The **least significant bit** (**LSB**) of a binary string is the last (furthest to the right) bit of the string, the one representing the unit's value. The least significant

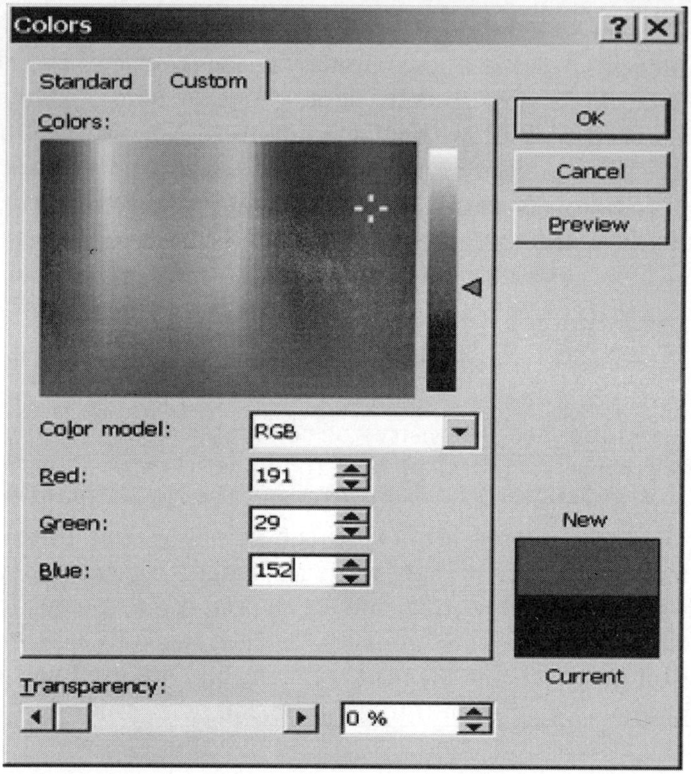

FIGURE 10.2. Color triples.

bit insertion method takes the binary representation of the hidden data and overwrites the least significant bit of the cover image with the LSB of the hidden image. Especially if we use 24-bit color, the change is imperceptible to the human eye.

Example 1: Three color pixels of the cover image are represented by the three 24-bit RGB words (9 bytes) below:

10010101 00001101 11001001
10010110 00001111 11001010
10011111 00010000 11001011

Now let's hide the payload or secret image 101101101 in the above cover image by overlaying them on the LSB of the cover image. For example the second of the nine is 00001101 which with the overlay of the second bit of 101101101 changes the last 1 to a 0. We get the following nine bits of data. The bits that have been changed are in bold.

10010101 0000110**0** 11001001
1001011**1** 0000111**0** 1100101**1**
10011111 00010000 11001011

We have hidden nine bits by changing only four of the LSB's.

FIGURE 10.3. Cover image (left). Extracted image (right).

Example 2: The following nine bytes of data are in the cover image:

00100111 11101001 11001000
00100111 11001000 11101001
11001000 00100111 11101001

Suppose the binary value for the letter `A` in the hidden message is (10000011). Inserting the binary value of `A` into the three pixels, starting from the left byte, results in the changed bytes shown in bold:

00100111 1110100**0** 11001000
0010011**0** 11001000 1110100**0**
1100100**1** 00100111 11101001

For example, if the cover image is the image of the trees, adding the hidden message "A" to every pixel yields a new image of a cat (see Figure 10.3). The cat is observable only to the receiver, who knows to drop all but the last two bits of each color component in each pixel and enhance the brightness 85 times.

There are numerous ways to conceal data in images. **DES** encryption algorithms are used by encrypting groups of 64 message bits using keys that are also 64-bits long. Every 8th key bit is ignored in the algorithm making the key size actually 56 bits.

One can also use **Sudoku puzzles** to conceal data by using a key to hide the data within an image. There are as many keys as there are possible solutions to the puzzle. There are $6.71 \cdot 10^{21}$ possible solutions to the standard 9×9 Sudoku puzzle. This is equivalent to around 70 bits, making it much stronger than the DES method which uses a 56-bit key.

Masking and filtering are used to hide images within images. We can manipulate the luminance of particular areas of the image in order to

FIGURE 10.4. Effect of changing luminance.

encode data. The contribution of each pixel is varied to achieve desired effects. Consider for example the image in Figure 10.4. We display an adjusted luminance, where the letters AMS appear throughout.

Another WWII technique that is still used today to hide images is **microdots**. The microdot technique uses photographs the size of a printed period that contain the information of an 8×11 printed page. The message is hidden by its small size and needs not be encrypted. The recipient merely needs to know the message is coming and that microdots are used. The recipient has to have a way of enlarging the period to create the image.

10.2.2. Audio and Video Embedding

Audio signals are encoded using sampling rates (frames) and a frequency range. The human ear can only distinguish from 20Hz to 20KHz. However, when there are two sounds created simultaneously with one louder than the other, generally only the louder sound is heard. Most digital audio is created by sampling the signal and quantizing the sample with a 16-bit quantizer. The rightmost bit, or low order bit, of each sample can be changed from 0 to 1 or 1 to 0. This modification from one sample value to another is not perceptible by most people and the audio signal still sounds the same. This is called low bit coding.

Phase coding, on the other hand, relies on the relative insensitivity of the human auditory system to phase changes and substitutes the initial phase of an audio signal with a reference phase that represents the data. It is more complex than low bit coding, but it is much more robust and less likely to distort the signal that is carrying the hidden data. Video encoding uses the human eye, which can only perceive approximately 24 frames per second. You can increase the frame rate and hide the payload in the superfluous frames. Since this may be detected visually, you can also encode each frame using each frame as a still image.

10.2.3. Digital watermarking

Watermarking has taken on a new importance in the digital era. Still images, video, music, text, and software are all easily copied and illegally distributed, causing the authors to lose out on considerable income in royalties. Digital watermarking is one other form of steganography used today to essentially copyright digital information. A message, which is just an identifier, is hidden in an image so that its source can be tracked or verified. Digital watermarking is used to protect against piracy or to identify an image's owner. Unfortunately, not all watermarking software is created equal. Some watermarks are easily removed or destroyed by manipulating various characteristics of the file, so companies are looking for more and more ways of securing the watermarks.

10.2.4. Steganography and Email, Chaffing and Winnowing, and Avatars

What was once thought of as junk email that got sent to your spam folder may be actually much more. This junk email may indeed be a message that is sent out, where only the recipient knows it is a message. The sender covers his tracks using techniques called chaffing and winnowing, rather than encryption.

In **chaffing and winnowing**, Alice, the sender, wants to send a message to Bob, the receiver. Alice lists the bits in her message and sends out each bit in a separate packet. Each packet contains the bit's serial number in the message, the bit itself (both unencrypted), and an identification code (**MAC**), whose secret key Alice shares with Bob. Charlie, the transmitter of Alice's packets to Bob, interleaves in a random order the packets with corresponding bogus packets (which are called **chaff**) with corresponding serial numbers, the bits inverted, and a random number in place of the MAC. Charlie does not need to know the key to transmit the message. Bob uses the MAC to find the authentic messages and drops the chaff. Bob's part of the process is called **winnowing**, thus the terminology chaffing and winnowing.

Now an eavesdropper located between Alice and Charlie can easily read Alice's message. But an eavesdropper between Charlie and Bob would have to tell which packets are bogus and which are real (i.e., to winnow, or "separate the wheat from the chaff"). That is infeasible if the MAC used is secure and Charlie does not leak any information on packet authenticity.

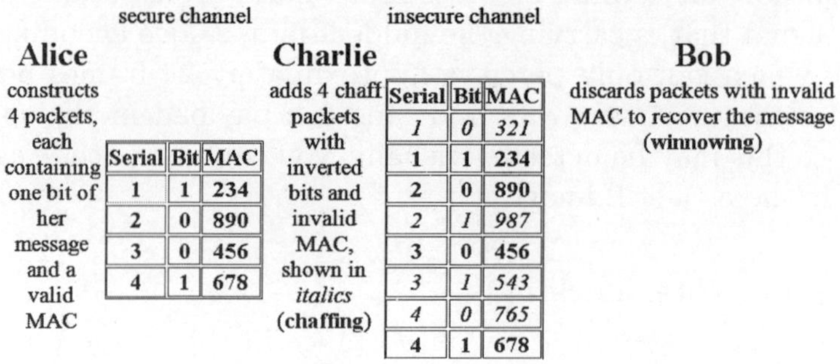

FIGURE 10.5. An example of chaffing and winnowing. (Image from Wikipedia Commons.)

An example is shown in Figure 10.5, where Alice wishes to send the message 1001 to Bob. Assume that all even MACs are valid and all odd ones are invalid.

When an adversary requires Alice to disclose her secret key, she can argue that she used the key simply for authentication. If the adversary cannot force Alice to disclose an authentication key, thereby allowing them to forge messages from Alice, then her message remains confidential. Charlie, on the other hand, does not even possess any secret keys that he could be forced to disclose.

An **avatar** in computing is any 2-dimensional or 3-dimensional graphical representation of a person, or object that represents the user. Avatars are used extensively in computer games, in internet forums, internet chats and blogs, in internet messaging, or even in artificial intelligence. These avatar graphical representations themselves can contain hidden messages. For example, the image in Figure 10.6 contains the message: "The invasion will take place December 12 at noon", using RedandBlueWing34 as a password. Be sure to do Exercise 10.6.9 to create your own avatar with hidden message.

The image shown in Chapter 1 of the cat was an avatar image also created using `mozaiq.org`.

10.3. Quantum Cryptography

> Nobody understands quantum theory.
>
> –Richard Feynman, Nobel Prize-winning physicist.

Quantum cryptography is based on quantum physics whose foundation is the **Heisenberg Uncertainty Principle**. In essence this principle says that it is impossible to know both an object's position and velocity at the same time. For cryptography, this uncertainty is used to secure the communication of messages. Once the secure key is transmitted using quantum

FIGURE 10.6. An avatar image constructed using `http://mozaiq.org/encrypt`.

cryptographic techniques, coding and encoding using the normal secret-key method can take place.

How is the secret key transmitted? Photons are used to transmit the key. You may have heard of **qubits**, the quantum computing analog of a bit. A qubit is a unit of quantum information. Each photon carries a qubit of information. A **bit** can take on the value of 0 or 1, but a qubit can take on the value of 0 or 1 ***or*** the superposition of the two. To create a photon, quantum cryptographers use light emitting diodes, (**LEDs**), as a source of unpolarized light to create one photon at a time. A string of photons are created through this process. Using polarization filters, we can force the photon to take one state or another, or polarize it (superposition of 0 and 1).

Spins and filters: Information is attached to the photon's spin as it emerges from a polarizing filter. Once a photon is polarized, it can't be accurately measured again, except by a filter like the one that initially produced its current spin. So if a photon with a vertical spin $|$ is measured through a diagonal filter $\times$, either the photon won't pass through the filter (we use ( ) to denote this), or the filter will affect the photon's behavior, causing it to take a diagonal spin (either $\diagup$ or $\diagdown$). In this sense, the information on the photon's original polarization is lost, and so too is any information attached to the photon's spin.

The questions become the following: How does a photon become a key? How do you attach information to a photon's spin? Binary codes come into play. Each type of a photon's spin represents one piece of information, usually a 1 or a 0, for a binary code. As we saw before, a binary code uses strings of 1's and 0's to create a coherent message. So a binary code can be assigned to each photon—for example, a photon that has a vertical spin $|$ can be assigned a 1.

Let's consider what is happening. Alice can send her photons through randomly chosen filters and record the polarization of each photon. She will then know what photon polarizations Bob should receive. When Alice sends Bob her photons using an LED, she'll randomly polarize them through either of two filters, so that each polarized photon has one of four possible states: $|$, —, $\diagup$ or $\diagdown$. As Bob receives these photons, he decides whether to measure each with either his + or × filter; he can't use both filters together. Keep in mind, Bob has no idea what filter to use for each photon, he's guessing for each one. After the entire transmission, Bob and Alice have a nonencrypted discussion about the transmission.

The reason this conversation can be public is because of the way it's carried out. Bob calls Alice and tells her which filter he used for each photon, and she tells him whether it was the correct or incorrect filter to use. Their conversation may sound a little like this: "Bob: Plus; Alice: Correct. Bob: Plus; Alice: Incorrect. Bob: ×; Alice: Correct." Since Bob isn't saying what his measurements are, only the type of filter he used, a third party listening in on their conversation can't determine the actual photon sequence.

For example, Alice sent one photon as a $\diagup$ and Bob says he used a + filter to measure it. Alice will say "incorrect" to Bob. But, if Bob says he used an × filter to measure that particular photon, Alice will say "correct." A person listening will only know that that particular photon could be either a $\diagup$ or a $\diagdown$, but not which one definitively. Bob will know that his measurements are correct, because a — photon traveling through a + filter will remain polarized as a — photon after it passes through the filter. After their odd conversation, Alice and Bob both throw out the results from Bob's incorrect guesses. This leaves Alice and Bob with identical strings of polarized protons. It might look like this: $— \diagdown | | | \diagdown — — | | | — \diagdown | \diagup \cdots$ and so on. To Alice and Bob, this is a meaningless string of photons. Now if the binary code is applied, the photons become a message. Bob and Alice can agree on binary assignments, say 1 for photons polarized as $\diagdown$ or — and 0 for photons polarized as $\diagup$ or $|$. This means that their string of photons is equivalent to 110001110001100, which can, in turn, be translated into English, Spanish, Navajo, prime numbers, or anything else Bob and Alice use as codes for the keys used in their encryption.

The goal of quantum cryptology is really to thwart attempts by a third party to eavesdrop on the encrypted message; we call the eavesdropper Eve. In modern cryptology, Eve can passively intercept Alice and Bob's encrypted message; she can get her hands on the encrypted message and work to decode it without Bob and Alice knowing she has their message. Eve might accomplish this by wiretapping their phones, or reading emails thought to be secure, hacking into their computers, or countless other ways.

Now the Heisenberg Uncertainty Principle comes into play when Eve makes her own eavesdrop measurements; it makes quantum cryptology the first cryptographic scheme to safeguard against passive interception.

Let's go back to Alice and Bob and bring Eve into the picture. If Alice sends Bob a series of polarized photons, and Eve has set up a filter of her own to intercept the photons, Eve is in the same boat as Bob—neither of them has any idea what are the polarizations of the photons Alice sent. Like Bob, Eve can only guess which filter orientation (for example, an $\times$ filter or a $+$ filter) she should use to measure the photons. After Eve has measured the photons by randomly selecting filters to determine their spin, she will pass them down the line to Bob using her own LED with a filter set to the alignment she chose to measure the original photon. She does this to cover up her presence and the fact that she intercepted the photon message. By measuring the photons, Eve inevitably altered some of them.

Now suppose Alice sends to Bob one photon polarized to a — spin, and Eve intercepts this photon. Eve incorrectly chose to use an $\times$ filter to measure the photon. If Bob randomly (and correctly) chooses to use a $+$ filter to measure the original photon, he will find it's polarized in either a $/$ or $\diagdown$, or it doesn't pass through, and he records ( ). Bob will believe he chose incorrectly until he has his conversation with Alice about the filter choice. After all of the photons are received by Bob, and he and Alice have their conversation about the filters used to determine the polarizations, discrepancies will emerge if Eve has intercepted the message. In the example of the — photon that Alice sent, Bob tells her that he used a $+$ filter. Alice tells him this is correct, but Bob knows that the photon he received didn't measure as — or $|$. Because of this discrepancy, Bob and Alice know that their photon has been measured by a third party, who inadvertently altered it. This is a powerful advantage of quantum cryptography over classical methods; a third party *must* leave evidence upon intercepting the signal.

Alice and Bob can further protect their transmission by discussing some of the exact correct results after they've discarded the incorrect measurements, by using a parity check on them. If the chosen examples of Bob's measurements are all correct, meaning the pairs of Alice's transmitted photons and Bob's received photons all match up, then their message is secure. Bob and Alice can then discard these discussed measurements and use the remaining secret measurements as their key. If discrepancies are found, they should occur in 50% of the parity checks. Since Eve will have altered about 25% of the photons through her measurements (half the time she guesses right and thus doesn't change anything, while for the other half of the time it passes through the incorrect filter and gets the correct orientation half of the time), Bob and Alice can reduce the likelihood that Eve has the remaining correct information down to a one-in-a-million chance by conducting 20 parity checks.

Not all is perfect, however, in quantum cryptography. The distance that the key can be transmitted is a technical limitation, approximately 67 kilometers. The distance possible with quantum cryptology is short because of interference. A photon's spin can be changed when it bounces off other particles, and so when it's received, it may no longer be polarized the way it was originally intended to be. This means that a 1 may come through as a 0, which is really the probability consideration in quantum physics. As the distance a photon must travel to carry its binary message is increased, so, too, is the chance that it will meet other particles and be influenced by them. Single photon detection is hard, and the technology to process more than a few photons is expected to be developed in the future. Numerous companies are doing the research to develop technologies for quantum cryptography, largely because they see quantum cryptography as a technique to combat hacking.

10.4. Cryptography and Terrorists at Home and Abroad

Chapter 1 talked about the history of encryption through the end of World War II. In this section, towards the end of the book, we consider various applications and possible applications of cryptographic techniques since 2000. For security reasons, the use of cryptographic techniques is often only hinted at in the press or in federal reports, such as the 2006 *Federal Plan for Cyber Security and Information Assurance Research and Development.*

10.4.1. Steganography as a Terrorist Tool

Steganography as a terrorist tool appeared in the news as early as February 5, 2001, in *USA Today* in two different articles, "Terrorist instructions hidden online" and "Terror groups hide behind Web encryption", and then in July of the same year, in an article title, "Militants wire Web with links to jihad". The *USA Today* articles were written by veteran foreign correspondent Jack Kelley, who in 2004 was fired after allegations emerged that he had fabricated stories and sources, so the validity of the information in these articles has been called into question. Later, after 9/11, *The New York Times* published an article claiming that al Qaeda had used steganography to encode messages into images, and then transported these via e-mail to prepare and execute the September 11, 2001, terrorist attack. The *Federal Plan for Cyber Security and Information Assurance Research and Development* of April 2006, highlighted the U.S. fears:

- "...immediate concerns also include the use of cyberspace for covert communications, particularly by terrorists but also by foreign intelligence services; espionage against sensitive but poorly defended data in government and industry systems; subversion by insiders, including vendors and contractors; criminal activity, primarily involving fraud and theft of financial or identity information, by hackers and organized crime groups...." (pp. 9–10)

- "International interest in R&D for steganography technologies and their commercialization and application has exploded in recent years. These technologies pose a potential threat to national security. Because steganography secretly embeds additional, and nearly undetectable, information content in digital products, the potential for covert dissemination of malicious software, mobile code, or information is great." (pp. 41–42)
- "The threat posed by steganography has been documented in numerous intelligence reports." (p. 42)

It is likely that steganography techniques are in constant use by both sides in the war against terrorism today.

10.4.2. Visual Cryptography

Visual cryptography, a modern variant of steganography, is a method developed by Naor and Shamir in 1995 of encrypting information by breaking up an image into a given number of ciphertext sheets, called **transparencies**. The hidden message is revealed by stacking the transparencies and viewing the result. (Recall the activity with the stack of cards—this is a computer version of that one.) Decryption is performed directly by the human eye with no special calculations required, so even those with no cryptographic experience or means of cryptographic computation can make use of this system. In a basic visual cryptography scheme in which the original message is split into two shares, each share on its own provides no information on the content of the other, thus ensuring perfect secrecy. This guarantee extends to any situation in which an image is divided into n shares, of which k must be assembled to reveal the image; any combination of $k - 1$ shares will contain no information on the appearance of the missing transparency. Further research has yielded more advanced methods of visual cryptography that permit the sender to specify qualified and forbidden subsets of n participants for image reconstruction, maximize the contrast of the image, and even conceal the existence of a hidden message by designing each transparency such that it exhibits some meaningful image rather than a random collection of pixels.

The most basic method of visual cryptography involves the division of a black and white image into two ciphertext transparencies. The original message is separated into its component black and white pixels, and each pixel is further divided into smaller sections. The system of pixel subdivision can be done in various ways, including a simple split into two rectangular halves, division into four blocks or replacement as a divided circle. Whatever the method, each combination that forms a full pixel always contains an equal

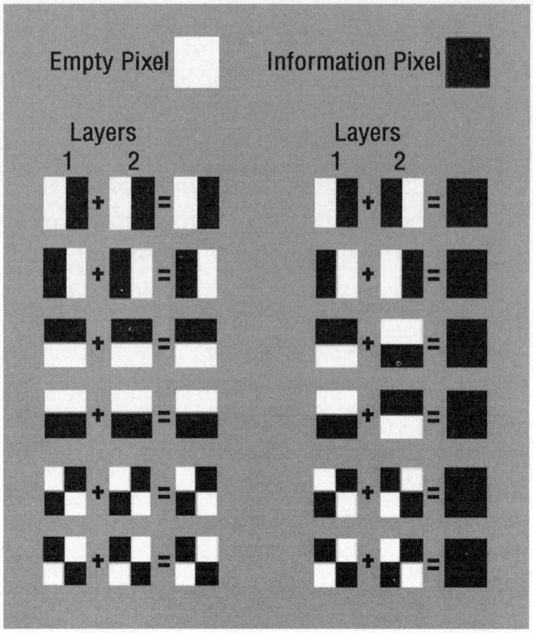

FIGURE 10.7. Blocks of pixels.

number of black and white (or, more precisely speaking, transparent) blocks. For example, a pixel divided into four square blocks can have six different states, each of which consists of two transparent blocks and two black blocks. With the image divided into pixels and a system of subdivision in place, the first transparency is made by assembling an image equal in size to that of the original message that consists of randomly selected subdivided pixels. The second layer is identical to the first except that those pixels that are black in the original image have an opposite state to that of the corresponding pixel in the first transparency, such that stacking the two layers yields an all-black pixel, and therefore the original message becomes visible. If the pixel states of layer one were selected at random, then both the empty and the information pixels in the second layer will have random states. Anyone who intercepts either of the images by itself will be unable to distinguish the character of the message by examination of its component pixels, so if the intended recipient of the message has been given a randomly generated transparency by the sender, the information layer can be transmitted publicly. This technique can be seen as a visual form of a one-time pad system; if the image on the first transparency contains truly random pixels, then the system guarantees near perfect security.

But while basic visual cryptography promises absolute secrecy, it inevitably comes with a few drawbacks. Like any one-time pad system, this method requires that all parties interested in communicating by these means must meet beforehand to exchange the randomly created image and that the key is never reused, which makes it a poor choice for groups who want to communicate without meeting or are interested in extended exchanges of information. Additionally, encryption by visual cryptography necessarily reduces the contrast of the reconstructed image, since each empty pixel in the final image will be visually interpreted as a shade of grey. In the case described above, an empty pixel in the final image will be composed of two black blocks and two transparent blocks, so there will be a 50% loss of contrast.

An important tool in ferreting out the "bad guys" is commonly called **text extraction**. Millions of emails and other messages are transmitted daily between people, some of which contain important information about the future plans of terrorists, some hidden in text and some out in plain sight. One of the most important software tools to be developed in the early part of the twenty-first century can process a million pages of text and extract what is the "most important" information, which helps prevent attacks in the U.S. and abroad. These programs, such as **Jigsaw**, provide a visual analytical approach to text extraction.

10.5. Summary

This chapter focused on cryptography in the twenty-first century, primarily on various forms of digital steganography and the "hope of the future", quantum cryptography. Quantum cryptography enjoys many advantages over classical methods, one of the most important being that the sender and receiver can detect when a third party has viewed the message. Even if the attacker cannot decode it, it is often very important to know that someone is on the prowl. This is not the end of the interplay between quantum mechanics and cryptography. An active area of research is **quantum computing**. These quantum computers work on qubits, and have the potential of solving many problems significantly faster than classical computers. An important example is factorization. When implementing RSA, it was essential that factorization is hard; if it were easy, then the encryption would not be secure. There exist extremely fast factorization algorithms built for quantum computers; all that remains is to actually build these machines! This leads to the subject of the final chapter. As these quantum computers are still years away, we are forced to use classical computers to attack the factorization problem. In Chapter 11 we describe *some* of the many factorization algorithms.

10.6. Problems

• From §10.1: Steganography—Messages You Don't Know Exist

EXERCISE 10.6.1. *Construct an example of plaintext that decodes to "A battalion will invade the island on Sunday."*

EXERCISE 10.6.2. *Using a null cipher, encode a message that decodes to "A battalion will invade the island on Sunday."*

EXERCISE 10.6.3. *Give an example of how you could encode and decode the statement "A battalion will invade the island tomorrow" using horizontal line shifting, and another using vertical line shifts.*

EXERCISE 10.6.4. *Give an example of how you could encode and decode the statement "A battalion will invade the island tomorrow" using horizontal word shifting, and another using vertical word shifts.*

EXERCISE 10.6.5. *Create an example of how to use feature coding to encode and decode the statement "A battalion will invade the island tomorrow."*

• From §10.2: Steganography in the Computer Age

EXERCISE 10.6.6. *The following nine bytes of data are in the cover image:*

$$\begin{array}{ccc} 11010011 & 00001111 & 00110011 \\ 10101010 & 00110010 & 01010101 \\ 10010010 & 00010001 & 10010011 \end{array}$$

Hide the payload, which is given by 110101001, *in the cover image. How many bits did you have to change?*

EXERCISE 10.6.7. *The following nine bytes of data are in the cover image:*

$$\begin{array}{ccc} 11010011 & 00001111 & 00110011 \\ 10101010 & 00110010 & 01010101 \\ 10010010 & 00010001 & 10010011 \end{array}$$

Hide the payload, which is given by 110101000, *in the cover image. How many bits did you have to change?*

EXERCISE 10.6.8. *Instead of a* 3×3 *image, imagine now you have an* $n \times n$ *image with* n *a large number. You want to hide a payload in it. If the payload is* n^2 *digits, approximately how many of the entry's* n^2 *final digits in the image do you expect to have to change?*

EXERCISE 10.6.9. *Use* `http://mozaiq.org/encrypt` *to create your own message and an avatar image using your own password and share it. Note you can do it directly on your own computer.*

EXERCISE 10.6.10. *Write a short report on the software Jigsaw for visual analytics. What is the mathematics behind Jigsaw?*

• From §10.3: Quantum Cryptography

EXERCISE 10.6.11. *Research quantum computing. When was it first proposed, and by whom?*

EXERCISE 10.6.12. *Research the Heisenberg Uncertainty Principle, and write a short note on it.*

- **From §10.4: Cryptography and Terrorists at Home and Abroad**

EXERCISE 10.6.13. *Research the program Carnivore, and write a short note on it.*

Chapter 11

Primality Testing and Factorization

This chapter is independent of the rest of the book. Though encryption provides a powerful *motivation* to study primality testing and factorization, these problems are interesting in their own right. Below we review some of the many primality tests and factorization algorithms, culminating in a description of the first provably fast, deterministic primality test.

11.1. Introduction

We studied RSA in Chapter 8, which depends on a fascinating disparity between our ability to test whether an integer is prime as well as our ability to factor integers. We saw that a fast factorization algorithm would allow one to break the RSA cryptography system, as we can deduce someone's private key from their public key. On the other hand, using RSA only depends on the ability to check numbers for primality in the key-generation process. In some sense this seems paradoxical, since we might expect primality testing and factoring to be closely related problems; however, it appears that factoring is a much harder problem. There are very fast algorithms which determine whether a number is prime, but all known factoring algorithms are extremely slow. Thus, although a modern computer can determine whether a 300-digit number is prime in under a second, it would take over a million years to factor it. Without this dichotomy, RSA would not work as a practical, secure system. We would either be unable to generate keys in the first place, if primality testing couldn't be done quickly, or we could break RSA by figuring out private keys from public ones, if factoring *could* be done quickly.

On the other hand, there is one clear connection between the problems of factoring and primality testing. If we can efficiently factor any number N, then we get for free a primality test (simply see whether or not any factor is strictly between 1 and N); the converse, however, need not be true. We'll see examples of algorithms that determine whether or not a number N is prime, but if N is composite, then they give us no information on its factors! At first this seems absurd. Every integer is either a prime or a product of primes; thus if our number isn't prime, surely we must know a divisor! Fortunately for the workability of RSA, this is not the case. It turns out that there are certain properties that prime numbers satisfy that composite numbers do not, and thus we can learn that a number is composite without actually knowing any factors!

We've been in this situation before. Let's revisit **Euclid's proof of the infinitude of primes** (see §7.5). Over 2000 years old, it's certainly one of the oldest proofs still taught in classes. The argument, a proof by contradiction, goes as follows. Assume there are only finitely many primes, say 2, 3, 5, 7, 11, ..., p_n. Consider the product $2 \cdot 3 \cdot 5 \cdot 7 \cdot 11 \cdots p_n + 1$. This number cannot be divisible by any prime on our list, as each leaves a remainder of 1. Thus our new number is either prime or else it's divisible by a new prime not in our list. In both cases, we see our list of primes was incomplete, contradicting the assumption that we had listed all the primes. In other words, there are infinitely many primes, *but* our proof doesn't give an infinite list of primes. The situation is similar for many primality tests, where we can learn a number is composite without knowing any of its factors.

In this chapter we explore primality testing and factorization. The material here is not directly needed to understand how to implement RSA, but it is essential for a conversation on its security, and thus we concentrate on the importance of efficiency for cryptographic applications. The example below looks at the sequence of primes that can be found by Euclid's proof. In addition to going through the details of applying Euclid's idea, it's a nice way to see how simple questions can lead to fascinating and complicated answers. In particular, while it's very easy to state Euclid's method and to discuss it theoretically, in practice it's exceptionally difficult to implement. Sadly, this can be the case for many algorithms in cryptography. We need to be able to factor large numbers to solve problem like the one that follows, and this quickly becomes a challenge. To date only the first 40 or so terms are known, and this example serves as a reminder of the need to find efficient algorithms.

EXAMPLE 11.1.1. *It's fascinating to look at the sequence of primes found by Euclid's method. We start with* 2. *We then add one and get* 3, *the next prime on our list. We then look at* $2 \cdot 3 + 1$; *this is the prime* 7, *which becomes the third element of our sequence. Notice that we've skipped* 5*— will we get it later? Let's continue. The next number comes from looking at* $2 \cdot 3 \cdot 7 + 1 = 43$, *which is again prime. The next number arises from* $2 \cdot 3 \cdot 7 \cdot 43 + 1 = 1807 = 13 \cdot 139$; *thus we can add* 13 *to our list. Note this is the first time Euclid's method gave a composite number, and is also the first time we got a smaller prime than the largest prime in our list to date. The sequence meanders around. The first few terms are*

2, 3, 7, 43, 13, 53, 5, 6221671, 38709183810571, 139, 2801, 11, 17, 5471, 52662739, 23003, 30693651606209, 37, 1741, 1313797957, 887, 71, 7127, 109, 23, 97, 159227, 643679794963466223081509857, 103, 1079990819, 9539, 3143065813, 29, 3847, 89, 19, 577, 223, 139703, 457, 9649, 61, 4357. *We obtained these from the* On-line Encyclopedia of Integer Sequences, *or the OEIS, entry* A000945 (`http://oeis.org/A000945`). *This website is a great resource to learn more about interesting sequences. There are many open questions about this sequence, the most important being:* Does it contain every prime? *We see it does get* 5; *does it get* 31?

We end this introduction with one an important warning. Factorization is thought to be hard because, despite hundreds of years of research by many intelligent people, no one has found a fast way to do it in general. Of course, failure to find a solution does not mean that a solution does not exist. For a long time, there was no known, provably fast primality test. What this means is that there were tests that ran quickly in practice, but to *prove* they would always run quickly depended on some of the deepest conjectures in number theory. Though there were many reasons to believe these conjectures, it was always possible that they might be false. Everything changed in 2002, when Agrawal, Kayal, and Saxena [**2**] devised an elementary fast primality test; we discuss their algorithm in §11.6. Thus, it's always possible that in a few years someone will have an "aha moment", and suddenly factorization becomes easy. One possible approach is through quantum computing (see for example [**16, 27**]).

11.2. Brute Force Factoring

We've seen that prime numbers play a key role in many cryptosystems. One reason is the **Fundamental Theorem of Arithmetic**, which asserts any number can be written uniquely as a product of powers of primes. Here "uniquely" means up to re-ordering, so we consider $2^2 \cdot 3$ to be the same as $3 \cdot 2^2$ or $2 \cdot 3 \cdot 2$. In fact, the desire to have a statement like this be true is why mathematicians have declared that 1 is not a prime. If 1 were prime, then unique factorization would fail. We could write 12 as either $2^2 \cdot 3$ or $1 \cdot 2^2 \cdot 3$, or even worse $1^{2013} \cdot 2^2 \cdot 3$. While the proof of this important result is not beyond the scope of this book (you don't get the label "Fundamental Theorem" lightly in mathematics!), it is surprisingly involved for such a familiar property of numbers. It uses the Euclidean algorithm, and its proof would take us too far afield. See [**34**] for details.

The security of RSA rests on the assumption that it is difficult to *quickly* factorize large numbers. The keyword, of course, is *quickly*, as there is a trivial algorithm to factor any number: brute force! While many problems in cryptography and other sciences can be solved by brute force, frequently the amount of time required is so large that such methods are essentially useless.

Brute force factorization: Let N be a positive integer greater than 1. Keep dividing N by all integers up to $N-1$ (i.e., divide N first by 2, then by 3, then by 4, and so on) until one of two things happens:

- if some a divides N, then N is composite, and a is a proper divisor;
- if no a from 2 through $N-1$ divides N, then N is prime.

If N is prime, then stop and our factorization is complete. If N is not prime, rerun the algorithm with N/a instead of N, and continue to run until we reach an input that is declared prime. Collecting the factors gives a factorization of the original number N.

This method is guaranteed to work. If N is composite, eventually we'll reach one of its factors; if N is prime, then no integers other than 1 and N divide N, and thus our algorithm will return prime. If N is composite and a divides N, we have a partial factorization. It's not immediately apparent, but a must be a prime number. The reason is we stop when we reach the first a that divides N; if $a = bc$ were composite, then we would have stopped at b or c, first, not a. Thus $N = aM$ for some M, and we have a partial factorization into primes. To complete the factorization, we simply run the algorithm on $M = N/a$.

For example, consider $N = 1776$. We first try 2, which is a factor as

$$1776 = 888 \cdot 2.$$

We can now conclude that 1776 is not prime, though we do not have its complete factorization. To complete the factorization of 1776, we need only factor 888 and then include the factor of 2 we have just found in our list of factors. Note that, once *one* factor is found, we have a new factorization problem but with a smaller number. We find that 2 divides 888, so 2^2 divides our original number 1776. Continuing in this manner, we see that 2^4 divides 1776 ($1776 = 16 \cdot 111$) but 2^5 does not, as

$$\frac{1776}{2^5} = 111.5.$$

Thus

$$1776 = 2^4 \cdot 111.$$

We've counted and removed all the powers of 2 from our number and now proceed to factor 111. Note that this problem is easier than the original problem of factoring 1776 for two reasons. First, our number is smaller, and second, we know that 2 cannot be a factor, as we have removed all powers of 2. We thus try the next number, 3, and find

$$111 = 3 \cdot 37.$$

As 3 does not divide 37, we have found all the powers of 3 that divide 111, and thus all the powers of 3 that divide 1776. Our factorization has improved to

$$1776 = 2^4 \cdot 3 \cdot 37.$$

We now try to factor 37. We do not need to try 2 or 3 as a factor, as we have already removed all of these. What should our next attempt be? There are two ways to proceed. One possibility is to try 4. The reason it is natural to try 4 is that it is the next integer after 3; however, clearly 4 cannot be a factor as 4 is just 2^2, and we have already removed all powers of 2. Thus the next number we try should be the next prime after 3, which is 5. Trying 5, we see that it is not a factor of 37 as

$$\frac{37}{5} = 7.4.$$

Continuing in this manner, we keep trying all the prime numbers up to 37, and find that none of them work. Thus 37 is prime, and the complete factorization of 1776 into products of powers of primes is

$$1776 = 2^4 \cdot 3 \cdot 37.$$

Remark: Before reading further, think about the example above. Do we really need to try *all* integers up to 37, or is there a significantly earlier cut-off point than 37 so that it suffices to look for prime factors up to that number?

From the standpoint of correctness, the brute force algorithm is perfect. It is guaranteed to work. It will tell if a number is prime, and if it's composite it will provide a complete factorization. Unfortunately, it's essentially worthless for real world applications due to its incredibly long run-time. To see this, consider the case when N is prime. If we use the naive approach of trying all numbers, we have N numbers to try; if we instead only try prime numbers as potential factors, by the prime number theorem we have essentially $N/\ln N$ numbers to check (the **prime number theorem** says the number of primes at most x is approximately $x/\ln x$). For example, if $N = 10^{100}$, then $\log 10^{100}$ is about 230. This means that there are about $10^{100}/230$ primes to check, or more than 10^{97} candidate factors! Alternatively, it means that around .4% of all integers up to 10^{100} are prime.

Can the brute force approach be salvaged? With a little thought, we can improve our algorithm enormously. We've already seen that instead of checking every number, clearly we need only look at prime numbers up to N. A much better observation is to note that it also suffices to just test primes up to $\sqrt{N}$. The reason is that if $N = ab$, then either a or b must be at most $\sqrt{N}$. If both a and b were larger than $\sqrt{N}$, then the product would exceed N:

$$\begin{aligned} a &> \sqrt{N}, \\ b &> \sqrt{N}, \\ a \cdot b &> \sqrt{N} \cdot \sqrt{N} = N. \end{aligned}$$

Thus for a given number N, either it is prime *or* it has a prime factor at most $\sqrt{N}$.

For example, consider $N = 24611$. As $\sqrt{N} \approx 156.879$, this means that either N is prime *or* it has a prime factor at most 156. Checking all primes up to 156 (the largest such prime is 151), we see that none of them divide 24611. Thus 24611 is prime, and we can show this by confining our search to primes at most 156. There are only 36 such primes, which is far fewer than all the primes at most 24611 (the largest is 24593, and there are 2726 such primes).

While it's an enormous savings to only have to check candidates up to $\sqrt{N}$, is it enough of a savings so that brute force becomes useful in practice? Let's analyze the problem in greater detail. Consider the situation of RSA, where an eavesdropper needs to factor a large number N which is the product of two primes. Let's say the primes p and q are around 10^{200} so that N is around 10^{400}. How long would it take us to try all possible divisors by brute force? The universe is approximately 13 billion years old, and is believed to contain less than 10^{90} subatomic items. Let's overestimate a lot, and say the universe is 1 trillion years old and there are 10^{100} subatomic items, each of which is a supercomputer devoted entirely to our needs and capable of checking 10^{40} prime numbers per second (note this is *magnitudes* beyond what the current best computers are capable of doing). How long would it take us to check the 10^{200} potential prime divisors of N? We first translate our age into seconds:

$$1 \text{ trillion years} = 10^{12} \text{ years} \cdot \frac{365.25 \text{ days}}{1 \text{ year}} \cdot \frac{24 \text{ hours}}{1 \text{ day}} \cdot \frac{3600 \text{ seconds}}{1 \text{hour}} < 10^{20}.$$

The number of seconds needed is

$$\frac{10^{200}}{10^{100} \cdot 10^{40}} = 10^{60};$$

the universe hasn't even existed for 10^{20} seconds, let alone 10^{60}!

The purpose of the above computation is to drive home the point that, just because we have an algorithm to compute a desired quantity, it doesn't mean that algorithm is useful. In our investigations in cryptography we encountered two variants of this problem. The first was a search for efficient algorithms, so we do not have to wait significantly longer than the universe has existed for our answer! The second is the opposite; for security purposes, we *want* problems whose answers take an incredibly long time to solve, as these should be secure. Of course, it's often necessary for us to solve these problems as well. This led to studying **trap-door functions** and problems. These are problems where the answer can be obtained significantly faster with an extra piece of information that is not publicly available. For instance, let's return to the factorization problem, and let's assume that the password is the largest prime factor of N, where $N = pq$ is the product of two primes. We've seen that if N is of the order 10^{400}, then it takes too long to compute the prime divisors by brute force; however, if we are given p, then we can immediately determine q from the simple division $q = N/p$.

As we mentioned in the introduction to this chapter, one of the greatest dangers in cryptography is that a problem which is believed to be difficult might in fact be easy. For example, just because no one knows a fast way to factor large numbers does not mean no such method exists, it just means we do not know how to do it. It's possible someone has discovered a fast way to factor large numbers, and is keeping it secret precisely for these reasons! In cryptography, the fear is that there might be a hidden symmetry or structure that an attacker could exploit, and the underlying problem might not be as difficult as we thought. Let's revisit our factorization problem. Recall the goal is to find the factors of N, with N a product of two primes. If we don't know either prime, we saw that even if we had incredible computer resources at our disposal, the life of the universe isn't enough time to make a dent in the problem. If we know the smaller of the two factors, the problem is almost instantaneous. What if we are foolish, and accidentally choose the two primes to be equal? In this case, the problem is again trivial. Even though we don't know p, we can quickly discover it by noting that $\sqrt{N}$ is an integer. For many cryptosystems, the danger is that there is something subtle like this lurking, unknown.

In the next sections we discuss various algorithms for primality testing and factorization. Some of these work only for numbers of certain, special form; others work for any input. While many of these algorithms are significantly faster than brute force, to date there is no known, fast way to factor a number, though there are known, fast ways to determine whether a number is prime.

11.3. Fermat's Factoring Method

As remarked above, one of the central difficulties of cryptography is that much of it is based on the belief that certain problems are hard and thus cannot be solved without extra information that is hidden from an attacker. How do we know these problems are hard? We know because no one has solved them yet! Of course, this sounds like circular logic. There could be an elementary approach that has just been missed for years, and the problem might actually be easy after all. We'll see an example of this in §11.6, where we give a provably fast way to determine if a number is prime. It is precisely because this proof was missed for so many years that we've decided to include its proof in this book, as a warning that the difficulty of the problem may not be as bad as we thought.

We now describe some interesting approaches to factoring numbers. The first is **Fermat's method**. We might as well assume our integer N is odd, as it's very easy to check whether or not it is divisible by 2. Further, we can assume that N is not a perfect square, as it's very easy to test if $N = n^2$ for some n. To do this, we simply approximate the square-root of N and then check the integers immediately below and above to see if either squares to N.

We therefore content ourselves with trying to factor odd N that are not perfect squares. Imagine that we can write N as the difference of two squares, so

$$N = x^2 - y^2.$$

If, somehow, we can find such an x and a y, then we can factor N as

$$N = (x-y)(x+y);$$

so long as $x - y$ and $x + y$ are neither 1 nor N we have factored N as a product of two nontrivial terms. For example,

$$2007 = 116^2 - 107^2,$$

so

$$2007 = (116+107)(116-107) = 223 \cdot 9.$$

For a more interesting example, consider

$$12069 = 115^2 - 34^2 = 149 \cdot 81,$$

or even more amazingly

$$\begin{aligned} 123456789987654321 &= 355218855^2 - 52188552^2 \\ &= 407407407 \cdot 303030303; \end{aligned}$$

we couldn't use 12345678987654321 as an example because of the entertaining fact that it equals 111111111^2.

Several questions immediately present themselves.

- Can N be written as a difference of two squares if N is an odd, composite number?
- *How do we find these squares if N can be written as a difference of two squares?*

Let's start with the first question about the difference of two squares. It turns out that it is always possible to write an odd, composite N that is not a perfect square as a difference of two squares as

$$\left(\frac{N+1}{2}\right)^2 - \left(\frac{N-1}{2}\right)^2 = \frac{N^2+2N+1}{4} - \frac{N^2-2N+1}{4} = N.$$

How did we arrive at this?

We see that our first question was poorly phrased. It's not enough to write N as a difference of two squares; we want the factors $x - y$ and $x + y$ to be neither 1 nor N. Clearly this cannot be possible if N is prime. What if N is composite, say $N = rs$ (where of course r and s are odd as N is odd)? We then need $(x-y)(x+y) = rs$. If we let $x - y = r$ and $x + y = s$, then adding the two gives $2x = r + s$ or $x = \frac{r+s}{2}$; note x will be an integer since r and s are odd so their sum is even. Similarly, subtracting the two equations gives $2y = s - r$ so $y = \frac{s-r}{2}$. In other words, if $N = rs$, then we have

$$N = \left(\frac{r+s}{2}\right)^2 - \left(\frac{s-r}{2}\right)^2.$$

Our discussion above gives us some cause for hope. We've shown that if N is composite, then we can write it as a difference of two squares in a nontrivial way, and if we can write it as a difference of two squares, then we can obtain a factorization.

The next question is whether or not we can *easily find* a difference of two squares equaling N. At first glance, there is a very obvious reason to worry about the practicality of Fermat's method: what if $N = x^2 - y^2$ but x and y are much larger than N? If this were the case, then it might take a long time to find the factors. We clearly need some control over how large x and y can be in order to bound how long Fermat's algorithm will take. This is readily done by noting again that $N = x^2 - y^2 = (x - y)(x + y)$. For a nontrivial factorization, $x - y \geq 2$ and thus $x + y \leq N/2$ (as $x + y = \frac{N}{x-y}$). In other words, x and y are at most $N/2$. Compared to our original attempt to factor by brute force, this is the same order of magnitude of how many numbers we must check; however, it's much worse than our refined brute force approach. There we saw it sufficed to check all numbers up to $\sqrt{N}$, whereas here we are still checking on the order of N numbers.

We can now explicitly state Fermat's method to factor N.

Fermat's method:

(1) If N is even, then 2 is a factor and apply Fermat's method to $N/2$.

(2) If N is a perfect square, then apply Fermat's method to $\sqrt{N}$.

(3) If N is an odd integer which is not a perfect square, let s equal the smallest integer greater than $\sqrt{N}$, and let $x = s, s + 1, \dots, \frac{N-3}{2}$. If for any of these choices of x we have $x^2 - N$ is a square (say y^2), then N is composite and factors as $N = (x - y)(x + y)$; if for each of these x's the number $x^2 - N$ is not a square, then N is prime.

While we know that it suffices to check x and y that are at most $N/2$, if we had to check all such integers, then Fermat's method would be too slow to be of practical use. Unfortunately, there are times when we would have to check all such numbers, for example, if N were prime. Fortunately, there is a large class of numbers where Fermat's method works quite well, namely those N which are the product of two primes each of which is near $\sqrt{N}$.

For example, consider $N = 327,653$; N is odd, and as $\sqrt{N}$ is approximately 572.41 we see N is not a perfect square. Letting s be the smallest integer at least $\sqrt{N}$, we have $s = 573$. Thus we must see if any of $s^2 - N$, $(s+1)^2 - N$, ... are perfect squares. We're in luck, as the very first case is

$$s^2 - N \;=\; 573^2 - 327653 \;=\; 328329 - 327653 \;=\; 676 \;=\; 26^2.$$

Rearranging gives

$$573^2 - 26^2 \;=\; (573 - 26)(573 + 26) \;=\; 547 \cdot 599.$$

For a more interesting example, consider $N = 223,822,733$. A straightforward computation shows that N is odd and not a perfect square. As $\sqrt{N} \approx 14960.7$, $s = 14961$. Thus we must check to see if any of $s^2 - N$, $(s+1)^2 - N$, and so on are perfect squares. We have

$$\begin{aligned} 14961^2 - N &= 8788 &\approx& \ 93.74^2, \\ 14962^2 - N &= 38711 &\approx& \ 196.75^2, \\ 14963^2 - N &= 68636 &\approx& \ 261.99^2, \\ 14964^2 - N &= 98563 &\approx& \ 313.95^2, \\ 14965^2 - N &= 128492 &\approx& \ 358.46^2, \\ 14966^2 - N &= 158423 &\approx& \ 398.02^2, \\ 14967^2 - N &= 188356 &=& \ 434^2. \end{aligned}$$

Thus

$$N = 14967^2 - 434^2 = (14967+434)(14967-434) = 15401 \cdot 14533;$$

though it is not obvious at all, the two factors above happen to be prime, and we have thus factored $N = 223,822,733$. While Fermat's method does not work well for all numbers, it is amazing how well it captures the factors of N whenever we can write N as rs with r, s close to $\sqrt{N}$. We were able to show a nine digit number is composite by just doing seven loops through the algorithm.

REMARK 11.3.1. Looking at Fermat's method, a natural question emerges: How many numbers can be written as a product of two numbers of approximately the same size? For example, consider

$$N = 7789357 \cdot 10354024466273 = 80651192954534856461.$$

Neither factor is particularly close to $\sqrt{N}$, which is approximately $9 \cdot 10^9$; however, our number has the alternative factorization

$$N = 8015447633 \cdot 10061969917,$$

and here the two factors are close to $\sqrt{N}$. There are many ways to group factors—for Fermat's method to work, all we need is for *one* grouping to have both terms near $\sqrt{N}$. Letting p_n denote the nth prime, we see

$$N = p_{15} \cdot p_{16}^2 \cdot p_{17} \cdot p_{231424} \cdot p_{231425};$$

the first factorization corresponds to

$$\left(p_{15} \cdot p_{16}^2 \cdot p_{17}\right) \cdot \left(p_{231424} \cdot p_{231425}\right),$$

while the second corresponds to

$$\left(p_{15} \cdot p_{16} \cdot p_{231424}\right) \cdot \left(p_{16} \cdot p_{17} \cdot p_{231425}\right).$$

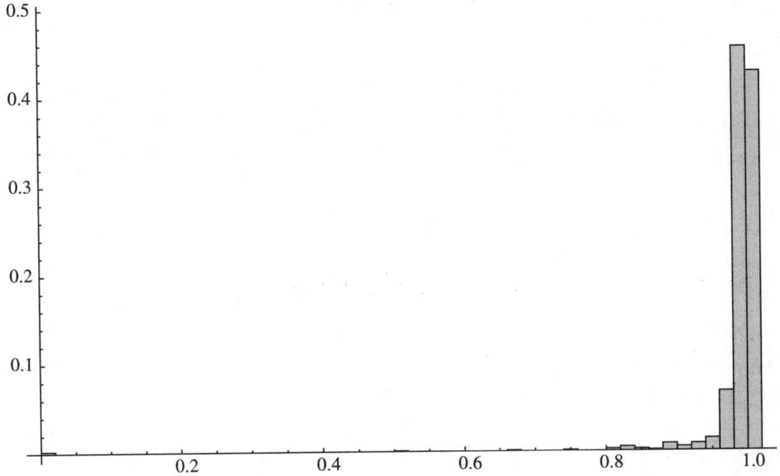

FIGURE 11.1. Histogram of percentage of witnesses in range 1000 to 2000. The x-axis is the percentage of witnesses (among the integers relatively prime to a given n), the y-axis is the percentage of integers in $[1000, 2000]$.

11.4. Monte Carlo Algorithms and FℓT Primality Test

In §7.6 we discussed a primality test based on Fermat's little Theorem. The Fermat primality test worked like this:

Fermat's primality test: Given an input $n > 1$:
- Step 1: Choose a random integer a satisfying $1 < a < n$.
- Step 2: Compute $a^n \pmod{n}$.
- Step 3: If $a^n \not\equiv a \pmod{n}$, then output "not prime".
- Step 4: Otherwise, if $a^n \equiv a \pmod{n}$, then output "maybe prime".

Fermat's little Theorem implies that n must be composite if a^n is not congruent to a modulo n. We call such an a a *witness* to the compositeness of n. In §7.6 we proved that for any composite number n that has at least one witness, at least half of the numbers a between 1 and n are witnesses. In fact, empirically speaking, most numbers seem to have far more witnesses than that, as shown in Figure 11.1. The problem with Fermat's test is that there are some composite numbers which essentially have no witnesses (the small bump at 0 in Figure 11.1 corresponds to these numbers). We need to use the word essentially as we don't care if it passes the test for an a dividing n, as in that case we can tell it's composite! In other words, these numbers pass the test for all a relatively prime to themselves, but are composite. We call these **Carmichael numbers**.

Carmichael number: A composite number n is a Carmichael number if for every $a \in \{1, 2, \ldots, n-1\}$ that is relatively prime to n we have $a^n \equiv a \pmod{n}$. Carmichael numbers exist; the first ten are 561, 1105, 1729, 2465, 2821, 6601, 8911, 10585, 15841, and 29341.

These numbers cause a lot of problems. No matter how many times we try Fermat's test on one of those numbers, it will never tell us that our number is not prime. However, suppose that what we want is not an algorithm to identify prime numbers, but an algorithm to identify numbers which are either prime or Carmichael numbers. Let's say a number has **the property $\mathcal{PC}$** if it is either a prime number or a Carmichael number. To check the property $\mathcal{PC}$ we would use exactly the same algorithm:

Fermat's text for $\mathcal{PC}$: Given an input $n > 1$:

- Step 1: Choose a random integer a satisfying $1 < a < n$.
- Step 2: Compute $a^n \pmod{n}$.
- Step 3: If $a^n \not\equiv a \pmod{n}$, then output "not $\mathcal{PC}$".
- Step 4: Otherwise, if $a^n \equiv a \pmod{n}$, then output "maybe $\mathcal{PC}$".

This algorithm isn't perfect. If it says n is not $\mathcal{PC}$, then we can trust it, and we know the status of n; however, if it says that n may have the property $\mathcal{PC}$, we don't know for sure whether or not it does. What we know from §7.6 is that if n doesn't have the property $\mathcal{PC}$, there is at most a 50% chance that Fermat's test will say that it may have the property. This algorithm thus qualifies as a **Monte Carlo algorithm** for testing the property $\mathcal{PC}$.

Monte Carlo algorithm for property checking: A Monte Carlo algorithm for checking whether an input has a certain property P is a probabilistic algorithm with the following properties:

(1) The *running time* of the algorithm has an upper bound which is determined just by the size of the input (i.e., the running time is nonrandom).
(2) The two possible outputs of the algorithm are: "my input does not have property P", or "my input may have property P".
(3) If the algorithm's input doesn't have property P, then it returns "my input may have property P" with probability at most 50% (regardless of the input!)

We see that Fermat's test for $\mathcal{PC}$ satisfies this definition. Note that it satisfies condition 1 since how long it takes to compute $a^n \pmod{n}$ can be bounded in terms of the size of $\log_2 n$, using the fast modular exponentiation algorithm discussed in §7.4. Fast modular exponentiation means that Fermat's test can be performed extraordinarily quickly; a modern computer can run it hundreds of times on numbers with hundreds of digits in less than a second!

Why are Monte Carlo algorithms useful? Up to half the time they may return an indeterminate answer (such as the number "may have the property P"). This indeterminacy is fundamental to the method, and is responsible for its name. It's a reference to the famous Le Grand Casino in Monte Carlo, Monaco. Considering all that is at stake in modern cryptography, do we really want to be "gambling" every time we run a primality test?

The important point is that by running a Monte Carlo algorithm over and over again, we can reduce our uncertainty to an arbitrarily small probability. For example, if we test the number 13571235581243 for the property $\mathcal{PC}$ with the Fermat test with one randomly chosen value for a, and find the algorithm returns that our number "may have the property $\mathcal{PC}$", we shouldn't be too confident that it really has this property. Remember there's a high probability, possibly as high as 50%, that our algorithm would return the same output of "maybe" even if our number doesn't have the property $\mathcal{PC}$.

But suppose we ran the algorithm ten times? Note that this involves independent choices of the random number used in each case, since the first step of the algorithm is to choose a random number, and so that step gets repeated too. If any of those times the algorithm returns that our number doesn't have the property $\mathcal{PC}$, then we know for sure that it doesn't. What if it answers that it "may have" the property $\mathcal{PC}$, ten times in a row? What is the chance this would happen if it doesn't have the property? Since each time the probability is at most $1/2$, and these are independent runs of the algorithm, the probability that it would return that it may have the property when it actually does not must be at most $(1/2)^{10} = 1/1024$, which is quite small. If we run Fermat's test 50 times and get that it "may have the property $\mathcal{PC}$" each time we are extremely confident, since the probability of getting 50 consecutive "maybes" when our number does not have the property is at most

$$\left(\frac{1}{2}\right)^{50} = \frac{1}{1,125,899,906,842,624}.$$

Note: There is a big assumption above: we're assuming all the tests are independent. For a note on the dangers of making this assumption, see Exercise 11.7.17.

The problem, unfortunately, is that although Fermat's test is a Monte Carlo algorithm for the property $\mathcal{PC}$, it is not a Monte Carlo algorithm for testing primality. It violates condition 3 of the definition, since if the input is a Carmichael number, then the input doesn't have the property of primality but the algorithm returns that the input "may have the property of primality" 100% of the time. This is just the definition of Carmichael numbers.

Fortunately Carmichael numbers are very rare. If we're testing random numbers for primality, which we do when generating RSA keys, we may be willing to live with a small probability of taking a composite number, but what if we're not? What if we want certainty? Remember, we can't reduce our probability of being wrong to an arbitrarily small number, as we could with a Monte Carlo algorithm. When testing primality with Fermat's test, there's always a chance that our input is a Carmichael number, and in that case running the algorithm over and over again won't make it any more likely that we find out the truth. In the next section, we'll discuss a Monte Carlo test for primality (the Miller–Rabin test), which allows us to very quickly determine primality to an arbitrary confidence level. We then end the chapter with the Agrawal–Kayal–Saxena Primality Test. While it is slower than the Miller–Rabin test, it is not probabilistic and leaves us with certainty.

11.5. Miller–Rabin Test

In this section we describe the **Miller–Rabin test**, which is a Monte Carlo algorithm for primality testing. This improves upon the Fermat test, which could not distinguish between primes and Carmichael numbers. There are many similarities between these two tests. Both are consequences of facts about exponentiation and congruences. Fermat's test follows from Fermat's little Theorem, which says $a^{p-1} \equiv 1 \pmod{p}$ if p is prime and a is relatively prime to p. The starting point for the Miller–Rabin test is the following.

THEOREM 11.1. *If p is a prime number and $a^2 \equiv 1 \pmod{p}$, then either $a \equiv 1 \pmod{p}$ or $a \equiv -1 \pmod{p}$.*

Before proving Theorem 11.1 we first discuss a needed result: if a prime p divides a product xy, then p divides x or y. This fact is a consequence of the Fundamental Theorem of Arithmetic, which asserts every integer can be written uniquely as a product of primes. Why does this imply that if p divides xy, then p must divide x or y? If this claim were false, consider the factorizations for x and y. Neither of these contain p, but $xy = mp$ for some m since p divides xy by assumption. Note that mp *does* contain p in its factorization. This contradicts the factorization into prime powers must be unique, and therefore either p divides x or it divides y.

Proof. As $a^2 \equiv 1 \pmod{p}$, this means $a^2 = 1+np$ for some n. Equivalently, $a^2-1 = np$, so p divides a^2-1. Since $a^2-1 = (a-1)(a+1)$, we have p divides $(a-1)(a+1)$. From the consequence of unique factorization described above, if a prime divides a product, it must divide at least one of the factors. Thus, either p divides $a-1$ (which is equivalent to $a \equiv 1 \pmod{p}$) or p divides $a+1$ (which is equivalent to $a \equiv -1 \pmod{p}$), completing the proof. □

It seems surprising that this observation will serve as the basis of a powerful, widely used primality test. We use this to create a sequence that takes on one set of values if n is prime and another set of values if n is composite. Before stating the test, we describe the sequence.

Let's take a prime number p. As every integer can be written as an odd number times a power of 2, we can write

$$p - 1 = b \cdot 2^k,$$

where b is an odd number and k is an integer. Another way to view this is that we're using the Fundamental Theorem of Arithmetic to partially factor $p-1$; after pulling out all the powers of 2 from $p-1$, what remains must be odd. For example, if $p = 61$, then $p - 1 = 60$ and $b = 15$ and $k = 2$, since $60 = 15 \cdot 2 \cdot 2$. We call b the **odd part** of $p-1$ and note that 2^k is the largest power of 2 dividing $p-1$.

We now describe the key sequence for the Miller–Rabin test. Choose an a between 1 and p, and consider the sequence

$$a^b,\ a^{2\cdot b},\ a^{4\cdot b},\ a^{8\cdot b},\ \dots,\ a^{2^k\cdot b},$$

where each term is reduced modulo p. This sequence can be found quickly. We calculate the first term by using fast modular exponentiation and then reducing modulo p, and then square each term modulo p to find the successive terms. For example, if $p = 241$, then $p-1 = 240$ and $b = 15$ and $k = 4$, so the sequence for $a = 3$ is

$$3^{15},\ 3^{30},\ 3^{60},\ 3^{120},\ 3^{240},$$

which modulo 241 is

$$8,\ 64,\ 240,\ 1,\ 1.$$

Notice two properties of this sequence:

(1) it eventually becomes 1 (and once it is 1, it stays 1); and
(2) the term immediately before the first 1 is congruent to -1 modulo the prime p.

There's nothing special about the prime 241. These two properties hold for any a relatively prime to a prime p, so long as we say the two properties hold if all the terms in the sequence are 1, as in that case, there is no term immediately before the first 1.

Why is the first property true? Since $b \cdot 2^k = p - 1$, the last term of the sequence $a^{b\cdot 2^k}$ is the same as $a^{p-1} \pmod p$. By Fermat's little Theorem this term is 1.

What about the second property? We now work backwards. Each term in the sequence is the square of the previous term. In Theorem 11.1 we proved that if a number squares to 1 modulo p, then that number is either 1 or -1 modulo p. Thus, if p is prime, then the term immediately before a term congruent to 1 must be either 1 or -1; if the term under consideration is the *first* term to be 1, then the term right before it must be -1.

This suggests the following test for primality.

Miller–Rabin primality test: Let n be an integer greater than 1.

- Step 1: Choose a random integer a satisfying $1 < a < n$.
- Step 2: Find b and k such that $a = b \cdot 2^k$.
- Step 3: Calculate, modulo n, the sequence

$$a^b,\ a^{2 \cdot b},\ a^{4 \cdot b},\ a^{8 \cdot b},\ \ldots,\ a^{2^k \cdot b}.$$

- Step 4: If the sequence is eventually 1 and if the last term before the first term is -1, output **maybe prime**.
- Step 5: Otherwise, output **not prime**.

The most intensive step is Step 3, which we can do extremely rapidly through fast exponentiation and then repeated squaring modulo n. Based on our discussion above, if n is prime, then the algorithm returns "maybe prime", since we showed that if n is prime, then the sequence (11.1) must have both the properties required in Step 4. If the sequence does not end with a 1, then n can't be prime by Fermat's little Theorem. If the sequence ends in 1 but the term before the first 1 is not -1 modulo n, then n can't be prime by Theorem 11.1. We thus see that the algorithm never lies, which of course is good!

On the other hand, to consider this a Monte Carlo algorithm for primality testing, we must show that if n is not prime, then there is at most a 50% chance that the algorithm returns "maybe prime". This analysis is trickier than the corresponding analysis for the Fermat test, but it can be shown that the probability of saying "maybe prime" when n is composite is at most 25%.

The Miller–Rabin test gives an extremely fast way of checking primality, and since it is a Monte Carlo algorithm, we can rerun the test several times on our input to ensure that the probability of an incorrect error is made arbitrarily small. For example, if we run the Miller–Rabin test 50 times (which can be done in less than a second even for numbers with hundreds of digits as are used for cryptography), the probability that the Miller–Rabin test would return "maybe prime" for an input which is composite is at most

$$\frac{1}{4^{50}} = \frac{1}{1{,}267{,}650{,}600{,}228{,}229{,}401{,}496{,}703{,}205{,}376}.$$

This is so small that even if every living person used this procedure to check a composite number for primality every second of every day, we would expect it to take more than 8 trillion years before anyone had the test indicate "maybe prime" on each of 50 trials for a number! The probability of an error of this type is much smaller than having incorrect results as a consequence of computer hardware errors.

So, for all practical purposes, the Miller–Rabin test (and other Monte Carlo primality tests like it) are perfect primality tests: they are extremely fast, simple to implement, and their error rate can be so small, regardless of the input, as to be completely irrelevant to any practical use we might have for them. But, it's natural to ask now, Is the randomness in these algorithms really necessary? Is it possible to have an algorithm, which doesn't use randomness at all, but which nevertheless is "fast"? It *is* possible to remove the randomness; we present this crowning achievement in the next section.

11.6. Agrawal–Kayal–Saxena Primality Test

For mathematicians, a question of great theoretical interest was whether it was possible to have a primality testing algorithm which was **deterministic** (did not use randomness) and a **polynomial-time algorithm** (which means that the running time of the algorithm is bounded by a polynomial of the number of digits of the input number). Although they are not deterministic algorithms, the Monte Carlo algorithms we discussed in the previous section have running times which are polynomially bounded by the number of digits in the input number. Roughly speaking, this is because, as we showed in §7.4, the number of steps it takes to do fast modular exponentiation is at most 8 times the number of digits in the exponent (we also need to realize that each squaring and multiplication used in each step of fast modular exponentiation is polynomially bounded by the number of digits in the inputs).

The concept of a polynomial-time algorithm captures a theoretical notion of "fast". For example, the brute force algorithm from §11.2 checks the primality of a number n by dividing it by all numbers up to n. This takes roughly n steps. Since the number of decimal digits in n is roughly $\log_{10} n$, the running time is not polynomially bounded by the number of digits. This is because $n = 10^{\log_{10} n}$, and it gives a horrendous run-time that is *exponential* in the number of digits.

A polynomial-time deterministic primality test had been a dream of mathematicians for many years, but it eluded researchers until 2002 when Agrawal and his two students Kayal and Saxena discovered the test of this section [**2**]. The goal of this chapter is not to prove why their algorithm works; there are numerous papers and resources online (we give references below) which do this. Instead, our emphasis is on explaining the new terms and concepts so that the interested reader can follow these works. Additionally, their result serves as a powerful warning. Just because a problem seems hard does not mean it is; this is an important lesson to keep in mind when thinking about the security of our different cryptosystems.

Before we can state their test, we need to define two concepts and state one result from combinatorics. We start with the simplest concept we need, the **Euler totient function** (we discussed this function in §8.1 and guided the reader through many of its properties in Exercises 8.8.7 through 8.8.11). Denoted φ, we define it by setting $\varphi(n)$ equal to the number of integers in $\{1, 2, \ldots, n\}$ that are relatively prime to n. In other words, we count how many of these integers don't share any prime factors with n. We thus have $\varphi(1) = 1$, $\varphi(2) = 1$, $\varphi(3) = 2$, $\varphi(4) = 2$, and $\varphi(12) = 4$. For example, if

$n = 12$, then we have the numbers

$$\{1,\ 2,\ 3,\ 4,\ 5,\ 6,\ 7,\ 8,\ 9,\ 10,\ 11,\ 12\}.$$

Which of these are relatively prime to 12? We lose 2, 3, 4, 6, 8, 10 and 12, and are left with

$$\{1,\ 5,\ 7,\ 11\},$$

and hence $\varphi(12) = 4$.

The totient function is very well understood. If p is prime, then $\varphi(p) = p - 1$. To see this, consider the set

$$\{1,\ 2,\ 3,\ \dots,\ p-2,\ p-1,\ p\}.$$

We want to know how many of these numbers don't share a proper divisor with p. Since p is prime, the only numbers that divide p are 1 and p; thus p cannot divide $1, 2, \dots, p-1$, but it does divide p. We therefore see that exactly $p-1$ of the p numbers are relatively prime to p, so $\varphi(p) = p - 1$.

While $\varphi(n)$ is the number of integers in $\{1, \dots, n\}$ that are relatively prime to n, frequently it matters *which* integers are relatively prime. These integers have special properties relative to n, and thus merit a name. We write $(\mathbb{Z}/n\mathbb{Z})^*$ for the subset of integers of $\{1, \dots, n\}$ that are relatively prime to n; this notation should look a bit strange, but it is the standard notation (and, if you take a group theory class, you'll learn why).

For example,

$$(\mathbb{Z}/12\mathbb{Z})^* \ = \ \{1,\ 5,\ 7,\ 11\}$$

and

$$(\mathbb{Z}/p\mathbb{Z})^* \ = \ \{1,\ 2,\ \dots,\ p-1\}$$

for any prime p.

The Agrawal–Kayak–Saxena (AKS) primality test is based on modular arithmetic, which we've seen throughout this book. We recall the definition, which says that

$$a \equiv b \pmod n \quad \text{if and only if} \quad n|(a-b).$$

In other words, $a \equiv b \pmod n$ means there is an integer k such that $n \cdot k = a - b$. For AKS we need a generalization of this notion of modular equivalence from integers to polynomials with integer coefficients. We say

$$f(x) \equiv g(x) \bmod m(x) \qquad \text{whenever} \qquad m(x)|f(x) - g(x).$$

By this we mean that there is another polynomial $h(x)$ with integer coefficients such that $h(x)m(x) = f(x) - g(x)$. For example,

$$3x^2 + 7x + 4 \ \equiv \ x^2 + 2x + 1 \bmod x + 1$$

as

$$(3x^2 + 7x + 4) - (x^2 + 2x + 1) \ = \ (2x+3)(x+1).$$

We can even combine polynomial congruence and modular congruence, and we say $f(x) \equiv g(x) \bmod (n, m(x))$ if there is an $h(x)$ such that $f(x) - g(x) -$

$h(x)m(x) \equiv 0 \bmod n$. Obviously the more congruences we have, the harder it is. As an example, we show $9x^2 - 3x + 1 \equiv x \bmod (7, x+1)$. We have

$$(9x^2 - 3x + 1) - x \;=\; 9x^2 - 4x + 1.$$

Because we'll eventually look at everything modulo 7, we write $9x^2$ as $2x^2 + 7x^2$ and $-4x$ as $3x - 7x$. We thus have

$$(9x^2 - 3x + 1) - x \;=\; 9x^2 - 4x + 1 \;=\; (2x^2 + 3x + 1) + (7x^2 - 7x).$$

We can factor $2x^2 + 3x + 1$ as $(2x+1)(x+1)$; note $x+1$ is our modulus $m(x)$. We have thus shown

$$(9x^2 - 3x + 1) - x \;=\; (2x+1)(x+1) + 7(x^2 - x)$$

or

$$(9x^2 - 3x + 1) - x - (2x+1)(x+1) \;\equiv\; 0 \bmod 7.$$

The final concept we need in order to state the AKS primality test is the **order of an element** in $(\mathbb{Z}/n\mathbb{Z})^*$. It turns out that $(\mathbb{Z}/n\mathbb{Z})^*$ is a group under multiplication modulo n. One consequence of this is that if $x \in (\mathbb{Z}/n\mathbb{Z})^*$, then there is an integer k such that $x^k \equiv 1 \bmod n$. For example, if $n = 12$, then we've seen $(\mathbb{Z}/n\mathbb{Z})^* = \{1, 5, 7, 11\}$, and we have

$$1^2,\ 5^2,\ 7^2,\ 11^2 \;\equiv\; 1 \bmod 12.$$

For another example, consider

$$(\mathbb{Z}/7\mathbb{Z})^* \;=\; \{1,\ 2,\ 3,\ 4,\ 5,\ 6\}.$$

A little calculation shows that each element is equivalent to 1 when raised to an appropriate power. The case of 1 is obvious; for the others, we have $2^4 = 8 \equiv 1 \bmod 7$, $3^6 = 727 \equiv 1 \bmod 7$, $4^2 = 8 \equiv 1 \bmod 7$, $5^6 = 15625 \equiv 1 \bmod 7$, and $6^2 = 36 \equiv 1 \bmod 7$. We denote the order of x modulo n by $\mathrm{ord}_n(x)$.

The needed combinatorial result is the following (see §3.2 of Chapter 3 for a detailed discussion). Recall that the **binomial coefficient** $\binom{n}{k}$ is defined by

$$\binom{n}{k} \;=\; \frac{n!}{k!(n-k)!}$$

when n is a positive integer and $0 \le k \le n$ is an integer; by convention we set $\binom{n}{0} = 0$. There is a nice combinatorial interpretation to these numbers; they are the number of ways of choosing k objects from n objects when order does not matter. (Thus we should view $\binom{n}{0}$ as saying that, mathematically, there is but one way to do nothing!) The key result for us is that

$$\binom{n}{k} \equiv 0 \bmod n \text{ for all } k \in \{1, \dots, n-1\} \text{ if and only if } n \text{ is prime.}$$

Note that we must exclude $k = 0$ and $k = n$, as $\binom{n}{0} = \binom{n}{n} = 1$ for all n.

Let's check the claim by looking at some examples. If we take $n = 4$, then since 4 is composite we expect at least one binomial coefficient not to be equivalent to 0 modulo 4. We have

$$\binom{4}{1} = 4, \quad \binom{4}{2} = 6, \quad \binom{4}{3} = 4;$$

while the first and last are congruent to 0 modulo 4, the middle is not (it's congruent to 2). If instead we take the prime $n = 5$, then we have

$$\binom{5}{1} = 5, \quad \binom{5}{2} = 10, \quad \binom{5}{3} = 10, \quad \binom{5}{4} = 5;$$

note all of these are equivalent to 0 modulo 5. We give a hint for proving one direction of the claim in Exercise 11.7.46.

We can now state the AKS primality test.

AKS primality test: Let $N > 1$ be a positive integer.

(1) Test to see if N is a perfect k^{th} power for a $k \geq 2$. If it is, then N is composite and stop, else proceed to Step 2.
(2) Find the smallest prime r such that the order of N modulo r is greater than $\ln^2 N$; in other words, for all $r' < r$ if k is the smallest integer such that $N^k \equiv 1 \bmod r'$, then $k \leq \ln^2 N$.
(3) If any of the numbers in $\{2, 3, \ldots, r\}$ have a nontrivial common factor with N (this means they share a divisor between 2 and $N - 1$), then N is composite and stop, else proceed to Step 4.
(4) If $N \leq r$, then N is prime and stop, else proceed to Step 5.
(5) For each positive integer a that is at most $\sqrt{\varphi(r)} \ln N$, check and see if $(x + a)^N \equiv x^N + a \bmod (x^r - 1, N)$. If there is such an a such that the equivalence fails, then N is composite; if the equivalence holds for all such a, then N is prime.

Remark: Note that if the AKS primality test terminates in either Step 1 or 3, then not only do we learn that N is composite, but we also find a factor (we find the factor in Step 3 by applying the Euclidean algorithm to r and N to find the greatest common divisor). Sadly, this is not the case if the program ends in Step 5.

There are numerous expositions describing both why the AKS primality test works, as well as why it will always terminate in polynomial time. The paper [**2**] is available online. Other good sources are an FAQ webpage [**47**] and a great survey article by Granville [**21**]. We refer the reader to these papers for the full details, and content ourselves with giving a rough analysis of some of the steps and then doing some illustrative examples.

It turns out that although AKS answered a crucial theoretical question about the existence of a polynomial-time algorithm for primality testing, it is not actually useful in practice. This is because, although the the running time is bounded as a polynomial of the size of the input number, the polynomial bound is large! Indeed, to test even just one 300-digit number for primality could take months or years (or more!) with AKS, depending on the implementation. This is the great value of randomness for primality testing; it allows us to have algorithms that finish in seconds instead of years.

Even though AKS is too slow for practical purposes, it's still extremely important as it does answer one of the central questions in the field. Since it is slow, it becomes natural to ask *why* it is slow. In particular, which steps take a long time?

Let's consider the first step. How difficult is it to determine if N is a perfect k^{th} power? First off, we should figure out how large k might be. Let's say $N = n^k$ for some n and k. Clearly the larger n is the smaller k is; thus k is largest when n is smallest. The smallest we may take n to be is 2, and thus the largest possible value of k, which we'll denote $k_{\max}$, must satisfy $2^{k_{\max}} \leq N$. If we take logarithms base 2 of both sides we find

$$\log_2 2^{k_{\max}} \leq \log_2 N.$$

We now use the power rule, which says $\log_b x^y = y \log_b x$, and find

$$k_{\max} \leq \frac{\log_2 N}{\log_2 2} = \log_2 N$$

(since $\log_2 2 = 1$). We've shown that Step 1 is very fast. The number of k to check is at most $\log_2 N$, which is at most $4 \log_{10} N$ (we used the change of base rule of logarithms, which in this case gives $\log_2 N = \log_{10} N / \log_{10} 2$). In other words, the number of k to check is bounded by a polynomial in the number of digits of N, which is our criterion for "fast".

Step 2 could take a long time for two reasons. One reason is that it might be hard to compute the order of N modulo r, and the second is that we might need to take r large before we find an r such that the order of N modulo r is at least $\ln^2 N$. Fortunately, results from number theory tell us that we can find an r without going too high. Specifically, we'll find an r among the first $\ln^5 n$ numbers, and thus r is small enough that computing the orders won't take too long.

As r is not too large, Step 3 is fairly fast. We just have to run the Euclidean algorithm to find the greatest common divisor of n and $a \leq r$, and the Euclidean algorithm is very fast.

Step 4 is the simplest of all to analyze. It's just a simple comparison, which takes no time.

We are thus left with Step 5. For most large numbers, almost all of the run-time is due to this step. It isn't pleasant to implement polynomial modular arithmetic, though this can be done in environments ranging from Java to Mathematica to Sage.

Let's look at some representative examples. For our first test, consider $N = 21$. Of course it's absurd to use the AKS primality test here, but the purpose is to highlight the method. Going through the algorithm, we see N is not a perfect k^{th} power. As $\ln^2 21 \approx 9.26$, we need to find the smallest prime r such that the order of N modulo r is at least 10. The smallest such r is 19, and the multiplicative order is 18. In other words, $21^{18} \equiv 1 \bmod 19$, and no smaller power of 21 is equivalent to 1 modulo 19. For example, if we took $r = 17$, we would have found 21 has multiplicative order of 4, as $21^4 = 194481 \equiv 1 \bmod 17$. We now move to Step 3 and look at the greatest common divisors of 21 and all $a \leq 19$; we're in luck as we very quickly discover that 21 and 3 have a common factor, and thus 21 is composite.

Let's do one more example. Consider now $N = 20413$, whose compositeness isn't as obvious. A quick check shows that N is not a perfect k^{th} power. We have $\ln^2 N \approx 98.4$, so we must find a prime r such that the multiplicative order of N modulo r is at least 99. The smallest such r is $r = 101$, and the order of N is 100. We now look at the greatest common divisors of N with $a \in \{2, \dots, 101\}$, and unfortunately all of these numbers are relatively prime to N. We thus continue to Step 4. As $N > r$, we move on to Step 5. We now have to deal with the polynomial congruences. Fortunately, this example isn't too bad; taking $a = 1$ shows that the congruence fails and thus N is composite! Explicitly, we have

$$(x+1)^{20413} \not\equiv x^{20413} + 1 \bmod (20413, x^{101} - 1).$$

To see the failure, we write $(x+1)^{20413} - (x^{20413} + 1) \bmod (20413, x^{101} - 1)$ below:

$$\begin{aligned}
18358 &+ 13974x + 12056x^2 + 7124x^3 + 19263x^4 + 16714x^5 \\
&+ 16714x^6 + 19263x^7 + 7124x^8 + 12056x^9 + 13974x^{10} \\
&+ 18358x^{11} + 11645x^{12} + 2603x^{13} + 19830x^{14} + 19591x^{15} \\
&+ \cdots + 19830x^{98} + 2603x^{99} + 11645x^{100},
\end{aligned}$$

which clearly is not zero!

11.7. Problems

• From §11.1: Introduction

EXERCISE 11.7.1. *Apply Euclid's method, but instead of starting with* 2 *instead start with* 3. *Do you obtain the same sequence as before? Calculate the first few terms. What are the first few terms if we start with* 11?

EXERCISE 11.7.2. *Look up a few interesting sequences on the OEIS, and learn how to use the website. Here are some suggestions: Recamán's sequence* (A005132), *the Busy Beaver problem* (A060843), *the Catalan numbers* (A000108), *the prime numbers* (A000040), *Mersenne primes* (A000043, A000668), *the Fibonacci numbers* (A000045), *and of course both the first sequence* A000001 *and the sequence often shown when you reach the webpage,* A000055.

EXERCISE 11.7.3. *There are* many *proofs of the infinitude of the primes. Read a few of them and write them up. Particularly fun ones are Furstenberg's proof, and a proof through Fermat numbers.*

• **From §11.2: Brute Force Factoring**

EXERCISE 11.7.4. *Using the brute force algorithm, determine whether or not* 1701 *is prime. If it is not prime, factor it.*

EXERCISE 11.7.5. *We assumed the amount of time it took to test whether or not a given n divided N was independent of n. In practice this is of course absurd. Consider the following two problems: let N be a* 10,000 *digit number, let x be a one digit number, and let y be a* 100 *digit number. Using long division, approximately how many digit multiplications are needed to divide N by x? To divide N by y? We thus see it is more "expensive" to divide N by y than it is to divide by x.*

EXERCISE 11.7.6. *We defined what it means for a graph to be connected in Exercise* 7.8.12. *Give a brute force, inefficient way to determine if a graph with finitely many vertices is connected.*

EXERCISE 11.7.7. *We defined the adjacency matrix of a graph in Exercise* 7.8.13. *Show that if A is the adjacency matrix of a graph G, then the matrix A^k encodes the number of paths of length k between two vertices; specifically, the element in the ith row and j^{th} column of A^k is the number of paths of length k starting at vertex i and ending at vertex j, where each step is along an edge in the graph connecting two vertices.*

EXERCISE 11.7.8. *Building on the previous problem, show a graph on n vertices is connected if and only if its adjacency matrix A is such that all the entries of $(A+I)^n$ are nonzero. We added I to A to allow us the option of staying at a vertex in a walk; if we didn't do this, we would have to add powers of adjacency matrices to consider all paths of length at most n.*

EXERCISE 11.7.9. *Prove an integer x is divisible by* 2 *if and only if its ones digit is a* 0, 2, 4, 6, *or* 8, *and prove x is divisible by* 5 *if and only if its ones digit is* 0 *or* 5.

EXERCISE 11.7.10. *Let x be an integer, and let $a_0, a_1, \dots, a_n$ be the digits of x, so $x = \sum_{k=0}^{n} a_k 10^k$ (so if $x = 314$, then $a_0 = 4$, $a_1 = 1$, and $a_2 = 3$). Prove x is divisible by* 3 *if and only if the sum of the digits, $a_0 + a_1 + \cdots + a_n$, is divisible by* 3*; similarly show x is divisible by* 9 *if and only if the sum of the digits is divisible by* 9.

EXERCISE 11.7.11. *Using the notation in the previous problem, show x is divisible by* 11 *if and only if the alternating sum of the digits is divisible by* 11. *In other words, x is divisible by* 11 *if and only if* 11 *divides $a_0 - a_1 + a_2 - a_3 + \cdots + (-1)^n a_n$.*

• From §11.3: Fermat's Factoring Method

EXERCISE 11.7.12. *Use Fermat's factoring method to factor* 7811.

EXERCISE 11.7.13. *Use Fermat's factoring method to factor* 361567.

EXERCISE 11.7.14. *Randomly take some large numbers and look at their factorizations. Experiment with grouping the factors and see how often you can find two factors near the square-root of the number. For definiteness, look at say* 1000 *numbers starting at say* 34225523532.

EXERCISE 11.7.15. *Let N be an odd number. Can every composite N be written as a difference of two cubes, say $N = x^3 - y^3$? If not, are there any natural conditions that N must satisfy in order to have a chance of having such a decomposition?*

EXERCISE 11.7.16. *How do you think Fermat's factoring method will work on a number that is the product of three primes all approximately the same size? Do you think it will find a factor faster or slower than the main case? Explain your reasoning. There are methods that are great at producing factors of special integers. For more information, read about Pollard's $p-1$ algorithm. This algorithm shares a feature with the Pohlig–Hellman algorithm mentioned in Exercise* 8.8.45, *namely that it requires a "smoothness".*

• From §11.4: Monte Carlo algorithms and the FℓT primality test

EXERCISE 11.7.17. *Here's a new probabilistic factorization test, which can be repeated to factorize a number. Let n be a positive integer greater than* 1*. Randomly choose a positive integer a and calculate the greatest common divisor d of n and a. If $d > 1$, then n is composite and d is a proper divisor; if, however, $d = 1$, then n* may *be prime. The more times we run this algorithm the more certain we are that n is prime. For a "typical" n, which do you think is a better choice of a :* 199, *or* 210*? Why? After running the test with $a = 210$ we next try $a = 189$, 280, and* 3780*. What is wrong with these choices? Are the results of these tests independent of the results from the $a = 210$ test? Why or why not?* Hint: The probability the algorithm finds a proper divisor is a function of n *and* the number of factors of a.

EXERCISE 11.7.18. *There are many questions one can ask about Carmichael numbers. Are there only finitely many, or are there infinitely many? If there are only finitely many (and not too many at that), then perhaps we don't have to worry about them in practice. It turns out that there are infinitely many, though amazingly this wasn't known until the work of Alford, Granville, and Pomerance* [**1**] *in* 1994*. If we let $\pi(x)$ denote the number of primes at most x and $C(x)$ denote the number of Carmichael numbers at most x, then $\lim_{x\to\infty} C(x)/\pi(x) = 0$ (this follows from a result of Erdős* [**17**] *and the Prime Number Theorem). In other words, if a number passes the Fermat test for all a relatively prime to n, it is probably a prime. Write a short note describing the paper* [**1**].

In many of the problems below the following theorem is useful; see [**44**] for a proof (as well as for more information about Carmichael numbers).

THEOREM 11.7.19 (**Korselt [24]**). *A positive composite integer n is a Carmichael number if and only if it is square-free and whenever p divides n, then $p-1$ divides $n-1$.*

EXERCISE 11.7.20. *Find the first Fermat witness for the compositeness of* 1905.

EXERCISE 11.7.21. *Prove that all Carmichael numbers are odd.*

EXERCISE 11.7.22. *Prove that if $6k+1$, $12k+1$, and $18k+1$ are all prime, then the product $(6k+1)(12k+1)(18k+1)$ is a Carmichael number (this result is due to Chernick,* 1939*). Find a few values of k such that each of the three numbers is prime.*

EXERCISE 11.7.23. *Prove that all Carmichael numbers have at least three distinct prime factors.*

EXERCISE 11.7.24. *Prove that* 561 *is a Carmichael number. (If you don't find a quick way of checking all the congruences, this problem will be very painful!)*

EXERCISE 11.7.25. *As a witness a to n must be relatively prime to n, we are left with a problem: Should we first compute $a^n - a$ and then, if this is not zero modulo n, check to see if a is relatively prime to n, or should we first test to see if a and n are relatively prime? Which do you think is more efficient?* Hint: Use the Euclidean algorithm to see if a and n are relatively prime.

EXERCISE 11.7.26. *Research the history and some of the applications of the Monte Carlo method, and write a short report.*

- **From §11.5: Miller–Rabin Test**

EXERCISE 11.7.27. *The Euclidean algorithm gives us the following: if x and y are two positive integers, then there are integers a and b such that $ax + by = \gcd(x, y)$. Show that this gives another proof of the key property we needed in our preparations for the Miller–Rabin test, namely that if a prime p divides nm but not m, then p divides n.*

EXERCISE 11.7.28. *Consider the number* 180*. The only primes that divide it are* 2, 3, *and* 5*. Show that whenever we write* 180 *as $x \cdot y$ that* 2 *always divides x or y (or both), and of course the same is true for* 3 *and* 5*.* Hint: There are nine ways to write 180 as $x \cdot y$.

EXERCISE 11.7.29. *Let $p = 11$. Show that the only $a \in \{1, 2, \ldots, 10\}$ that square to* 1 *modulo* 11 *are* 1 *and* 10 *(and, of course, $10 \equiv -1 \pmod{11}$).*

EXERCISE 11.7.30. *Use the Miller–Rabin test to determine if* 11 *is prime.*

EXERCISE 11.7.31. *Use the Miller–Rabin test to determine if* 15 *is prime.*

EXERCISE 11.7.32. *Use the Miller–Rabin test to determine if* 21 *is prime.*

EXERCISE 11.7.33. *There are other probabilistic primality tests. Look up and write a short explanation of the Solovay–Strassen primality test, which has mostly been replaced by the Miller–Rabin test.*

EXERCISE 11.7.34. *If the Generalized Riemann Hypothesis is true, then there is a deterministic version of the Miller–Rabin test. Specifically, it suffices to test all* $a \le 2\ln^2 n$. *Look up and write a short explanation of the Riemann hypothesis and the generalized Riemann hypothesis.*

- **From §11.6: Agrawal–Kayal–Saxena Primality Test**

EXERCISE 11.7.35. *Compute* $\varphi(15)$, $\varphi(21)$, $\varphi(33)$, *and* $\varphi(35)$.

EXERCISE 11.7.36. *Notice each of the numbers in the previous problem is the product of two primes. Find a pattern between* $\varphi(pq)$, $\varphi(p)$, *and* $\varphi(q)$ *for the four numbers of the previous problem, where* p *and* q *are distinct primes. Based on your success, conjecture a formula for* $\varphi(n)$ *when* n *is the product of two primes. Prove your claim.*

EXERCISE 11.7.37. *Compute* $\varphi(4)$, $\varphi(9)$, $\varphi(25)$, *and* $\varphi(49)$.

EXERCISE 11.7.38. *Based on the results from the previous exercise, guess a formula for* $\varphi(p^2)$ *where* p *is a prime. Prove your result.* Hint: If you think the answer is a polynomial in p, you can figure out the coefficients and then try to prove if you're right. Clearly, $\varphi(p^2) < p^2$, so if it is a polynomial it must be of the form $ap + b$. By looking at two choices for p, you can figure out what a and b are and then prove your claim.

EXERCISE 11.7.39. *Find a formula for* $\varphi(p^3)$ *for* p *primc by trying out various* p. *Prove your claim.* Hint: Argue as in the previous exercise to try to find the formula. What's the highest power of p that can occur if our answer is a polynomial in p?

EXERCISE 11.7.40. *Find a formula for* $\varphi(pq)$ *when* p *and* q *are distinct primes. Prove your claim.* Hint: If the answer is a polynomial in p and q, the only possibility is $apq + bp + cq + d$, though an argument is required to eliminate the possibility of terms like p^2 or q^2.

EXERCISE 11.7.41. *Is* $x^2 + 2x + 1 \equiv 2x + 1 \bmod x + 1$*?*

EXERCISE 11.7.42. *Is* $x^2 + 7x + 5 \equiv x^2 + 3x + 1 \bmod x + 1$*?*

EXERCISE 11.7.43. *Is* $x^2 + x + 2 \equiv x^2 + 1 \bmod (3, x + 1)$*?*

EXERCISE 11.7.44. *Find the orders modulo* 9 *for* $(\mathbb{Z}/9\mathbb{Z})^*$ *and the orders modulo* 11 *for* $(\mathbb{Z}/11\mathbb{Z})^*$. *Note that if* $y \equiv -x \bmod n$, *then the orders of* x *and* y *are related; if* $\mathrm{ord}(x)$ *is even, then* $\mathrm{ord}(y) = \mathrm{ord}(x)$, *while if* $\mathrm{ord}(x)$ *is even, then* $\mathrm{ord}(y) = 2\mathrm{ord}(x)$. *This observation can save a lot of time in computing orders!*

The next three problems involve the following claim:

$$\binom{n}{k} \equiv 0 \bmod n \text{ for all } k \in \{1, \dots, n-1\} \text{ if and only if } n \text{ is prime.}$$

EXERCISE 11.7.45. *Verify the claim for $n = 6$ and $n = 7$. Note that it suffices to just look at the binomial coefficients with $k \leq n/2$ as $\binom{n}{k} = \binom{n}{n-k}$.*

EXERCISE 11.7.46. *One direction of the claim isn't too bad. Consider $\binom{n}{k} = \frac{n!}{k!(n-k)!}$. Because this has the combinatorial interpretation of being the number of ways of choosing k objects from n objects when order does not matter, we know it must be an integer. Imagine now that n is a prime. Show that n cannot divide any term in the denominator if $1 \leq k \leq n-1$, and thus $\binom{n}{k}$ must be divisible by n as claimed.*

EXERCISE 11.7.47. *For the brave: Prove the other direction of the claim, namely that if n is composite, then $\binom{n}{k}$ is not divisible by n for all $1 \leq k \leq n-1$.* Hint: If n is composite, we must have $n = ab$ for some $a, b \geq 2$. Try and keep track of how often powers of a and b divide the numerator and the denominator. Try looking for a good choice of k.

EXERCISE 11.7.48. *Step 1 of the algorithm asks us to make sure that N is not a perfect k^{th} power. Show that it suffices to check for k prime. For example, while 2176782336 is a 12th power, it is also a perfect square.*

We used properties of the **logarithm** in analyzing the AKS primality test. The next few exercises collect some of their most important properties. Recall $\log_b x = y$ means $x = b^y$; here $b > 1$ is the **base** of the logarithm and $x > 0$.

EXERCISE 11.7.49. *Prove $\log_b(b) = 1$ and $\log_b(1) = 0$.*

EXERCISE 11.7.50. *Why do we want to avoid the base b equaling 1? Would logarithms make sense base $1/2$?*

EXERCISE 11.7.51. *Prove $\log_b(x_1x_2) = \log_b x_1 + \log_b x_2$ and $\log_b(x_1/x_2) = \log_b x_1 - \log_b x_2$. In other words, the logarithm of a product is the sum of the logarithms, and the logarithm of a quotient is the difference of the logarithms.*

EXERCISE 11.7.52. *Prove $\log_b(x^r) = r\log_b(x)$. Note that this plus the result for the logarithm of a product implies the result for the logarithm of a quotient, as $\log_b(x_1/x_2) = \log_b(x_1x_2^{-1})$.*

EXERCISE 11.7.53 **(The Change of Base Formula)**. *If $b, c > 1$, prove $\log_c(x) = \log_b(x)/\log_b(c)$. This is one of the most important of the logarithm laws, as it allows us to compute logarithms base b if we know them base c. In other words, it suffices to have just* one *table of logarithms. If you've taken a probability or a statistics course, this is similar to standardizing random variables and having just a look-up table for the standard normal.*

Chapter 12

Solutions to Selected Problems

12.1. Chapter 1: Historical Introduction

Exercise 1.5.1: (a) 32, 15, 33, 13, 34, 32, 24, 33, 22, 21, 43, 34, 32, 44, 23, 15, 43, 34, 45, 44, 23.

Exercise 1.5.3: (a) C, Y, C, M, A, D, E, T.

Exercise 1.5.11: The message says, "Do not fire until you see the whites of their eyes."

Exercise 1.5.13: There are six potential possibilities.

12.2. Chapter 2: Classical Cryptography: Methods

Exercise 2.11.5: The percentage that has two letters switched and the other 24 letters sent to themselves is

$$\frac{325}{26!} \cdot 100\% \;=\; \frac{1}{1240896803466478878720 0}\%.$$

Exercise 2.11.7: 25!.

Exercise 2.11.11: MEET LATER.

Exercise 2.11.17: There is no added security.

Exercise 2.11.19: (a) $f(2n)$ equals $(2n-1)\cdot f(2n-2)$.

Exercise 2.11.21: Neither 1776 and 1861 or 1701 and 35 are pairs of congruent numbers modulo 26.

Exercise 2.11.23: Reducing by the modulus gives

$$\begin{aligned} 11 \bmod 3 &= 2,\\ 5 \bmod 3 &= 2,\\ 29 \bmod 26 &= 3,\\ 19 \bmod 26 &= 19. \end{aligned}$$

Exercise 2.11.27: There is no x such that x is congruent to 2 mod 30 and 3 mod 72.

Exercise 2.11.29: 149.

Exercise 2.11.31: We have

$$1800 \;=\; 2^3 \cdot 3^2 \cdot 5^2.$$

Exercise 2.11.35: 11 yields a valid affine cipher.

Exercise 2.11.37: For an alphabet with p letters for a prime p, there are $p^2 - p$ valid affine ciphers.

Exercise 2.11.41: THERE ISNOP LACEL IKEHO ME.

Exercise 2.11.47: SEIZE THE DAY.

Exercise 2.11.53: All permutations of three elements: 123, 132, 213, 231, 312, 321.

Exercise 2.11.57: (a) We have

$$\begin{pmatrix} 218 & 28 \\ 53 & 14 \end{pmatrix}.$$

Exercise 2.11.61: (a) We have

$$\begin{pmatrix} 7 & 18 \\ 1 & 5 \end{pmatrix}.$$

Exercise 2.11.63: There are 288 matrices for modulo 6; 2880 for modulo 10; 12,096 for modulo 14; 79,200 for modulo 22.

Exercise 2.11.65: There are 96 invertible matrices modulo 4; 3888 modulo 9; 300000 for modulo 25.

12.3. Chapter 3: Enigma and Ultra

Exercise 3.7.1: When order doesn't matter it is 120; when it does, it's 720.

Exercise 3.7.3: 518,400.

Exercise 3.7.5: For three people, we get 6.

Exercise 3.7.7: $(n-1)!$.

Exercise 3.7.9: 1,947,792.

Exercise 3.7.11: For $n = 1$, 2, 3, and 4, the answers are 2, 4, 8, and 16.

Exercise 3.7.17: 2598960.

Exercise 3.7.19: The answer is approximately 0.00198079231.

Exercise 3.7.21: The answer is approximately 0.00394.

Exercise 3.7.25: 3003.

Exercise 3.7.33: $(2n)!! = 2^n \cdot n!$.

Exercise 3.7.35: 50% work in a two-letter alphabet.

12.4. Chapter 4: Classical Cryptography: Attacks I

Exercise 4.9.1: (a) When should I return. Shift of 7.

Exercise 4.9.5: These aren't the droids you're looking for. Shift of 19.

Exercise 4.9.7: The inverse function is $f(x) = 2x + 5$.

Exercise 4.9.9: The inverse is $g(y) = y^{\frac{1}{3}} + 5$.

Exercise 4.9.11: It is $g(y) = 9y + 7 \bmod 26$.

Exercise 4.9.19: The pairs (5,6) and (5,5) can be used for the affine cipher, the pairs (13,17) and (6,6) cannot.

Exercise 4.9.21: RLOIP HUVY.

Exercise 4.9.23: MEETA TBUST OP.

Exercise 4.9.25: $n^2 - n$.

Exercise 4.9.29: (a) The solution is $a = 5, b = 19$.

Exercise 4.9.31: TO BE OR NOT TO BE (this is too famous not to write with the proper spacing).

Exercise 4.9.35: 73007/73008.

Exercise 4.9.39: Do or do not—there is no try. Judge me by my size do you? Reckless is he. Now things are worse.

Exercise 4.9.41: Shakespeare.

Exercise 4.9.43: The length of the keyword is 4.

Exercise 4.9.45: ITWAS THEBE STOFT IMESI TWAST HEWOR STOFT IMESI TWAST HEAGE OFWIS DOMITWASTH EQGEO FFOOL ISHNE SSITW ASTHE EPOOCH OFBEL IEFIT WASTHEEPOC HOFIN CREDU LITYI TWAST HESEA SONOF LIGHT ITWAS THESE ASONO-FDARK NESSI TWAST HESPR INGEF HOPEI TWAST HEWIN TEROF DESPA IR.

12.5. Chapter 5: Classical Cryptography: Attacks II

Exercise 5.3: 24.

Exercise 5.5: A man, a plan, a canal, Panama.

Exercise 5.17: BWGQAXASPJNIEBDDIJHQYEFDJOZVWKWTGJB-XIZEHNSLPYQSSGO

Exercise 5.19: It's more likely that "E" would be involved in a good position than "Q".

Exercise 5.23: "Jane update please."

Exercise 5.25: The message is "Karla Needs Out".

12.6. Chapter 6: Modern Symmetric Encryption

Exercise 6.11.1: (a) We have $1011_2 = 1(8) + 0(4) + 1(2) + 1(1) = 11$.

Exercise 6.11.3: We have (a) $14 = 8 + 4 + 2 = 1110_2$.

Exercise 6.11.7: It is HELLOSTUDENT.

Exercise 6.11.9: Signed binary representations of numbers are not unique, with the exception of 0.

Exercise 6.11.11: Every number has a unique restricted signed ternary representation.

Exercise 6.11.17: **HI** becomes 00111 01000.

Exercise 6.11.19: ?**SY** becomes 11100 10010 11000.

Exercise 6.11.33: ☺CHDIU_.

Exercise 6.11.41: (a) We have

D	X	M	P
0 0 0 1 1	1 0 1 1 1	0 1 1 0 0	0 1 1 1 1

Exercise 6.11.45: The muddled block is 0001.

12.7. Chapter 7: Introduction to Public-Channel Cryptography

Exercise 7.8.1: The degrees of the vertices in the graph in Figure 1 are: A:1, B:3, C:4, D:5, E:4, F:3, G:3, H:2, I:3.

Exercise 7.8.7: $d + 1$ vertices.

Exercise 7.8.17: Take $a = 2$ and $b = (n^2 + 1)/2$ for n odd.

Exercise 7.8.21: We have

$$\begin{aligned} M &= 5(4)-1=19, \\ e &= a'M+a=3(19)+5=62, \\ d &= 6(19)+4=118, \\ n &= [62(118)-1]/19=385, \\ \text{Decryption} &= md \bmod n \;=\; 27\cdot 118 \bmod 385 \;=\; 106. \end{aligned}$$

Exercise 7.8.29: It takes five steps.

Exercise 7.8.31: The answers are (a) 19, (b) 73, (c) 92, which in binary is 1011100_2.

Exercise 7.8.33: Base 8 expansion of 25: $3\cdot 8^1+1=31_8$. Therefore, $25_{10}=31_8$.

12.8. Chapter 8: Public-Channel Cryptography

Exercise 8.8.1: We have $n=pq=11\cdot 23=253$. Thus,

$$\varphi(n)=(p-1)(q-1)=10\cdot 22=220.$$

We can see that $13=e$ is relatively prime to 220. So, $d=e^{-1} \bmod \varphi(n))=17$. We may thus take our public key to be $(13,253)$ and our private key to be $(17,253)$.

Exercise 8.8.21: Since you are decrypting the message, you use the public key, $(3,55)$. According to the RSA algorithm, you compute $4^3 \bmod 55$, which is 9.

Exercise 8.8.25: The ToyHash value is 4, and we sign with 49.

Exercise 8.8.43: We have

$$g^{ab} \bmod n \;=\; 1030^{1001\cdot 843} \bmod 1967 \;=\; 701.$$

Exercise 8.8.57: Sketch of the proof: Without loss of generality assume $m=kp$ for some $k<q$, and let $c=m^3 \pmod n$. Compute c^d modulo p and q and combine using the Chinese Remainder Theorem.

12.9. Chapter 9: Error Detecting and Correcting Codes

Exercise 9.9.11: The message could also be the codeword 1, followed by 1, followed by 0, followed by 1. It could alternately be 1, followed by 10, followed by 1.

Exercise 9.9.13: The minimum distance is 2.

Exercise 9.9.15: The minimum distance is 1. Consider the code words 0000000000 and 0000000001.

Exercise 9.9.19: The first could be the intended code, but not the second.

Exercise 9.9.21: There are 15 possible messages that can be received that contain exactly one error.

Exercise 9.9.27: The Hamming distance, $d(0110011, 0100101)$ is 3.

Exercise 9.9.29: Using essentially one digit for information, we can have $n+1$ ones or $n+1$ zeros.

Exercise 9.9.33: The intended message was 101011000.

Exercise 9.9.37: This message is a valid Hamming (7, 4) codeword.

Exercise 9.9.41: (1) This is encoded incorrectly.

Exercise 9.9.45: (1) This lattice will have points at all coordinates that have integers as the x and y values.

12.10. Chapter 10: Modern Cryptography

Exercise 10.6.7: Bolding the changes, we have:

11010011000011110011001**1**
1010101**0**0011001001010101
100100100001000**1**1001001**1**

12.11. Chapter 11: Primality Testing and Factorization

Exercise 11.7.1: Starting the sequence with 3 we get: 3, 2, 7, 43, 13, 53, 5, ... Switching the first two terms gives us the same sequence as when starting with 2.

Exercise 11.7.13: $N = (604 - 57)(604 + 57) = 547 \cdot 661$.

Exercise 11.7.17: The better choice is 210 for a.

Exercise 11.7.29: We have

$$\begin{aligned}
1^2 &\equiv 1 \mod 11,\\
2^2 &\equiv 4 \mod 11,\\
3^2 &\equiv 9 \mod 11,\\
4^2 &\equiv 16 \equiv 5 \mod 11,\\
5^2 &\equiv 25 \equiv 3 \mod 11,\\
6^2 &\equiv 36 \equiv 3 \mod 11,\\
7^2 &\equiv 49 \equiv 5 \mod 11,\\
8^2 &\equiv 64 \equiv 9 \mod 11,\\
9^2 &\equiv 81 \equiv 4 \mod 11,\\
10^2 &\equiv (-1)^2 \equiv 1 \mod 11.
\end{aligned}$$

Exercise 11.7.31: For $n = 15$, $a = 2$, $b = 7$, $k = 1$, the sequence is 8, 4. Therefore, n is not prime.

Exercise 11.7.35: We have

$$\begin{aligned}
\phi(15) &= 8,\\
\phi(21) &= 12,\\
\phi(33) &= 20,\\
\phi(35) &= 24.
\end{aligned}$$

Exercise 11.7.37: We have

$$\begin{aligned}
\phi(4) &= 2,\\
\phi(9) &= 6,\\
\phi(25) &= 20,\\
\phi(49) &= 42.
\end{aligned}$$

Exercise 11.7.39: $\phi(p^3) = p^3 - p^2$.

Exercise 11.7.41: $x^2 + 2x + 1 \not\equiv 2x + 1 \mod (x+1)$.

Exercise 11.7.43: Yes.

Bibliography

[1] W. R. Alford, A. Granville and C. Pomerance, *There are infinitely many Carmichael numbers*, Annals of Mathematics **139** (1994), 703–722. Available online at `http://www.math.dartmouth.edu/~carlp/PDF/paper95.pdf`.

[2] M. Agrawal, N. Kayal and N. Saxena, *PRIMES is in P*, Ann. of Math. (2) **160** (2004), no. 2, 781–793. Available online at `http://annals.math.princeton.edu/wp-content/uploads/annals-v160-n2-p12.pdf`.

[3] G. Ateniese, C. Blundo, A. de Santis, and D. Stinson, *Visual cryptography for general access structures*, Information and Computation **129** (1996), no. 2, 86–106.

[4] K. R. Babu, S. U. Kumar and A. V. Babu, *A survey on cryptography and steganography methods for information security*, International Journal of Computer Applications **12** (2010), no. 3, 13–17, published by the Foundation of Computer Science.

[5] W. Barker (editor), *The History of Codes and Cipher in the United States Prior to WWI*, Aegean Park Press, Laguna Hills, CA, 1978.

[6] W. Barker (editor), *The History of Codes and Cipher in the United States, Part II*, Aegean Park Press, Laguna Hills, CA, 1989.

[7] D. J. Baylis, *Error Correcting Codes: A Mathematical Introduction*, Chapman Hall/CRC Mathematics Series, 1997.

[8] M. Campbell, *Uncrackable codes: The Second World War's last Enigma*, New Scientist, magazine issue 2813, May 30, 2011.

[9] R. D. Carmichael, *Note on a new number theory function*, Bull. Amer. Math. Soc. **16** (1910), no. 5, 232–238.

[10] J. H. Conway and N. J. A. Sloane, *Lexicographic codes: error-correcting codes from game theory*, IEEE Trans. Inform. Theory **32** (1986), no. 3, 337–348.

[11] J. H. Conway and N. J. A. Sloane, *Sphere Packings, Lattices and Groups*, third edition, Springer-Verlag, New York, 1998.

[12] A. D'Agapeyeff, *Codes and Ciphers—A History of Cryptography*, Blackfriars Press, 1949.

[13] H. Davenport, *Multiplicative Number Theory*, 3rd edition, revised by H. Montgomery, Graduate Texts in Mathematics, Vol. 74, Springer-Verlag, New York, 2000.

[14] Daily Mail Online, *Al-Qaeda planned to hijack cruise ships and execute passengers, reveals "treasure trove of intelligence" embedded in PORN video*, retrieved 5/1/2012: `http://www.dailymail.co.uk/news/article-2137848/Porn-video-reveals-Al-Qaeda-planns-hijack-cruise-ships-execute-passengers.html`

[15] S. Droste, *New results on visual cryptography*, In Advances in Cryptology—CRYPTO '96, pp. 401–415, Springer, 1996.

[16] A. Ekert and R. Jozsa, Richard, *Quantum computation and Shor's factoring algorithm*, Rev. Modern Phys. **68** (1996), no. 3, 733–753.

[17] P. Erdős, *On pseudoprimes and Carmichael numbers*, Publ. Math. Debrecen **4** (1956), 201–206. Available online at `http://www.renyi.hu/~p_erdos/ 1956-10.pdf`.

[18] W. F. Friedman, *History of the Use of Codes*, Aegean Park Press, Laguna Hills, CA, 1977.

[19] J. Gallian, *Contemporary Abstract Algebra*, seventh edition, Brooks Cole, Belmont, CA, 2009.

[20] M. J. E. Golay, *Notes on digital coding*, Proc. I.R.E. **37** (1949), 657.

[21] A. Granville, *It is easy to determine whether a given integer is prime*, Bull. Amer. Math. Soc. (N.S.) **42** (2005), no. 1, 3–38. Available online at `http://www.dms.umontreal.ca/~andrew/PDF/Bulletin04.pdf`.

[22] M. Kanemasu, *Golay codes*, MIT Undergraduate Journal of Mathematics **1** (1999), no. 1, 95–99. Available online at `http://www.math.mit.edu/ phase2/UJM/vol1/MKANEM~1.PDF`

[23] Klagenfurt University, *The Breakthrough of Frequency Analysis*, Universitat Klagenfurt, Aug. 2005.

[24] A. Korselt, *Probléme chinois*, L'intermédiaire des mathématiciens **6** (1899), 142–143.

[25] J. Leech and N. J. A. Sloane, *Sphere packings and error-correcting codes*, Canad. J. Math. **23** (1971), 718–745. Available online at `http://cms.math.ca/cjm/v23/cjm1971v23.0718-0745.pdf`.

[26] R. Lewin, *Ultra Goes to War*, Pen and Sword, Barnsley, United Kingdom, 2008.

[27] S. Loepp and W. K. Wootters, *Protecting Information: From classical error correction to quantum cryptography*, Cambridge University Press, 2006.

[28] M. Marayati, Y. Alam and M. H. at-Tayyan, *Al-Kindi's Treatise on Cryptanalysis*, vol. 1, Riyadh, KFCRIS & KACST, 2003. Print. Ser. on Arabic Origins of Cryptology.

[29] R. McCoy, *Navajo code talkers of World War II*, American West **18** (1981), no. 6, 67–74.

[30] W. C. Meadows, *They Had a Chance to Talk to One Another...: The Role of Incidence in Native American Code Talking*, Ethnohistory **56** (2009), no. 2, 269–284.

[31] W. C. Meadows, *The Comanche code talkers of World War II*, University of Texas Press, Austin, 2002.

[32] A. R. Miller, *The Cryptographic Mathematics of Enigma*, NSA Pamphlet, 2001. `http://www.nsa.gov/about/\_files/cryptologic\_heritage/publications/wwii/engima\_cryptographic\_mathematics.pdf`

[33] S. J. Miller, *The Probability Lifesaver*, Princeton University Press, to appear.

[34] S. J. Miller and C. E. Silva, *If a prime divides a product...*, preprint. `http://arxiv.org/abs/1012.5866`

[35] S. J. Miller and R. Takloo-Bighash, *An Invitation to Modern Number Theory*, Princeton University Press, Princeton, NJ, 2006, 503 pages.

[36] M. Naor and A. Shamir, *Visual cryptography, advances in cryptology*, Eurocrypt '94 Proceeding LNCS (1995), 950, 1–12.

[37] National Science and Technology Council, *Federal Plan for Cyber Security and Information Assurance Research and Development*, April 2006. `http://www.cyber.st.dhs.gov/docs/Federal%20R&D%20Plan%202006.pdf`

[38] T. Nicely, *The pentium bug*, `http://www.trnicely.net/pentbug/pentbug.html`.

[39] T. Nicely, *Enumeration to* 10^{14} *of the twin primes and Brun's constant*, Virginia J. Sci. **46** (1996), 195–204.

[40] D. Nicholas, *Lucky break*, History Today **57** (2007) no. 9, 56–57.

[41] R. Nichols, *Lanaki's Classical Cryptography Course*, Lecture 6, Part II: "Arabian Contributions to Cryptology", American Cryptogram Association, Jan. 1996. Accessed from the web February 9, 2013. `http://www.threaded. com/cryptography6.htm`.

[42] L. Savu, *Cryptography role in information security*, in Proceedings of the 5th WSEAS international conference on Communications and information technology (CIT11), N. Mastorakis, V. Mladenov, Z. Bojkovic, F. Topalis and K. Psarris editors. World Scientific and Engineering Academy and Society (WSEAS), Stevens Point, Wisconsin, USA, pp. 36–41.

[43] B. R. Roshan Shetty, J. Rohith, V. Mukund, R. Honwade and S. Rangaswamy, *Steganography Using Sudoku Puzzle* (2009), 623–626. doi:10.1109/ARTCom.2009.116.

[44] J. Silverman, *A friendly introduction to number theory*, Pearson Prentice Hall, 2006.

[45] S. Singh, *The Code Book: The Science of Secrecy from Ancient Egypt to Quantum Cryptography*, Anchor Books (a division of Random House), New York, 1999.

[46] S. Singh, *Arab Code Breakers*, SimonSingh.net, 2012, accessed February 14, 2013. `http://simonsingh.net/media/articles/maths-and-science/arab-code-breakers`.

[47] A. Stiglic, *The PRIMES is in P little FAQ*, September 22, 2008, `http://www. instantlogic.net/publications/PRIMES\%20is\%20in\%20P\%20little\%20FAQ.htm`

[48] T. M. Thompson, *From error-correcting codes through sphere packings to simple groups*, The Carus Mathematical Monographs, Number 21, the Mathematical Association of America, 1983.

[49] United States Department of Justice, *Criminal complaint by Special Agent Ricci against alleged Russian agents*, June 2010. `http://www.justice.gov/opa/documents/062810complaint2.pdf`.

[50] University Klagenfurt, *People behind information*, node on "The Breakthrough of Frequency Analysis" in Virtual Exhibitions Informatics, accessed February 27, 2013. `http://cs-exhibitions.uni-klu.ac.at/index.php?id=279`.

[51] E. R. Verheul and H. C. A. van Tilborg, *Constructions and properties of k out of n visual secret sharing schemes*, Design Codes and Cryptography **11** (1997), no. 2, 179–196.

[52] B. Watson, *Jaysho, moasi, dibeh, ayeshi, hasclishnih, beshlo, shush, gini*, Smithsonian (1993), 2434.

[53] Wikipedia, User 'SilverMaple', *Al-Kindi*, Wikimedia Foundation, accessed February 14, 2013. `http://en.wikipedia.org/wiki/Al-Kindi`.

[54] J. Wilcox, *Solving the Enigma—History of the Cryptanalytic Bombe*, NSA Pamphlet, 2001. `http://www.nsa.gov/about/\_files/cryptologicvheritage/publications/wwii/solving\_enigma.pdf`

[55] W. R. Wilson, *Code talkers*, American History **31** (Jan/Feb 97), no. 6, 16–21.

[56] F. W. Winterbotham, *The Ultra Secret*, Dell, 1975.

[57] F. B. Wrixon, *Codes, Ciphers, Secrets, and Cryptic Communication*, Black Dog and Leventhal Publishers, New York, 1998.

Index

Published Titles in This Series

29 **Margaret Cozzens and Steven J. Miller,** The Mathematics of Encryption, 2013
28 **David Wright,** Mathematics and Music, 2009
27 **Jacques Sesiano,** An Introduction to the History of Algebra, 2009
26 **A. V. Akopyan and A. A. Zaslavsky,** Geometry of Conics, 2007
25 **Anne L. Young,** Mathematical Ciphers, 2006
24 **Burkard Polster,** The Shoelace Book, 2006
23 **Koji Shiga and Toshikazu Sunada,** A Mathematical Gift, III, 2005
22 **Jonathan K. Hodge and Richard E. Klima,** The Mathematics of Voting and Elections: A Hands-On Approach, 2005
21 **Gilles Godefroy,** The Adventure of Numbers, 2004
20 **Kenji Ueno, Koji Shiga, and Shigeyuki Morita,** A Mathematical Gift, II, 2004
19 **Kenji Ueno, Koji Shiga, and Shigeyuki Morita,** A Mathematical Gift, I, 2003
18 **Timothy G. Feeman,** Portraits of the Earth, 2002
17 **Serge Tabachnikov, Editor,** Kvant Selecta: Combinatorics, I, 2002
16 **V. V. Prasolov,** Essays on Numbers and Figures, 2000
15 **Serge Tabachnikov, Editor,** Kvant Selecta: Algebra and Analysis, II, 1999
14 **Serge Tabachnikov, Editor,** Kvant Selecta: Algebra and Analysis, I, 1999
13 **Saul Stahl,** A Gentle Introduction to Game Theory, 1999
12 **V. S. Varadarajan,** Algebra in Ancient and Modern Times, 1998
11 **Kunihiko Kodaira, Editor,** Basic Analysis: Japanese Grade 11, 1996
10 **Kunihiko Kodaira, Editor,** Algebra and Geometry: Japanese Grade 11, 1996
9 **Kunihiko Kodaira, Editor,** Mathematics 2: Japanese Grade 11, 1997
8 **Kunihiko Kodaira, Editor,** Mathematics 1: Japanese Grade 10, 1996
7 **Dmitri Fomin, Sergey Genkin, and Ilia V. Itenberg,** Mathematical Circles, 1996
6 **David W. Farmer and Theodore B. Stanford,** Knots and Surfaces, 1996
5 **David W. Farmer,** Groups and Symmetry: A Guide to Discovering Mathematics, 1996
4 **V. V. Prasolov,** Intuitive Topology, 1994
3 **L. E. Sadovskiĭ and A. L. Sadovskiĭ,** Mathematics and Sports, 1993
2 **Yu. A. Shashkin,** Fixed Points, 1991
1 **V.M. Tikhomirov,** Stories about Maxima and Minima, 1991